AF815088

Handbuch der Aromaforschung

H. MAARSE · R. BELZ

Isolation, separation and identification of volatile compounds in aroma research

Handbuch der Aromaforschung

Unter Mitarbeit namhafter Wissenschaftler des In- und Auslandes entstandene Sammlung von Monographien auf dem Gebiet der Aromaforschung

herausgegeben von
MANFRED ROTHE
Potsdam-Rehbrücke

Isolation, separation and identification of volatile compounds in aroma research

By Dr. H. MAARSE and Drs. R. BELZ

Division for Nutrition and Food Research TNO
Institute CIVO-Analysis, Zeist, The Netherlands

With cooperation by

J. M. H. BEMELMANS
J. H. DHONT
L. J. VAN GEMERT
P. J. GROENEN
C. VAN INGEN
M. C. TEN NOEVER DE BRAUW
J. SCHAEFER
S. VAN STRATEN

With 90 figures and 86 tables

AKADEMIE-VERLAG · BERLIN
1985

Erschienen im Akademie-Verlag, DDR-1086 Berlin, Leipziger Straße 3—4
© Akademie-Verlag Berlin 1985
Lizenznummer: 200 · 100/486/85
Printed in the German Democratic Republic
Gesamtherstellung: VEB Druckerei „Thomas Müntzer“, 5820 Bad Langensalza
Lektor: Karl Abel
LSV: 3695
Bestellnummer: 762 520 0 (5984/5)
05800

Foreword

In 1973 Dr. M. ROTHE approached Dr. C. WEURMAN — at the time head of the Flavour Section, Central Institute for Nutrition and Food Research TNO (CIVO TNO) — on whether he would be willing to compose a series of monographs on techniques applied in aroma research. Dr. WEURMAN and his co-workers accepted the proposal, but the actual writing was considerably delayed because of Dr. WEURMAN's death in January 1975.

After some hesitation his colleagues decided to nevertheless carry the work out. This, however, proved to be far more time consuming than expected.

At CIVO TNO work on aroma problems already began in 1956. The original objectives were the development of analytical methods and instruments for the chemical investigation of the aroma of food products and the application of such techniques in unravelling complex aromas at the request of interested industries. Although the emphasis of aroma research has shifted somewhat to what are called negative quality aspects, such as off-flavours, toxicants and odour pollution of air, the relevant techniques do not essentially differ from those used in food aroma research in the strict sense.

After the initial years, during which the purely instrumental facilities were built up, sensory analysis has become an integrated part of aroma research and of off-flavour and odour pollution studies.

Prof. Ir. B. KROL,
Director, Division for
Nutrition and Food Research TNO

Preface

This monograph has been written by research workers of the Flavour Section and the Section of Instrumental Analysis of Institute CIVO-Analysis TNO.
Being editors as well as co-authors, it was our task to try and integrate the relevant contributions and seeing to it that the whole subject was covered and no overlapping occurred. Co-authors and other colleagues rendered valuable assistance in critically perusing the manuscripts. In this connection we would specially mention Dr. Ir. H. T. Badings of the Netherlands Institute of Dairy Research at Ede.
We are also grateful to many collaborators of the Flavour Section and the Section for Instrumental Analysis who assisted in selecting the illustrations, arranging lists of references, etc. The co-operation of the TNO translation units that took an active part in checking the manuscripts on the correct use of the English language is also much appreciated.
Finally, we would like to mention the enthusiasm with which Mrs. J. C. de Beauveser organized and supervised the typing of the manuscripts.

September 1978

The Editors
H. Maarse
R. Belz

Contents

1. Introduction

H. Maarse

One of the aims of aroma research is to identify the compounds that play a part in the aroma of a food product. Because of the low concentrations in which the majority of these compounds occur, identification was difficult to achieve before the beginning of the era of modern instrumental aroma research some twenty years ago. From the mid-fifties to the mid-sixties emphasis was laid on identifying as many compounds as possible. The development of gas chromatography in those years and its ensuing use in combination with mass spectrometry (GC-MS) tremendously accelerated the output rate in aroma research (fig. 1—1). It was rather a disappointment that many newly identified compounds proved to be unimportant for the odour of a product.

From about the late sixties, identification work became more systematic and was increasingly directed to the really important compounds by proper use of sensory evaluation. Those who tried to set up a complete survey of all volatile compounds ever encountered in foods, as we did (1), will agree that impressive results have been obtained. Figure 1-1 shows considerable increase of the number of identified compounds over the years.

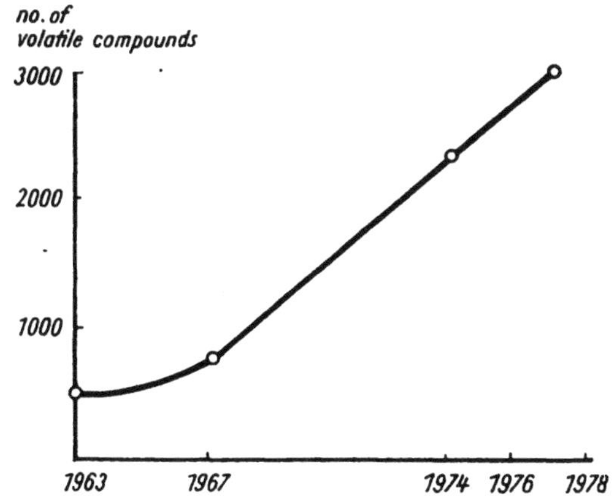

Fig. 1—1
Number of volatile compounds reported to occur in food products (1).

Many of the techniques and methods described in this book were specially developed for the study of the odorous components of food products. The peculiar nature of this field of application becomes evident from the following facts:

- The number of volatile substances in foods is extremely high.
- There is great variety in these substances, both in physical and chemical properties and in the concentrations in which they occur, which range from picogrammes to milligrammes per kilogramme product.
- Components contributing most significantly to odour may well occur at the low end of the concentration range.
- Quantitative ratios among aroma components are often more important than the effects of individual ones.
- Many odorous components become unstable when they are isolated from their original environment. Some foods must be regarded as live systems in which the qualitative and quantitative composition of odorous compounds is regulated by biochemical processes; interference with these processes, as occurs in isolation procedures, e.g. maceration, generally induces changes in composition.
- In contrast to other fields of analysis, the concentrations in the vapour emanating from the product instead of those within the product are of primary importance.
- Volatile compounds can be identified and measured instrumentally, but only the human nose can estimate their actual contribution to the odour of a product.

The only handbook on methodology in aroma research appeared in 1971 (2). In its preface, the authours state inter al.: "Flavour research is an active and rapidly changing field. No doubt by the time this book reaches the reader many advances will be made". Seven years later it can indeed be stated that developments have continued ever since. In this period, some centres of attention have been: a) the mechanisms through which odorous compounds are formed; b) the relationship between concentrations of compounds within a food product and those in the vapour phase above it; c) the relation of sensory to instrumental analysis in the development of new methods for assessing the quality of a product. As regards chromatographic techniques, the increased use of glass instead of stainless steel capillary columns and the incipient use of high-pressure-liquid chromatography for aroma research are worth mentioning. The possibilities of GC-MS were further extended by the introduction of the computer as part of the combination.

In this book the following items will be treated successively:

- methods for the isolation and concentration of volatile compounds;
- separation methods;
- methods for the identification of volatile components;
- sensory analysis as a step-by-step control of the instrumental procedures.

Although this book is entirely devoted to the analysis of volatile compounds, it should be remembered that besides odour, taste is an equally important aspect of the quality of a food product.
Attention will be mainly focussed on recent developments and applications, while only a summary account of the theoretical part is given. For the fundamentals, the reader will be referred to the handbooks. As a rule, illustrative material has been

derived from the analysis of odorous compounds; only if no appropriate examples were available has material from other research areas been used.
The subject of thin-layer chromatography (TLC) is treated more extensively than the rest, rendering the possibility to incorporate unpublished work from the author's laboratory.

Since a separate monograph on sensory analysis will be published in this series, the chapter on sensory analysis is limited to the aspects relevant to the step-by-step control of the instrumental analysis, beginning with the sampling stage right through to the final identification of purified components. The importance of such guidance by sensory methods can be illustrated by the following example. When one relies solely on instrumental analysis, sensorially significant components, which after gas chromatographic separation are overlaid by other components in much higher concentration, often escape detection.

The final chapter deals with some applications of methods, developed by aroma investigators, in other fields of research that are more or less related to aroma research proper, viz. off-flavours, mal-odours of the environment, pheromones, aroma and toxic aspects of tobacco smoke, volatile toxic chemicals, and medical (clinical) chemistry.

In the past, misuse by flavour chemists of terms from the sensorial field, such as "odour", "flavour" and "aroma", often led to confusion. Being primarily interested in the physical and chemical properties of substances, they do not always carefully choose the terms to describe the organoleptic properties of isolated components or fractions. Speaking the same "language" actually will be conducive to a good co-operation between instrumental and sensorial analysts. In the last few years, attempts have been made to give sharp definitions of these terms and to discourage the use of others. In anticipation of the results of international consultations, which in the meantime resulted in two international standards and one draft international standard (3), the British Standards Institute (B.S.I.) has issued a comprehensive list of terms (4). A few of these are given below:

to smell	— to test for sensation by use of the olfactory organ
odour	— 1) the sensation perceived via the olfactory organ from certain volatile substances 2) quality of this particular sensation due to these substances
aroma	— an odour with a pleasant connotation
taste	— 1) the sensation perceived via the taste buds resulting from the presence of certain soluble substances 2) quality of this particular sensation due to these substances
flavour	— the combination of taste and odour. It may be influenced by sensations of pain, heat, cold (e.g. spices, horseradish and menthol) and by tactile sensations
off-flavour	— an atypical flavour, usually associated with deterioration
taint	— a taste or odour foreign to the product.

In the following chapters we will try to adhere to these definitions.

References

1. Straten, S. van, Fl. de Vrijer and J. C. de Beauveser, Volatile Compounds in Food, 4th ed., Central Institute for Nutrition and Food Research TNO, Zeist, The Netherlands 1977, pp. 325 and first supplement to this 4th edition.
2. Teranishi, R., S. Hornstein, Ph. Issenberg and E. L. Wick, Flavor Research, Principles and Techniques, Marcel Dekker, Inc., New York 1971, pp. 315.
3. —, Sensory analysis, Vocabulary, Part I, International Standard ISO 5492/I-1977, Part II, International Standard ISO 5492/II-1978, Part III, Draft International Standard ISO/DIS 5492/III-1978
4. —, Glossary of terms relating to sensory analysis of food BS 5098: 1975

2. Isolation and concentration of volatiles from foods

2.1. Isolation and concentration from the product phase

J. M. H. Bemelmans

2.1.1. Introduction

Notwithstanding the spectacular development of analytical instruments in recent years, the human nose is as yet more sensitive to many odorous compounds than the detectors presently available for aroma analysis. This difference in sensitivity, the minute concentration of many odorous compounds in food and the amount of sample needed for their identification make it necessary to isolate the odorous compounds from the food and to concentrate them prior to the actual analysis.

An excellent review on methods for the isolation and concentration of food volatiles was given by Weurman (1). He pointed out that combinations of procedures used by some investigators to prepare odour concentrates were not always logical. Some isolation and concentration methods were also reviewed by Forss (2) and by Teranishi et al. (3).

This chapter will deal with distillation and extraction procedures used to isolate volatiles from foods directly. Their advantages and disadvantages in laboratory-scale experiments will be discussed. Attention will also be given to the various methods for preparing concentrates of the odorous compounds, followed by a discussion on some procedures to fractionate these concentrates.

Volatiles can also be isolated from the vapour phase in equilibrium with the food, or from gas flows led through the food. The relevant procedures are discussed in Chapter 2.2.

A major difficulty encountered in aroma investigations is that the amount of volatile compounds in foods is very small. Table 2—1 gives the concentration of the odorous compounds in some foods and shows that only minute quantities are present. Concentrations of individual compounds may be as low as 1 ppb or less, requiring large quantities of food to be processed in order to obtain sufficient material for analysis.

Table 2—1
Concentration of aroma compounds in various products according to WEURMAN (4)

product	concentration, ppm
banana	12—18
strawberry	2—8
grape	110
passion fruit	36
tomato	3—5
cocoa	100
beef	34
bread	6—10 (excl. ethanol)

Characteristic properties of food odours, summarized in Table 2-2, have always to be considered during isolation and concentration. Food odours usually contain between 50 and 250 different volatile compounds having a wide range of boiling points (7). Some products contain an even larger number: more than 700 have already been found in coffee.

Table 2—2
Characteristic properties of food odours are:*

1. the complexity of the mixture of odorous components present in most food products
2. the low concentration in which these components are generally present
3. the chemical reactivity and lability of the components
4. the polar character of many aroma compounds
5. the wide range of boiling points of the components
6. the large differences in the relative amount of the individual components
7. the presence of large quantities of volatile non-aroma compounds (a.o. water)
8. the strict relation between the quantitative composition of the mixture and the organoleptic properties
9. the differences in odour potencies of the different components

* According to WEURMAN and VAN LUNTEREN (5), but with additions from BADINGS (6)

Since some volatiles are heat sensitive or reactive, precautions have to be taken to ensure that they do not react during their isolation or concentration. Furthermore, it should be noted that foods themselves are complex matrices containing many chemical constituents, some of which might react and generate volatiles. The formation of new compounds (artifacts) should be prevented as much as possible, since it can lead to erroneous conclusions. This is also true for extraneous compounds (contaminants) introduced during analysis from e.g. solvents or glass-ware.

When isolating and concentrating food volatiles the above-mentioned factors must be taken into account; in this connection the following guide-lines are recommended:

1. The volatiles should be isolated from the food as completely as possible.
2. Artifacts and contaminants have to be kept down to a minimum: labile volatiles should be prevented from reacting or decomposing; no volatiles should be generated from the non-volatile food constituents.
3. Preferably, recovery data for the relevant methods should be ascertained.

In this chapter, the procedures are classified throughout as isolation and concentration methods, although with some of these the volatiles are simultaneously isolated and concentrated.

2.1.2. Selection and pretreatment of samples

Before analysis it should be clearly defined which odour characteristics of the product are to be studied. The sample of which the odour has to be analyzed must possess these odour notes as strongly as possible, while less desirable notes should be weak or absent. This implies that great care is required when selecting the sample which actually is to be submitted to analysis. Thus CHANG et al. (8) employed a panel for sample selection, using the profile method, and WOBBEN et al. (9) had their samples selected by experienced flavourists.

Usually the product under investigation has to be submitted to a pretreatment before the volatile constituents can be removed from the nonvolatile food matrix, unless it is a liquid of low viscosity. This pretreatment can comprise one or more of the following procedures: milling, homogenization, centrifugation, filtration or pressing. Solid products may have to be homogenized with water, so as to produce a slurry.

Formation of artifacts by chemical or biochemical reactions during pretreatment of a sample must be avoided. To minimize chemical reactions, it is usually necessary to ensure that the heat developed during milling or homogenization is dissipated (e.g. by cooling), so as not to expose the product to a high temperature. Also, exposure to light and/or oxygen must be excluded as much as possible.

Special procedures have to be applied for foods containing active enzymes. During maceration or homogenization, cell structures are disrupted, allowing enzymes and precursors to react. DRAWERT et al. (10) have shown that esters are hydrolyzed and that fast oxidative reactions occur in the presence of oxygen, as soon as the cell structure is disturbed. A large increase in the hexanal and hex-2-enal concentrations has been observed after homogenization of various fruit tissues (11). Similarly, FLEMING et al. (12) found a rapid formation of carbonyl compounds when cucumbers were homogenated in the presence of oxygen, while KAZENIAK and HALL (13) noted that the quantity of aldehydes in sliced tomatoes was markedly smaller than in homogenized ones. The latter phenomenon was confirmed by other investigators (14, 15). Cutting banana and potato tissues gave rise to an increased ethylene production (16).

When studying the volatiles of the undamaged tissue the enzymes have to be inactivated before or immediately after homogenization. DRAWERT et al. (17) have recommended that homogenization be carried out in the presence of methanol, thus inhibiting the enzyme activities; they followed this procedure for the isolation of apple volatiles. The same method was employed in studies of banana volatiles (18), and

grape aroma (19). In the latter study artifact formation was moreover avoided by crushing the grapes in a carbon dioxide atmosphere, followed by heat inactivation of the enzymes. In another investigation (20) fermentation of the juice was blocked by sodium fluoride, when crushing the grapes while oxidation was prevented by the presence of ascorbic acid.
The introduction of extraneous compounds, even if only small amounts are involved, will interfere with the analysis. The product itself may already be contaminated before it is submitted to analysis, e.g. by exhaust gases or constituents of the packaging material. Glassware and other equipment used for pretreatment and analysis may also be sources of contamination. BADINGS (6) and JUNK et al. (21) emphasized the necessity to use thoroughly cleaned utensils for which they described procedures. Contaminants may also be introduced by the water used to dilute or to homogenize a product, by extraction solvents and by anti-foaming agents. Therefore, it is most important to carry out blank experiments so as to ascertain whether or not contaminants are introduced into the odour concentrates. Corrections have to be made when the presence of contaminants cannot be prevented.

2.1.3. Isolation of volatiles

2.1.3.1. Introduction

The volatiles in the vapour phase above a food product are responsible for the odour perceived. Therefore, odour research should preferably be carried out by analysing the vapour phase. However, vapour samples, generally do not contain sufficient material for analysis. For many compounds the human nose is more sensitive than the instrumental detectors available at present, which implies that compounds contributing to an odour may escape instrumental detection. This disadvantage becomes even more evident when unknown compounds have to be identified, since in that case larger quantities are needed. Usually, the volatiles will have to be accumulated from large volumes of vapour phase. For this, methods have been developed which are applied in food odour analysis as well as in environmental studies (cf. Chapter 2.2.). A disadvantage of vapour phase analysis is that this accumulation is time consuming and that during the course of an investigation odour concentrates may have to be prepared several times. Therefore, it is often preferred to isolate the odorous compounds from the food product itself rather than from the vapour phase.
It should be noted that the concentration ratios among the volatiles in the vapour phase are not the same as in the product. Nor are the concentration ratios in an odour concentrate prepared by isolating the volatiles from the product phase, the same as those in the vapour phase. These ratio differences may affect the odour and should be taken into account in judging the odour quality of the concentrate.
Essentially two groups of methods are available to isolate volatiles directly from food products. The first one is based on the volatility of the compounds concerned and includes various distillation procedures to separate the volatile fraction from the non-volatile food matrix. The other group is based on a favourable distribution coefficient of the volatiles between an extractant and the food. Based on these two principles many procedures have been developed, the most important of which will be discussed.

First, however, an outline is given of the criteria for the selection of a method to be used for a particular problem.

2.1.3.2. Selection of method

The criteria that are important in selecting an isolation method are listed in Table 2—3.

Table 2—3
Criteria for selecting an isolation method are:

1. the volatilities and boiling points of the odorous compounds
2. the polarity of the compounds
3. the stability of the compounds at elevated temperatures, the influence of oxygen
4. the concentrations in the product and thus the quantity of product that has to be processed
5. the purpose of analysis: qualitative or quantitative
6. the distribution of volatiles throughout the product
7. the physical state and over all composition of the product

Usually, more than one method can be applied to isolate volatiles from a food product. The most appropriate one should be carefully chosen, since any losses of important compounds, caused by improper selectivity or low efficiency of the method, cannot be made up for in a later phase of the investigation.
In the isolation of volatiles from a food matrix it has to be borne in mind, as already said in the introduction, that they are present in large numbers, covering a wide range of boiling points and volatilities. When employing a distillation method, conditions should be selected on the basis of the *volatilities* of the odorous compounds of interest. Their *boiling points* have to be considered when selecting an extraction solvent, since good recoveries in the solvent removal step are only obtained when extractant and volatiles have sufficiently differing boiling points (cf. 2.1.4.2.).
The *polarity* of the volatiles is important in order to be able to choose a solvent when extraction is applied. The distribution coefficient should be biased towards the solvent, so that only comparatively small volumes of solvent are required.
The conditions for isolation have to be adapted to the *stability* of the volatiles, which means that low temperatures have to be applied if the volatiles are heat-sensitive and oxygen is to be excluded if they are sensitive to oxidation. In fact, this pertains not only to the volatile compounds, but also to the non-volatile fraction of the food.
Another criterion is the *concentration* of the volatiles, since this determines the amount of material to be processed in order to obtain sufficient quantities of the odorous compounds of interest. The concentration as well as the *purpose of analysis* should be considered when methods for the preparation of an odour concentrate are selected. "Large" amounts of volatiles have to be isolated if unknown compounds are to be identified. For this purpose, the isolation and concentration methods need not be quantitative and/or reproducible. If, on the other hand, the exact concentrations of the compounds in a food product are to be determined, the recovery of the method should be known. Often, food products are not homogeneous and the volatiles present are not evenly *distributed*. Isolation of an odour-bearing fraction from the product

(e.g. fat) prior to the analysis may be advantageous and will become a decisive factor in selecting a method.
It is obvious that the *physicial condition* of the food is an important criterion. In this respect food products can be divided in three groups:

1. solid products
2. liquid products on a water basis
3. fats and oils

These groups, which CHANG et al. (22) also mentioned, have no clearly defined boundaries; from solid products slurries can be made to be processed subsequently as a liquid.
Solid and liquid foods have to be dealt with according to different methods. Most equipment is designed for batch-wise operation and the capacity is the deciding factor with regard to the quantity of food that can be handled in a single operation. For liquid products, laboratory-scale equipment is available for continuous or semi-continuous operation, which allows relatively large amounts of product to be processed in one experiment.
Finally, the over-all *composition* of a product will to some extent determine the percentage of the volatiles recovered from a food matrix, since e.g. proteins and other macro-molecules are known to bind volatiles (23, 24). This composition also determines whether the product is likely to form an emulsion during extraction or will foam during distillation.

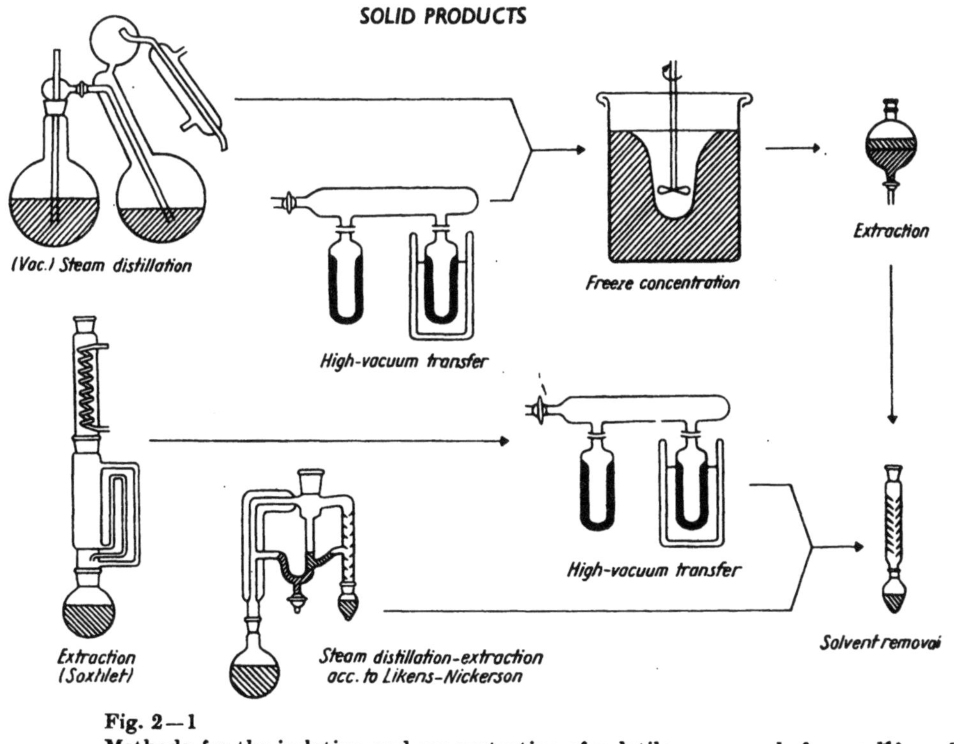

Fig. 2—1
Methods for the isolation and concentration of volatile compounds from solid products.

Usually, little information is available on the nature and the concentration of the volatiles in the product, so that criteria 1—4 are not of much help in selecting the appropriate method for their isolation. Then the isolation method is determined by the physical state and the composition of the product under investigation (criterion 7), taking into account the purpose of the analysis (criterion 5) and the distribution of the volatiles within the product (criterion 6). Information on the latter can be obtained by some rather simple experiments including the isolation of e.g. the fatty phase from the product.

Unfortunately, all isolation methods are selective in that the isolation of some compounds is favoured over that of others. This selectivity makes it impossible to isolate all compounds quantitatively in one single procedure. Therefore, sometimes more than one method will have to be applied. This was done by e.g. ARNOLD et al. (25) who isolated high-boiling volatiles by solvent extraction and low-boiling ones by an aeration technique.

Figures 2—1, 2—2 and 2—3 give an outline of the main methods as used at the author's institute to isolate and concentrate the volatiles from the three groups of food products. Figure 2—1 shows that the volatiles in solid products can be isolated by (vacuum-) steam distillation, high-vacuum transfer, extraction, or by a combined steam distillation-extraction method. The dilute aqueous solution obtained by steam distillation

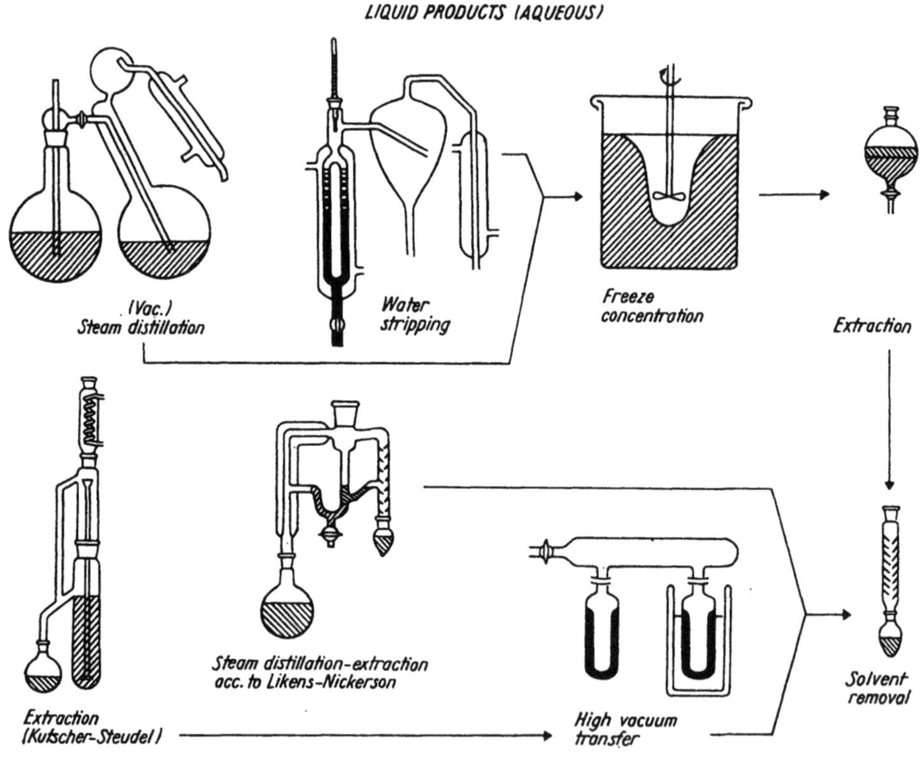

Fig. 2—2
Methods for the isolation and concentration of volatile compounds from aqueous liquid products.

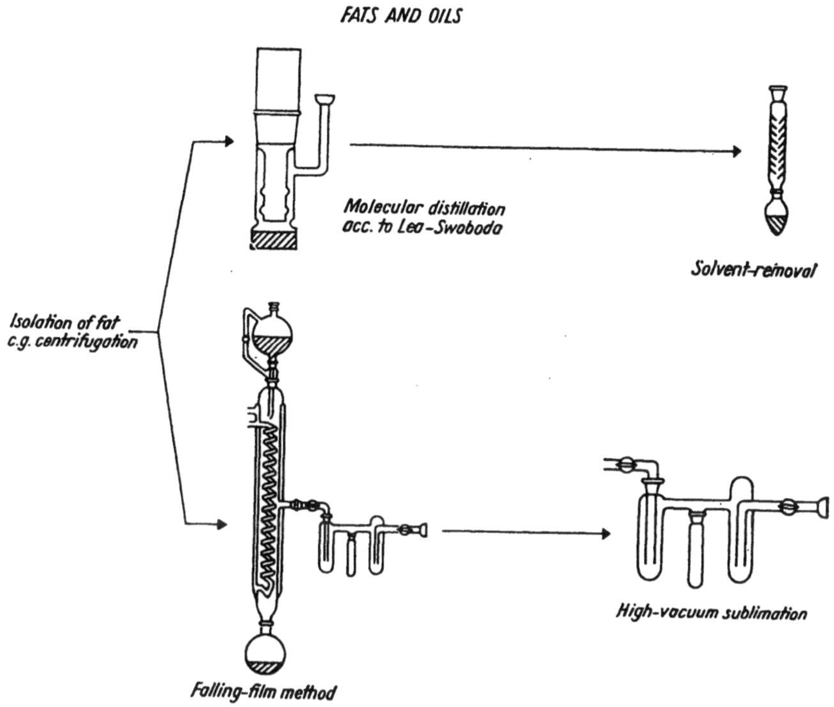

Fig. 2—3
Methods for the isolation and concentration of volatile compounds from fats and oils.

or high-vacuum transfer is reduced in volume by freeze-concentration and subsequently, the volatiles are recovered from the freeze concentrate by extraction. Odour concentrates are then prepared by partial removal of the extractant. Odorous compounds extracted directly from the food in a Soxhlet apparatus are concentrated by removing the bulk of the solvent, separated from the co-extracted non-volatile material by high-vacuum transfer, and further concentrated by removal of most of the remaining solvent. Finally, a LIKENS-NICKERSON (26) apparatus can be used, in which products are simultaneously steam-distilled and extracted. Here too, the extract is concentrated by distilling off the solvent.
Similar schemes are outlined for liquid foods on a water basis in Figure 2—2 and for fats and oils in Figure 2—3.
The methods depicted in these figures are discussed in detail.
Since the composition of foods varies considerably, as any relevant table shows, an odour research laboratory must be equipped with a wide variety of facilities for the isolation and concentration of volatiles.

2.1.3.3. Distillation

Introduction

In all distillation methods the volatile food constituents are separated from the non- or less-volatile ones. The advantage of distillation over extraction is that in

aroma isolates and concentrates obtained by distillation no non-volatile matter is present that might interfere with the ensuing chromatographic analysis. A drawback is, however, that by distillation dilute aqueous solutions are obtained from which the volatiles have to be isolated, which means an extra step in the analysis.
In this section the main distillation methods used in aroma research are divided in three groups, i.e. steam, high-vacuum and molecular distillation. This division is adopted for practical reasons and is rather arbitrary.
A rather special case presents itself when the isolation procedure is directed at a selected chemical group of volatiles. The isolation procedure can then be adjusted in order to obtain the maximum yield of the group in question. Several steam distillation procedures for such purposes are discussed in the section on thin-layer chromatography, see 3.4.5.1.

Steam distillation

For the isolation of food volatiles, steam distillation can be carried out at atmospheric or reduced pressure. At reduced pressure artifact formation is minimal, since the product is not exposed to a high temperature; a disadvantage is the lower yield of high-boiling components.
The equipment used for steam distillation can be similar to that used in organic chemistry, although several distillation devices have been developed specially for aroma research. Volatiles from small amounts (5—10 ml) of liquid products or slurries can be distilled at atmospheric pressure in the SCHMIDT apparatus (27). The sample is put in a glass tube that is placed in a round-bottomed flask with water (Fig. 2—4).

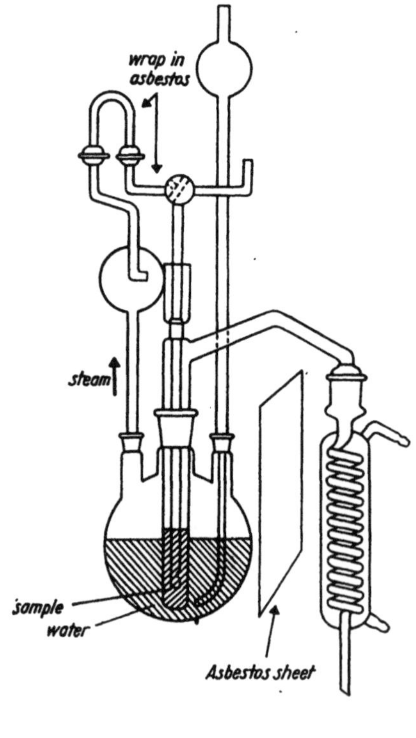

Fig. 2—4
Apparatus for the steam-distillation of small samples at atmospheric pressure according to SCHMIDT (27).

The latter is heated to generate the steam which is led through the sample via a three-way valve. The steam and the entrained volatiles are condensed in a cooler and collected in a flask. The apparatus has been used by MAARSE et al. (28) to isolate for example coumarin from alcoholic beverages.
Steam distillation at atmospheric pressure is frequently used to isolate essential oils from e.g. herbs and spices. The oil from *Achillea millefolium* and thyme was thus isolated with a CLEVENGER apparatus (29, 30). KÄMPF (31) compared the recoveries of a modified CLEVENGER device, recommended by the Swiss Pharmacopoeia, with those obtained using the apparatus prescribed by the European Pharmacopoeia (32). Yet another version, known as "Karlsruher Apparatur" (33), is recommended in the Dutch Pharmacopoeia (34). Steam distillation is not only used to isolate essential oils from herbs and spices, but also to recover volatiles from fats and oils. For the latter a spinning band apparatus was used (35).
In the literature, several appliances for steam distillation at reduced pressure have been described (8, 36). Of particular interest is that of FORSS and HOLLOWAY (37) in which samples of 0.1—1 kg can be processed (Fig. 2—5).

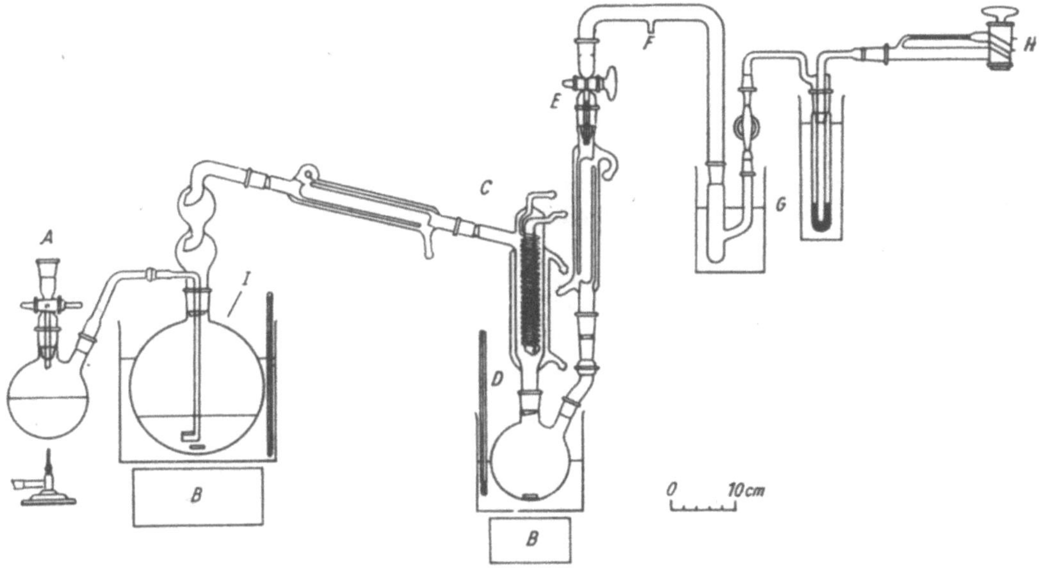

Fig. 2—5
Apparatus for steam distillation of volatiles from aqueous slurries at reduced pressure and for concentration of the distillate, as described by FORSS and HOLLOWAY (37). *A*. steam flask; *B*. magnetic stirrer and heater; *C*. condensers; *D*. collecting flask; *E*. stopcock with capillary; *F*. to vacuum gauge; *G*. liquid nitrogen containers; *H*. to manostat and mechanical pump; *I*. product flask.

In using this method, sufficient water should be added to the product, so as to make a thin slurry that can be stirred thoroughly during distillation. Foaming of the product must be prevented, preferably not by addition of an anti-foaming agent (possible introduction of contaminants), but rather by breaking the foam in its early stages by a sudden release of the vacuum, or by addition of magnesium sulphate (38, 39, 40). According to FORSS and HOLLOWAY (37), the recovery of C_3-C_9 n-alkan-2-ones from

butter oil is approximately 75—85%, but decreases considerably as the chain length increases. A disadvantage is, that cleaning the apparatus is rather time consuming.
The equipment discussed so far is operated batchwise. Products having a low viscosity, such as milk or dilute aqueous slurries, however, can be distilled continuously. Several investigators (41, 42, 43) described equipment for flash-evaporation of volatiles in which large volumes of product (10—100 l/day) can be processed continuously at reduced pressure. The product is introduced into a heat exchanger and heated rapidly to boiling-point temperature so that a flash-evaporation is effected. The evaporated water, together with the entrained volatiles, is condensed and collected in a series of cold traps. The stripped product and the condensate are removed from the system by pumps. A similar apparatus, in which the stripped product is removed by gravitational force, has been used by BADINGS (44) and is also being applied at the author's institute. It was used in a study of the volatiles of peanuts (45), tobasco pepper (46) and beef stew (47). Equipment of different design was used by HERZ and CHANG (48) and KALLIO (49).
Flash evaporation has the advantage that the product is heated only briefly, which reduces the danger of artifact formation. This does not hold for batch-wise distillation processes, especially if carried out at atmospheric pressure. PICKETT et al. (50) found that, depending on the pH of the medium, 30—90% of some terpene alcohol esters was lost during steam distillation at atmospheric pressure and that a variety of reaction products had been formed. Similar results were observed by AZZOUZ et al. (51), who compared cold-pressed and distilled lime oils by gas chromatography. A number of terpenes (e.g. α-thujene, neral, geranial) were lost almost completely during distillation, while other compounds (e.g. α-phellandrene, α- and β-terpineol) had been formed.

High-vacuum distillation

Volatiles can be isolated from a food product also by high-vacuum distillation. The vacuum to be achieved depends on the presence of water in the product. When water is absent, a vacuum of 10^{-3} mm Hg or less can be reached and a molecular distillation carried out (cf. next section). When water is present, a high-vacuum distillation can be carried out; the water vapour then acts as an "entrainment" gas, for the volatile molecules released from the product. For more information on high-vacuum distillation the reader is referred to PERRY (52).
MERRITT and co-workers (53) developed a closed system to achieve a "total take-off" distillation at high vacuum, thus separating the volatile from the non-volatile food compounds.
Equipment for high-vacuum distillation, often referred to as high-vacuum transfer, is simple and consists basically of a closed two-flask system connected to a vacuum pump (Fig. 2—6). A slurry is made of the product and shell-frozen in the sample flask, which is subsequently attached to the system and cooled with liquid nitrogen. The system is evacuated until the appropriate pressure has been reached and is then disconnected from the vacuum pumps. Distillation is started by transferring the cooling agent from the sample flask to the receiving flask and continues until "all" volatiles have been transferred to the cooled flask (the one on the left).
High-vacuum distillation was used to isolate volatiles from 50—200 g of fish. According to the authors (55), trimethylamine and n-butanol were completely recovered

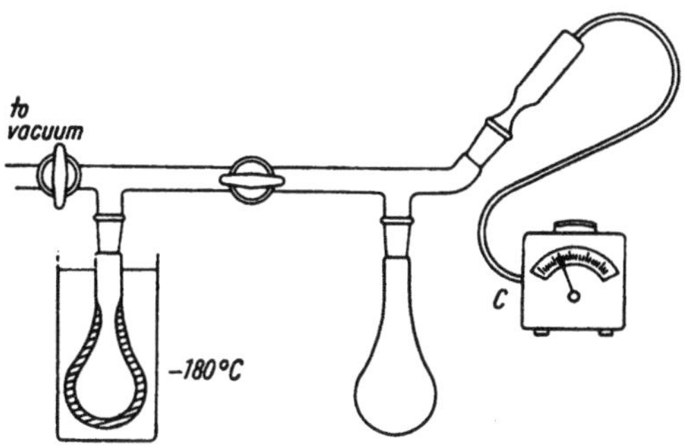

Fig. 2—6
Schematic drawing of a high-vacuum distillation system according to WEURMAN et al. (54).

from an aqueous solution. Similarly the C_2—C_5 fatty acids were isolated from biological tissues, but no recovery data were reported (56).
High-vacuum distillation is used not only for the isolation of volatiles from foods, but also to remove co-extracted non-volatile compounds from odour extracts and concentrates, preceding further analysis (Figs. 2—1 and 2—2). For this purpose the method was applied by MULDERS et al. (57) in preparing odour concentrates of bread suitable for chromatographic analyses. Recently, ITO (58) thus removed colouring agents from a molasses odour concentrate.
The advantages and drawbacks of this method have been discussed by WEURMAN et al. (54). Artifact formation is not likely to occur, since the product is kept at low temperatures. Moreover, the method can be used for all kinds of food, irrespective of their composition. Drawbacks are that the distillation takes quite some time, which appears to be extended when more polar non-volatile material is present in the product; as a result of this, handling large samples (5—10 kg) is time consuming. Furthermore, the recovery of the volatiles, which depends to some extent on the character of the non-volatile fraction, is incomplete; this can be deduced from the distinct odour emitted by the residue when rehydrated.
The main advantage of the high-vacuum distillation method, i.e. the minimum risk of artifact formation, is also offered by the lyophilization process described by WICK et al. (59) and used by HAYMON (46) to isolate the volatiles of tobasco peppers. During lyophilization the system is being evacuated continuously in order to maintain a low pressure.

Molecular distillation

In molecular distillation the more volatile components vaporize from a liquid sample surface and 'travel' relatively short distances to a cooled surface where they are condensed. The 'travelling' distance should be short compared to the 'mean free path' of the molecules, which requires a short distance between the two surfaces on the one hand, and a low pressure ($< 10^{-3}$ mm Hg) on the other. Such a low pressure can only be obtained with products that are free of water, e.g. fats and oils. The low pressures imply the use of low temperatures, which is an advantage.
Two types of molecular stills are available. Large volumes of product (0.1—2 l) can be processed in a 'flowing film' still, while small volumes are handled in a 'pot' still. In a flowing film still either gravitational force (falling-film still) or mechanical force (rota-film still) is used to obtain a thin film of the product.

BADINGS (6) used a rota-film molecular still to isolate butterfat volatiles at high vacuum. The temperature of the fat was kept below 55 °C to prevent the formation of artifacts. For a number of alkanals from butterfat the recoveries varied between 72 and 89% (Table 2—4). The method was used to isolate a catty odour from the fat fraction of tainted cheese (60). Similarly, BRANDT and JENSEN (61) found recoveries between 71 and 114% for compounds with boiling points ranging from 40—450 °C. The recovery studies were carried out for volatiles added to soybean oil in or below the ppm range. The recovery of volatiles with high boiling points increases as a function of the temperature to which the fat or oil is exposed. BILLS and SLOAN (62) determined the recovery of chlorinated insecticides from milkfat, also using a rota-film molecular still. The fat was brought to a temperature of 100, 150 and 200 °C respectively. The recoveries were nearly complete at 200 °C, but were poor at 100 °C, as is shown in Table 2—4.

Table 2—4
Recovery of a number of compounds isolated from fat with a rota-film molecular still

compound	type of fat	concentration (ppm)	Recovery (%)				Reference
			55 °C	100 °C	150 °C	200 °C	
pentanal	butter fat	0.4	80				BADINGS (6)
heptanal	butter fat	0.6	89				BADINGS (6)
nonanal	butter fat	0.8	85				BADINGS (6)
decanal	butter fat	0.8	72				BADINGS (6)
lindane	milk fat	40.0		24.4	71.4	96.0	BILLS and SLOAN (62)
heptachlor epoxide	milk fat	80.0		42.2	90.2	98.1	BILLS and SLOAN (62)
DDT	milk fat	120.0		49.3	90.6	94.7	BILLS and SLOAN (62)

As stated above, in the rota-film molecular still a mechanical device spreads the oil or fat in a thin film that is exposed to the vacuum. Several investigators used the force of gravity instead to obtain such a film. Their equipment is probably more reliable and also easier to clean than the rota-film still. Figure 2—7 shows a falling-film still as

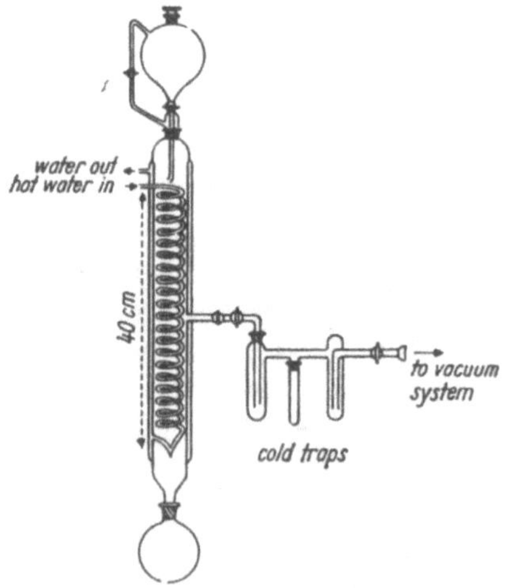

Fig. 2—7
Vacuum degassing system used at the author's institute for the isolation of volatiles from fats and oils.

used at the author's laboratory. It contains a vertical helix on which the oil-film is formed. PATTERSON (63) employed a similar device to isolate the compound that is associated with the boar taint in the fat of grown male pigs. The yield of that compound, 5α-androst-16-ene-3-one, which has a high boiling point, did not exceed 30%. This was improved by so redesigning the apparatus, that the distance between the heated surface covered with a film of fat and the cold surface on which the released volatiles were condensed was considerably shortened. However, for quantitative analysis, a pot-still technique was adapted and a recovery of 90% was achieved. The falling-film still was used also to isolate the volatiles from the oil pressed from roasted peanuts (64, 65).

With both the rota-film and the falling-film methods considerable amounts of oil (0.1—2 l) can be processed, whereas for small volumes (0.01—0.1 l) pot-still techniques are more suitable (66, 67). In these procedures, the fat is heated under reduced pressure while being stirred. The emanating volatiles are condensed on a liquid-nitrogen-cooled "finger". NAWAR et al. (68) determined the distillation efficiency of a modified version of the DE BRYUN and SCHOGT apparatus (66) for normal alkanes and alkenes from C_9 to C_{22}. Recovery for the C_9 to C_{17} compounds was between 85 and 99%, but decreased markedly for the longer-chain compounds. FORSS and HOLLOWAY (37), using equipment of their own design, compared the recoveries of alkanones and alkanols from butter oil by molecular distillation and high vacuum distillation respectively.

Recoveries can be increased by raising the temperature to which the fats and oils are exposed. However, temperatures critical for the formation of artifacts, should not be reached. During molecular distillation, extraneous volatiles may also be introduced by contamination. FERRETTI and FLANAGAN (69) identified benzothiazole in a mixture of non-enzymatic browning products. Its formation could not be rationalized and it was assumed that the compound originated from rubber tubing used as the connecting parts of the distillation apparatus. Furthermore, vacuum grease which is frequently used at the glass joints of the various parts of distillation devices, may release contaminants (70). Blank experiments always are required to determine whether contaminants interfere with the analysis.

2.1.3.4. Extraction

Introduction

Extraction methods are used in aroma research either to isolate volatiles directly from foods (Figs. 2—1 and 2—2), or to recover them from dilute aqueous solutions obtained via distillative isolation. The methods are based on a favourable distribution coefficient of the volatiles between a solvent phase and the food, or the aqueous distillate. If all volatile compounds of a food product have to be studied, the solvent should not be selective. Frequently, however, the analysis does not concern all volatiles. In that case, the solvent should facilitate the extraction of the compounds of interest and, if possible, exclude those that are not important or that interfere with the subsequent analysis. Solvents used for food aroma research have been listed by WEURMAN (1).

The efficiency of the extraction of an aqueous solution (e.g. a distillate) can be raised considerably by saturating the solution with salt. JOHNSON et al. (71) recommended that the extractant be added before saturating the solution; otherwise the salting-out process might result in losing volatiles in the atmosphere. Generally, NaCl or Na_2SO_4 is used; $(NH_4)_2SO_4$ is less suitable since it may decompose and liberate ammonia. According to WEURMAN (1), the selectivity of the extraction can be altered by saturation with salts, which will also prevent or reduce any formation of emulsions. Sometimes, compounds likely to interfere with the extraction can be eliminated, as was shown by EDELENYI et al. (72), who treated wines with a lead acetate solution prior to extraction.

Various devices are available for the extraction of foods. Selection of the apparatus is based inter al. on the physical state of the food (liquid or solid), the density and boiling point of the solvent to be used and the amount of food to be extracted. Foods having a medium moisture content can be processed as a solid product after mixing with a drying agent until a free-flowing powder is obtained.

Choice of solvent; recoveries

For the extraction of alcoholic beverages solvents are usually selected which discriminate against ethanol and other lower alkanols, since these compounds interfere with the analysis, but hardly contribute to the odour. This is effected by the use of less-polar solvents, such as iso-pentane (73, 74) and pentane-methylene chloride mixtures (75, 76). HARDY (77) used trichlorofluoromethane and determined the recovery of 1-alkanols, 2-alkanones and ethyl alkanoates by continuous extraction from dilute aqueous and aqueous-alcoholic solutions. The recovery of all compounds improved as the chain-length increased, whether they were extracted from water or from water-ethanol mixtures (Table 2—5). Since the lower alkanols were not extracted efficiently, trichlorofluoromethane was considered suitable for extracting odorous compounds from beverages and foods in which ethanol is a major volatile constituent, e.g. bread.

Table 2—5
Recovery (%) of alcohols, ketones, and esters from water or 10% ethanol by extraction with trichlorofluoromethane according to HARDY (77)

Carbon number	1-Alkanols		2-Alkanols		Ethyl alkanoates	
	Water	10% Ethanol	Water	10% Ethanol	Water	10% Ethanol
2	0					
3	0		0			
4	9		19		51	
5	68	47	65		72	
6	86	73	72	66	78	71
7	87	76	78	73	81	78
8	93	83	81	78	82	83
9			82	81	84	86
10	100	92	84	85	85	88
11			91	95		
12			95	105		

The extraction of alcohols and esters from aqueous solutions containing various amounts of ethanol was also studied by WILLIAMS and TUCKNOTT (78). They compared the efficiency of ether, pentane and trichlorofluoromethane for continuous extraction. A complete recovery of all compounds other than ethanol was obtained with pentane and trichlorofluoromethane. The authors reported that for exhaustive extraction no advantage is to be expected from adjusting the ethanol concentration of brandies prior to extraction, as is recommended by some investigators (74, 79).

SCHULTZ et al. (80) compared several solvents with regard to their suitability to extract volatiles from an apple essence. Iso-pentane and 1,2-dichlorotetrafluoroethane were equally effective, though less so than ether, which gave higher recoveries especially for lower alcohols. Extraction with liquid carbon dioxide gave similar results as ether. Despite the differences, all concentrates had the characteristic apple odour.

Freons, including those already mentioned, are often used for the extraction of volatiles, not only because of their selectivity, but also because they are essentially odourless. The latter property enables sensorial evaluation of the odour concentrate without interference from the solvent, which cannot be said of most other extractants. Freons also offer the advantage of being non-flammable.

ZENZ et al. (81) used carbon disulphide as the extractant in a study of beer volatiles. They found reproducible recoveries of alcohols and acids, but encountered difficulties in the quantitative determination of esters. The recoveries of the latter decreased as the ethanol content increased when extracted from aqueous solutions with carbon disulphide (82). The same effect was observed for higher alcohols. For quantitative analysis it was found necessary to adjust the ethanol content to 10 percent v/v before extraction. The latter is not in agreement with the results reported by WILLIAMS and TUCKNOTT (78).

Acetonitrile has been used succesfully in the extraction of volatiles from products having a high fat content. WONG and PARKS (83) packed a column with a mixture of shredded cheese and Celite, and eluted this with acetonitrile. The volatiles in the first drops of eluent were highly concentrated and could be chromatographed as such. This procedure was also followed to determine lactones in blue cheese (84).

Extractions are usually carried out with low-boiling solvents, and the extracted volatiles subsequently are concentrated by distilling off the solvent. If the investigator is particularly interested in volatiles with low boiling points, very low-boiling solvents, such as ethyl chloride (bp 12 °C), trichlorofluoromethane (bp 23 °C) or iso-pentane (bp 28 °C) may be used. Also liquid carbon dioxide (bp —78 °C) was applied as an extractant (85, 86), while recently an apparatus was developed for extraction of food slurries with dichlorodifluoromethane, which boils at —29 °C (87).

WEURMAN (1) disagreed with the use of low-boiling solvents, since the volatiles of interest still tend to get lost during solvent removal. It may instead be advantageous to use high-boiling solvents, such as glycerol (bp 290 °C) or propylene glycol (bp 188 °C). The volatiles are then recovered from the solvent by distillation at reduced pressure (88). This procedure was also followed by NELSON and HOFF (89), who macerated fruits in the presence of paraffin oil. The same authors also studied the partition coefficients for homologous series of aldehydes, ketones, alcohols and esters between paraffin oil and an aqueous sodium sulphate solution (90). An advantage of this method is that any solvent left in the concentrate does not interefere with the gas chromatographic analysis, as is often the case with low-boiling solvents. A drawback is, however, that it is difficult and time-consuming to adequately purify the high-boiling solvents and avoid the introduction of contaminants. This might explain why this method has not been applied more frequently.

Methods

Solid as well as liquid products can be extracted with a solvent if a direct contact can be established, e.g. the product plus the solvent can be put in an Erlenmyer flask, which is placed in a shaking apparatus. This method, though successfully used to extract substances such as pesticides from animal feed, is seldom applied in aroma research. The sample — solid or liquid — can also be mixed together with the solvent in a blender, thus ensuring close contact of sample and solvent (89). However, for the extraction of solid products, a Soxhlet apparatus is generally used, in which the solvent is continuously recycled. This device can handle large quantities of food (0.1 to 10 kg). Figure 2—8 shows a Soxhle-type apparatus, specially adapted for extraction with methyl chloride; MARTIN (91, 92) described a variety of such appliances.

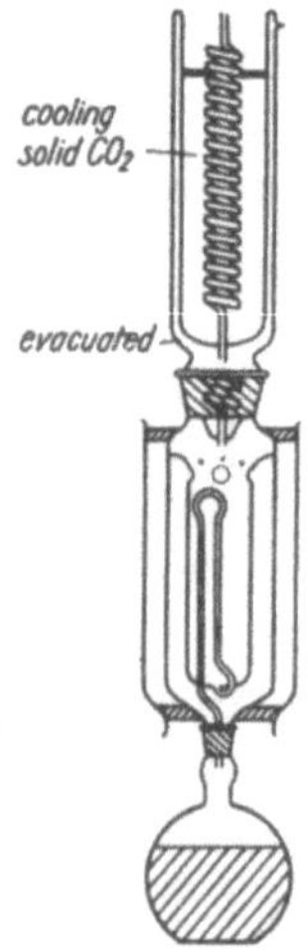

Fig. 2—8
Soxhlet-type apparatus for extraction of foods with methyl chloride according to WEURMAN (1).

Volatiles in liquid foods can be extracted batch-wise in a separating funnel. Centrifugal force or the addition of salt will break any emulsions formed. Some devices for continuous extraction of liquids are described by STAGE and GEMMEKER (93) and by MARTIN (91, 92). If the extractant is lighter than water, the KUTSCHER-STEUDEL apparatus (Fig. 2—9) may be used, in which the solvent is repeatedly percolated through the aqueous liquid to be extracted. The process is continuous, since the solvent is returned to the collection flask and reused after evaporation and condensation. AHNOFF and JOSEFSSON (94) designed an apparatus for continuous extraction of polluted water, which can be used both in the laboratory and outisde. The device might be useful for the isolation of high-boiling odour compounds.
A method for the isolation of short-chain fatty acids from very small sample volumes (0.5 ml of water or of a bacterial medium) was described by SCHWARZE et al. (95). Model experiments showed that the C_1—C_4 fatty acids (2.2—5.1 mmol/l) as well as lactic acid (21.5 mmol/l) could be recovered quantitatively. Recently methods were published to extract volatiles from small samples of beer (96) and body fluids (97).

Extraction with liquid carbon dioxide requires special equipment: it has to be carried out in a high-pressure system (85, 86). The carbon dioxide was evaporated continuously from the odour extract and recycled after condensation. SCHULTZ et al. (98) developed the method further and tested it on a pilot-plant scale. Also dichlorodifluoromethane was used to extract volatiles from fruits and fruit distillates; as in the case of carbon dioxide, the extraction was carried out at elevated pressure (99).

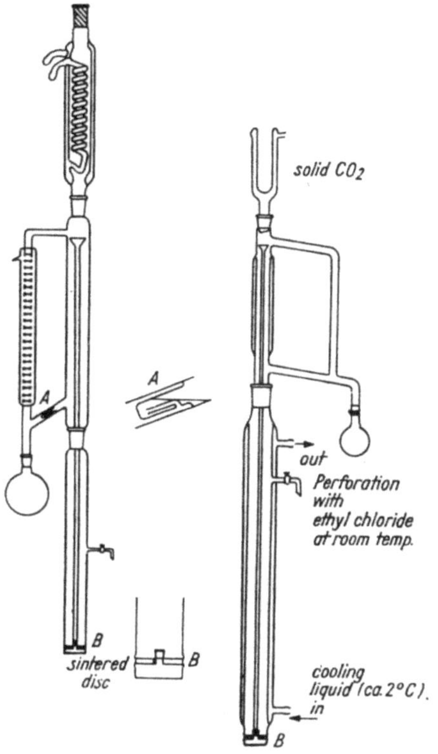

Fig. 2—9
Modifications of Kutscher-Steudel perforators used for isolation of volatiles by liquid-liquid extraction according to WEURMAN (1).

It should be emphasized here that extraction solvents must be pure to prevent undue identification of trace contaminants as food odour constituents (70). The need for purification is made clear in a report by LIBBEY and WALRADT (100). The antioxidant BHT, found in several odour isolates, was derived from the diethylether used for the extraction. Solvents are usually purified by distillation; several methods to this end have been described by BADINGS (6). After distillation the solvents should not be stored in plastic containers but in thoroughly cleaned glass vessels, to avoid contamination. It is strongly recommended to check the purity of solvents in blank experiments.

2.1.3.5. Simultaneous steam distillation/extraction

An ingenious device designed by LIKENS and NICKERSON (26) is frequently used to isolate food volatiles (Fig. 2—10; top left). An aqueous slurry of the product is put in a flask attached to the right side of the apparatus, while a flask connected to the other side holds the solvent (Fig. 2—10; top left). The flasks are heated separately, and steam and solvent vapour are condensed at the same time in the cooled central section. Subsequently, the water and the solvent phases, which are immiscible, are separated in the "U"-section of the apparatus and flow back to their respective flasks. This results in a continuous procedure during which distillation and extraction are carried

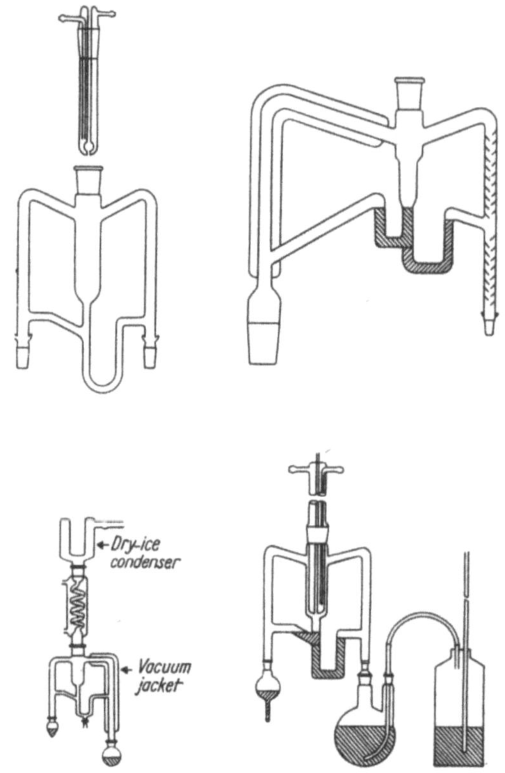

Fig. 2—10
Various Likens-Nickerson devices for the isolation of volatiles by simultaneous steam distillation extraction:

top left:	Original apparatus as designed by Likens and Nickerson (26)
top right:	Modification after Groenen et al. (40), suitable for extraction with 'heavy' solvents; solvent arm with Vigreux indentations
bottom left:	Modification after Maarse and Kepner (112) with vacuum jacket to minimize premature condensation and dry-ice condensor to reduce volatilization losses
bottom right:	Modification after Römer and Renner (39); steam is generated in a separate vessel.

out simultaneously and which requires only a small volume of solvent to extract large quantities of food. Originally, the apparatus was used to study hop oil components in beer (101); later applications were e.g. the isolation of volatiles from carrots (102), bell peppers (103), tomatoes (14, 104), olive oil (105), potato chips (106), peanuts (107), mushrooms (108), poultry (109, 110) and nuts (111).
Maarse and Kepner (112) isolated terpenes from Douglas fir needles with a modified apparatus: a vacuum jacket was added to reduce the premature condensation of

steam and a dry-ice condensor to minimize solvent losses (Fig. 2—10; bottom left). This equipment, without the dry-ice condensor, was used to study volatiles from cognac (113), from the oil of Origanum vulgare (114) and from tainted eggs (115). Römer and Renner (39) added an external steam generator to the original apparatus (Fig. 2-10; bottom right). MacLeod and Cave (116) adapted the apparatus so as to minimize the formation of emulsions when volatiles were isolated from eggs, and a dry-ice/acetone condensor was attached to prevent the loss of volatiles. Volatiles in irradiated boiled and roasted beef were recovered with a similar apparatus (117, 118).

Solvents lighter than water were used in all studies mentioned above. However, Groenen et al. (40) adapted the Likens-Nickerson device to isolate volatile nitrosamines from meat products with dichloromethane having a density greater than that of water (Fig. 2—10; top right). As a further modification, they recommended the use of Vigreux indentations in the solvent side of the apparatus.

Considering the frequent application of the Likens-Nickerson apparatus for the isolation of volatiles, surprisingly little is known about the recoveries obtained. In a model experiment Likens and Nickerson (26) found recoveries between 72 and 97% for a number of unspecified hop oil constituents. The recoveries proved to depend on the pH of the test solution. Maarse (114) recovered 96.7% of the monoterpene and 91.3% of the sesquiterpene fraction from the oil of Origanum vulgare, but only 57.1% of the thymol fraction. Farmer et al. (119) determined the recoveries from aqueous solutions of several homologous series of compounds (Table 2-6). The recoveries for the compounds having a short chain (less than 5 carbon atoms) were poor, but the longer ones were almost completely recovered.

Table 2—6
Recovery (%) of compounds from an aqueous solution with the Likens and Nickerson apparatus according to Farmer et al. (119)

carbon number	1-alkanols	2-alkanones	alkanals	alkanes
3	trace			
4	trace	trace	trace	
5	73	79	101	
6	97	104	91	64
7	101	101	101	94
8	102	94	94	103
9	99	97	83	94
10		102		90
11		101		94
12				104

The Likens-Nickerson apparatus is usually operated at atmospheric pressure, which implies that the product is exposed to a rather high temperature. Therefore, it is not surprising that artifacts are being formed. McGill and Hardy (120) reported the formation of cis-hept-4-enal from lipids when the volatiles in cod were thus isolated. Similarly, 2,4-nonadienal and 2,4-decadienal were found as artifacts in odour concentrates of nuts (111). To minimize artifact formation, the Likens-Nickerson apparatus was used at reduced pressure to isolate volatiles from vegetables (121) and papaya fruit (122).

A more complicated device for continuous steam distillation extraction of foods was described by WILLIAMS (123). In contrast with the LIKENS-NICKERSON apparatus, separate condensors were used for the water and the solvent vapour.

2.1.3.6. Interpretation of recovery data

The investigator is usually interested not only in the qualitative but also in the quantitative composition of food odours. The latter can only be determined when the respective recoveries of the isolation methods are known. Generally, the recovery of the various compounds will be different, as has been mentioned before..
Recovery studies can be carried out with model systems or, preferably, with a food matrix to which known amounts of the compound of interest have been added. Even the latter approach may give erroneous results, since the volatiles added generally will not become localized in the same part(s) or cells of the product as the volatiles which the food contains naturally.
It shoudld always be borne in mind that recovery data for one food may not be valid for another product of different composition. Therefore, the recovery data for some procedures given earlier in this chapter must be regarded with due care. Especially recovery data for volatiles isolated from aqueous model solutions may differ considerably from those for foods. Not only can their solubility in the food be quite different, but part of the volatiles can be adsorbed to macromolecules, such as proteins and carbohydrates. The adsorption might account for the difficulty to fully remove all odorous compounds from foods e.g. by high-vacuum distillation and to obtain essentially odourless residues.

2.1.4. Concentration of volatiles

2.1.4.1. Dilute aqueous solutions

Introduction

Distillative isolation of volatile compounds from foods often results in dilute aqueous solutions. The volatiles they contain have to be concentrated by one or more of the following procedures:

distillation
freeze concentration
zone melting
adsorption
extraction

Sometimes, one of these techniques may render an odour concentrate suitable for direct analysis, whereas others have to be used in combination; e.g. freeze concentration is usually followed by extraction, as Figures 2-1 and 2-2 show. The methods. listed above will now be discussed in some detail.

Distillation

Distillation can be applied to isolate volatiles from food products or to concentrate them from aqueous solutions. FORSS et al. (124) described an apparatus for the recovery of volatiles from aqueous solutions by distillation at reduced pressure, which can handle a wide range of volumes (5—5000 ml; Fig. 2—11). The solution is kept in flask C, and the distillate is collected in a cold trap (G, H or I). The distillation is continued until the volume of the distillate amounts to 1/1000—1/50 of the original volume. In a model test concentration factors of up to 400 were obtained. Alcohols, ketones and esters of intermediate chain lengths (C_5—C_9) gave the best recoveries; the shorter ones tended to remain in solution.

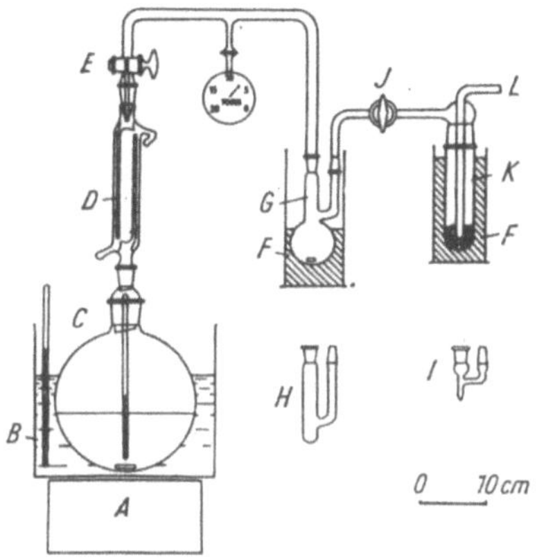

Fig. 2—11
Apparatus for concentration under reflux according to FORSS et al. (124); *A*. Magnetic stirrer and heater; *B*. water bath; *C*. 5 litre flask; *D*. Double-surface condensor; *E*. stopcock, 2 mm; *F*. Liquid-nitrogen vessels; *G*. collecting flask, volume 100 ml; *H*. collecting trap for volumes of 2 to 10 ml; *I*. collecting trap for volumes below 1 ml; *J*. stopcock; *K*. trap containing glass beads; *L*. connection to mechanical pump

PICKET et al. (36) enriched beer volatiles from an aqueous solution by distillation at reduced pressure. The volatiles were collected in a series of traps which were cooled at different temperatures. A low-temperature, fractional distillation technique was used by SHIPTON and WHITFIELD (125) to recover volatiles from aqueous distillates containing 5% ethanol

Freeze concentration

WEURMAN (1) strongly recommended that the volume of dilute aqueous solutions of volatiles be reduced by freeze concentration prior to extraction, so as to minimize the amount of extractant needed, and thus minimizing the introduction of contaminants.

In freeze concentration of aqueous solutions pure ice is frozen out, while the solutes are concentrated in the remaining liquid phase (126, 127). Both freeze concentration and zone melting (cf. next section) are non-selective concentration methods. The underlying theoretical aspects were discussed by GOUW (128). BAKER (129, 130) studied the effect of various physical, chemical and mechanical parameters on the freeze concentration process of aqueous solutions of organic solutes and also dealt with the relevant theoretical aspects. Several investigators stressed the necessity of continuously stirring the boundary layer between the solid and liquid phase during the concentration process, e.g. by stirring the solution (131). This is necessary since the rate of freezing is high compared with the rate of diffusion of the solute molecules at the water/ice interface.

Initially, freeze concentration was used to recover non-volatile solutes (132, 133, 134). SHAPIRO (135) was one of the first to use the technique for concentrating volatile compounds; he determined recoveries of acetone and isopropyl alcohol from dilute aqueous solutions. A more elaborate investigation was carried out by KEPNER et al. (136), who concentrated a number of alcohols, esters and alkanones with a SHAPIRO-type arrangement (Fig. 2—12).

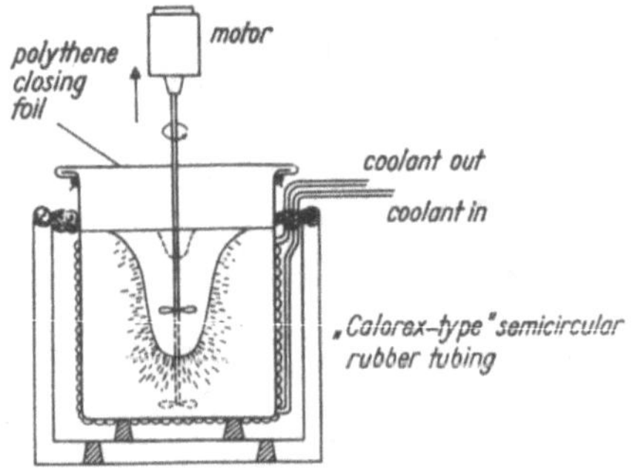

Fig. 2—12
Freeze concentration apparatus; Shapiro type arrangement as described by KEPNER et al. (136).

Volatilization losses were minimized by covering the vessel containing the solution with polythene foil. The recovery of all compounds was 90% or more, while a 5 to 40 fold concentration was obtained. Figure 2—13 gives the results of a freeze concentration

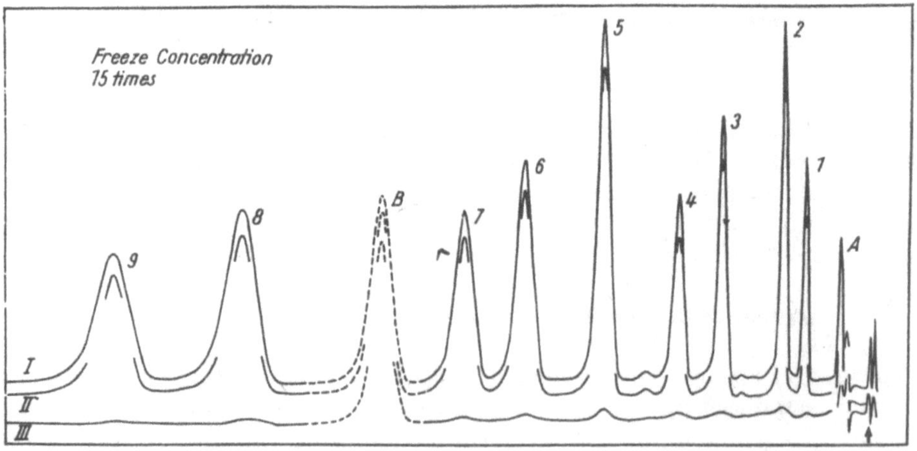

Fig. 2—13
Head space analyses of freeze concentrate fractions obtained with SHAPIRO-type arrangement by KEPNER et al. (136). A. Air peak, B. Standard (2-heptanone); 1. Ethyl formate, 2. Ethanol, 3. 3-Pentanone, 4 Ethyl butyrate, 5. 1-Butanol, 6. 4-Heptanone, 7. 3-Heptanone, 8. Ethyl caproate, 9. 3-Octanone.

experiment with an aqueous model solution, peak B being an added internal standard. Chromatogram I represents the headspace of the model solution. The freeze concentrate, diluted back to its original volume, gave almost the same chromatogram (II). Practically no volatiles were left in the ice phase (III).
Considering the fact that only simple equipment is needed and excellent recoveries can be obtained, it is rather surprising that freeze concentration has been used by only a few investigators.

Zone melting

In the zone melting process too, solutions of volatiles are enriched by freezing out the pure solvent, but the technique is quite different from that applied in freeze concentration. The solution to be concentrated is brought in a tube and subsequently frozen. A zone of the frozen solution is melted and kept molten by a heater, which is moving slowly along the tube, so the molten zone is 'travelling' through the frozen solution (127).
The technique in question was initially used to remove impurities from liquids. Later it was realized that the method could also be applied to concentrate trace compounds, e.g. volatiles from dilute solutions. Both SCHILDKNECHT (137) and SWOBODA (127) concentrated high-boiling volatiles using this technique. It was first used for odour research by HUCKLE (138), who extracted volatiles from raspberries with benzene; the volatiles were subsequently concentrated by zone melting the frozen benzene. The concentrate thus obtained was described as having a 'fresh' odour. In employing zone melting to concentrate fish volatiles (55), a 10—35 fold concentration was obtained.
SCHILDKNECHT (139) described a multi-stage zone melting apparatus that can handle 15—25 g substance and is suitable for the purification of solvents with melting points of 0—300 °C. A simple device applied by SCOTT (140) has the disadvantages that it must be placed in a refrigerator and that the refining process takes rather a long time. The necessary convection in the molten phase is usually effected by spinning the sample tube. An intensive stirring-effect was created by frequently changing the sense of rotation of the tube (141). The time needed to complete the procedure could be considerably reduced by the application of a fully automatic apparatus. ARKENBOUT (142) thus obtained a 300-fold concentration of the volatiles from 100 ml of an aqueous solution (Fig. 2—14).
Despite the high concentration factors that can be achieved, zone melting has not been widely applied in odour research, probably because of the complexity of the earlier equipment and the comparatively small volumes that can be handled. Nowadays, equipment is commercially available and can be used conveniently.

Adsorption

Adsorption is usually employed to accumulate volatiles from the vapour phase and, less frequently, from solutions. Chapter 2.2., which deals with the sampling of volatiles from the vapour phase, gives some information on the theoretical aspects of adsorption.

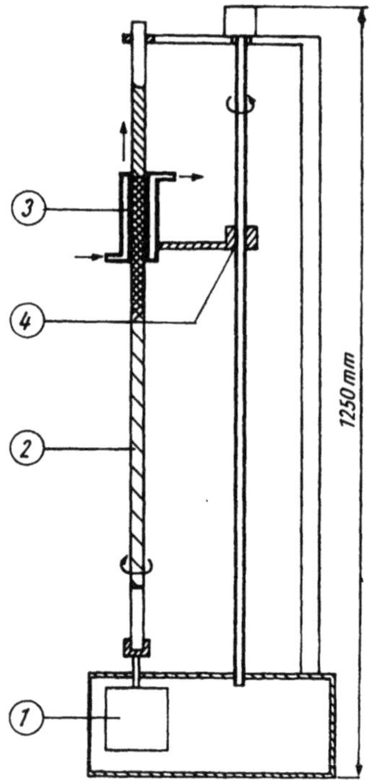

Fig. 2—14
Zone-melting apparatus for concentrating dilute aqueous solutions developed by ARKENBOUT (142); 1. Alternating rotation stirring mechanism; 2. Pyrex tube containing 100 ml aqueous solution 3. Heat exchanger connected to low temperature circulator; 4. Vertical transport mechanism.

Before synthetic resins became available, charcoal was the usual adsorbent. The advantages and drawbacks associated with the use of charcoal have been discussed by WEURMAN (1). To accumulate volatiles, dilute aqueous solutions can either be passed through a tube holding the adsorbent, or the adsorbent can be added to the solution. The former procedure was used to recover volatiles from an apple essence (80), while the latter was chosen to study the volatile phenols in alcoholic distillates from fermented grains (143).

In recent years polymers have been used as adsorbing agents. JUNK and co-workers (21) used a low-polarity styrene-divinylbenzene copolymer, known as Amberlite XAD-2, to study low (10—100 ppb) levels of organic compounds in water. The solutes were removed from the water by passing it through a column of XAD-resin. The compounds were subsequently desorbed by elution with diethyl ether. Recovery studies included alcohols, aldehydes, acids, phenols, esters and pesticides. The data presented in Table 2—7 demonstrate the high efficiency of the method. The authors, however, pointed out that the analysis included a number of critical steps and that serious discrepancies could occur if techniques and conditions were not properly applied. The method was used to study the contamination of water by polymer tubing (144).

LEONI et al. (145) showed Tenax to be suitable for the isolation of pesticides and polychlorinated biphenyls from water at the 1 ppb level. They observed fewer substances interfering with the gas chromatographic analysis than when liquid-liquid extraction was applied. Tenax may also be suitable to recover odorous volatiles.

As yet, only one report is known to have appeared about the use of resins to recover odorous volatiles from aqueous solutions: VITZTHUM and WERKHOFF (146) recovered coffee volatiles from a steam distillate by passing it through a column filled with Porapak Q.

Table 2—7
Recovery of solutes in water at the 10 to 100 ppb level* with the porous-polymer method according to JUNK et al. (21)

Compounds tested	Recovery (%)**	Compounds tested	Recovery (%)
Alcohols		*Acids****	
1-Hexanol	93	Octanoic acid	108
2-Octanol	100	Decanoic acid	90
1-Decanol	91	Oleic acid	100
Benzylalcohol	91		
		*Phenols****	
Aldehydes + ketones		Phenol	40
2,6-Dimethyl-4-heptanone	93	o-Cresol	73
Acetophenone	92	o-Chlorophenol	96
Benzophenone	93	2,4,6-Trichlorophenol	99
Benzaldehyde	101		
		Ethers	
Esters		Hexyl ether	75
Benzyl acetate	100	Benzyl ether	99
Diethyl phthalate	92	Anisole	87
Diethyl fumarate	86	2-Methoxynaphthalene	97
Methyl decanoate	95	Phenyl ether	91
Methyl octanoate	98		
Methyl palmitate	70		
		Alkyl benzenes	
		Ethylbenzene	81
		Cumene	93
		p-Cymene	92

* ppb = µg/kg
** The average reproducibility of these values is ±12%. With dedicated control of all variables the reproducibility limit may be decreased to ±5%.
*** The water was acidified by adding 5 ml of concentrated HCl prior to adding the organic solute. XAD-4 resin was used.

Extraction

Volatile compounds can also be recovered from dilute aqueous solutions by extraction. First, however, the volume of the solution should be reduced by freeze concentration to minimize the introduction of contaminants, as has been mentioned before.
Saturation of the solution with a salt could increase the efficiency of the extraction. Solvent selectivity can be applied advantageously to exclude the bulk of certain compounds from further analysis. Moreover, classes of compounds may be eliminated by choosing appropriate extraction conditions (cf. section 2.1.5.). For further information the reader is referred to section 2.1.3.4.

2.1.4.2. Dilute organic solutions

Introduction

The preparation of odour concentrates involves procedures which transfer the volatiles from a food product or an aqueous distillate to an organic solvent (Figs. 2—1 and 2—2). Usually, their concentration in the solvent is too low to permit direct analysis; therefore, the volatiles have to be concentrated first.
Most extracts contain traces of water that must be removed, either with drying agents or by freezing out. The drying agent must be inert and should not induce any chemical changes. In this connection, NURSTEN and WILLIAMS (147) studied the suitability of calcium and sodium sulphate and found calcium sulphate to give the best result. They recommended that large volumes of extract be dried first with sodium sulphate and subsequently with calcium sulphate. Sodium sulphate was also used by JUNK et al. (21), but SWAIN (148) considered this drying agent unsuitable, since it affected the concentration of compounds in the extracts. No data were given to support this statement. Residual water from diethyl ether fractions was also removed by brief immersion in a liquid nitrogen bath and by subsequently decanting the ether (21). This method was found as effective as the use of sodium sulphate and did not cause any losses.

Distillation

The concentration of volatiles from dried low-boiling solvents by distillative fractionation was extensively studied by JUNK et al. (21). An almost 100% solute recovery was found with both a VIGREUX and a three-cavity SNYDER column (Fig. 2—15), but the investigators gave several reasons for choosing the VIGREUX-column when concentrates of less than 0.6 ml were needed. Solute recoveries proved to be dependent on the shape of the concentration vessel. The maximum loss with vessel A (Fig. 2—16) was 6%; with vessels B and C 40—90%. The high losses were attributed to a less effective washing of the exposed vessel walls by the solvent. Table 2—8 gives recoveries of some rather

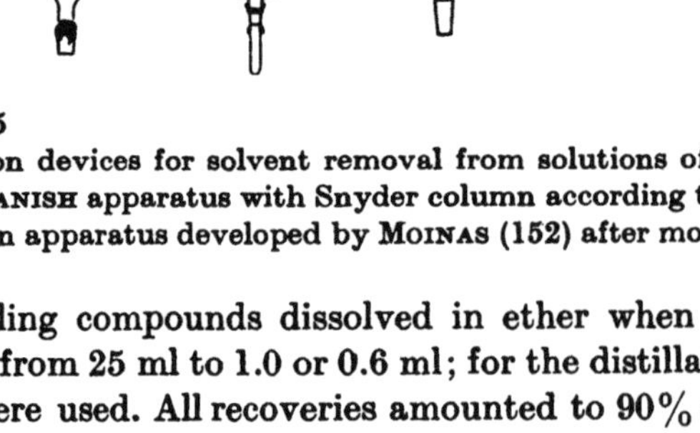

Fig. 2—15
Distillation devices for solvent removal from solutions of volatiles; A. SNYDER column; B. KUDERNA-DANISH apparatus with Snyder column according to FREHSE and NIESSEN (149); C. Microdistillation apparatus developed by MOINAS (152) after modification by BADINGS (154).

high-boiling compounds dissolved in ether when the volume of their solution was reduced from 25 ml to 1.0 or 0.6 ml; for the distillation a Snyder-column and a type A vessel were used. All recoveries amounted to 90% or more and no additional loss was

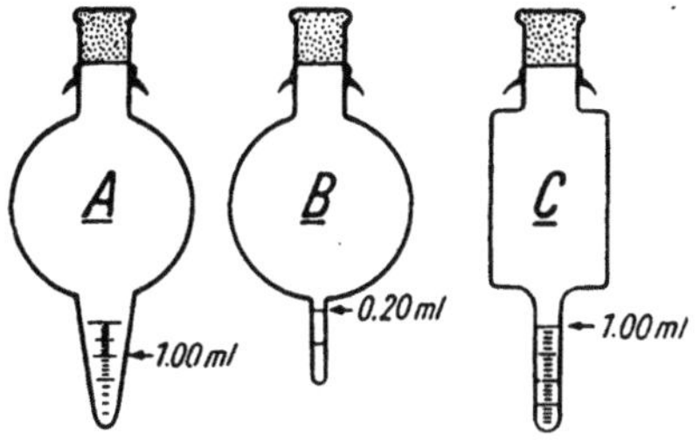

Fig. 2—16
Scale drawings of concentration vessels according to JUNK et al. (21): A is recommended, B is unsatisfactory and C is questionable.

observed when the volume was brought down from 1.0 to 0.6 ml; the latter measurement was made for a few compounds only.

Table 2—8
Recovery of solutes in diethyl ether during concentration from 25 ml to 1.0 and 0.6 ml according to JUNK et al. (21)

compound	Recovery (%)	
	25 ml to 1.0 ml	25 ml to 0.6 ml
Naphthalene	94	95
Methylnaphthalene	105	102
Dimethylnaphthalene	93	101
Acenaphthene	99	96
Biphenyl	101	97
Octane	96	
Dodecane	103	
Cumene	99	
p-Cymene	101	
Toluene	91	
Ethylbenzene	98	
Decalin	97	
Tetralin	101	

In a study on the volatiles of the oil of Origanum vulgare, MAARSE (114) compared the recoveries of mono- and sesquiterpenes obtained by making use of:

1. a vessel without VIGREUX-column
2. a vessel with VIGREUX-column
3. a vessel with VIGREUX-column and reflux unit (1:1)

The volume of pentane-ether (2:1) solutions was reduced from 100 to 2.0 ml. The results which are summarized in Table 2-9 show that a VIGREUX-column gives slightly higher recoveries.

The data given in Tables 2-8 and 2-9 concern rather high-boiling compounds present in concentrates with final volumes between 0.6 and 2.0 ml. However, in food odour analysis the concentration of the volatiles may still be too low for analysis, so that the volumes will have to be reduced further. This can be done with various types of distillation equipment, such as a column packed with FENSKE helices (150), a WIDMER column (49) and a KUDERNA-DANISH apparatus (151) as depicted in Figure 2-15 B.

Table 2–9
Recovery of compounds after reducing the volume of the solution from 100 to 2.0 ml according to Maarse (114)

compound*	Recovery (%)		
	vessel without Vigreux column	vessel with Vigreux column	vessel with Vigreux column and reflux unit (ratio 1:1)
α-pinene	95.0 ± 0.1	97.2 ± 0.3	95.8 ± 0.3
β-pinene	95.0 ± 0.2	97.4 ± 0.1	95.6 ± 0.2
3-carene	98.6 ± 0.2	98.9 ± 0.1	96.9 ± 0.2
limonene	98.4 ± 0.1	99.3 ± 0.2	96.6 ± 0.2
trans-β-ocimene	97.7 ± 0.3	99.3 ± 0.1	96.2 ± 0.1
terpinolene	96.2 ± 0.1	99.6 ± 0.1	96.2 ± 0.2
linalool	94.4 ± 0.0	97.2 ± 0.1	94.6 ± 0.1
cadinene	98.5 ± 0.1	99.0 ± 0.1	97.6 ± 0.2
citronellyl acetate	98.4 ± 0.1	98.8 ± 0.1	98.6 ± 0.1

* The concentration of the monoterpene hydrocarbons in the original solution was 120 ppm; the concentration of the other compounds was 200 ppm; the solutes were dissolved in a pentane-ether mixture (2 + 1).

Frequently, the final step in reducing the volume of an odour concentrate is achieved by purging with a gentle stream of nitrogen. Junk et al. (21) observed high losses during such a purging procedure and were therefore of the opinion that it should be eliminated. The micro-distillation apparatus applied by Moinas (152) makes this possible, i.e. the volume is reduced by distillation to 5–10 μl (Fig. 2-15C). Bemelmans and te Loo (153) determined recovery data for a number of compounds with this apparatus, modified by Badings (154). The results are summarized in Table 2-10, together with those obtained with a Vigreux-column, followed by careful evaporation with a nitrogen stream. The data show no substantial losses for most compounds when the volume was reduced to 50 μl with the micro-distillation device; however, they increased significantly when the volume was reduced to 15 μl, though the recoveries of most compounds were still considered to be acceptable. By distillation with a Vigreux-column followed by purging with nitrogen, the losses, after the volume was reduced to 50 μl, were greater than those found with the microdistillation apparatus.
Usually, fairly large volumes of solvent have to be removed, most of which can be distilled off rather rapidly without intensive refluxing, and with negligible losses. Distillation becomes critical as soon as the volume is reduced to 1 ml; further concentration has to be done very carefully, so as to ensure a high recovery. This implies that many devices are not suitable for the distillative removal of solvent, and that the final volume of the concentrate should be chosen deliberately.

2.1.4.3. Artifact formation

Weurman et al. (54) demonstrated that artifacts can be formed during the distillative concentration of odour extracts. They distilled a solution of ethanol, butanol and acetic acid in pentane in concentrations likely to be encountered during the analysis of food odours and subseqently analysed the concentrate by gas chromatography. In addition to the compounds mentioned, also 1,1-diethoxybutane was found, which had been

Table 2—10

Recoveries (%) obtained by concentrating 2 ml of a solution of volatiles in ether to volumes as indicated, using either a micro-distillation apparatus or a Vigreux column, according to BEMELMANS and TE LOO (153)

compound	boiling point (°C)	Concentration (ppm)*	micro-distillation apparatus			Vigreux***		
			100 µl**	50 µl**	15 µl**	100 µl**	50 µl**	15 µl**
ethyl formiate	54.5	3.1	22.0	15.5	10.3	27.4	15.9	13.5
ethanol	78	3.1	60.4	56.8	28.7	67.6	57.2	29.0
butanal	75.7	3.1	37.9	32.6	12.8	29.8	27.3	14.2
3-pentanone	101.7	3.7	81.7	76.1	30.9	70.0	61.7	31.4
ethyl butyrate	97.4	3.7	95.0	86.0	51.7	74.4	62.1	42.3
hexanal	128	3.7	97.0	84.5	53.5	72.2	58.6	52.5
1-butanol	117.3	4.3	94.2	86.5	53.0	63.8	53.1	45.3
3-heptanone	147	4.3	94.8	85.3	55.0	64.2	50.6	41.2
3-methylbutan-1-ol	128	5.0	94.7	85.2	53.7	67.2	55.2	46.7
3-octanone	167	5.0	91.2	82.9	48.7	68.2	53.5	43.7
ethyl caproate	168	5.0	91.0	81.1	48.3	68.2	52.4	44.4

* Concentration of original solution.

** Volume of concentrate.

*** Volume reduced to 500 µl by distillation; further reduction by purging with a stream of nitrogen.

formed during the concentration procedure and should therefore be considered an artifact. Similarly, esters are formed during concentration of a dilute solution of ethanol and aliphatic acids.
Formation of artifacts can also take place during storage of odour concentrates. BADINGS (6) reported that only at −40 °C could a concentrate of volatiles from cold-stored butter be kept unchanged for 12 hours. A similar observation was made by SWAIN (148) who found chromatographically that odour extracts of boiled ham changed during storage at 0 °C. On the other hand, it was reported that concentrates of beer volatiles could be stored for over a month without apparent deterioration (36). It is obvious that the formation of artifacts depends on the stability of the volatiles present in the odour cocentrate.

2.1.5. Group separation of volatiles

Odour concentrates are rather difficult to analyse because they contain many compounds with sometimes closely similar physico-chemical properties, making isolation of the compounds of interest rather laborious. The procedure can be shortened by dividing the concentrate into less complex fractions prior to gas chromatographic analysis. Such a separation can be based either on differences in functional groups or on different polarities of the various compounds.
Another reason for the investigator to effect a 'gross' fractionation prior to the isolation of individual compounds, might be that he wants to identify only those compounds that contribute to the odour investigated. To this end, the importance of the odorous fractions is determined by sensory evaluation, so that only the interesting ones need be analysed further.

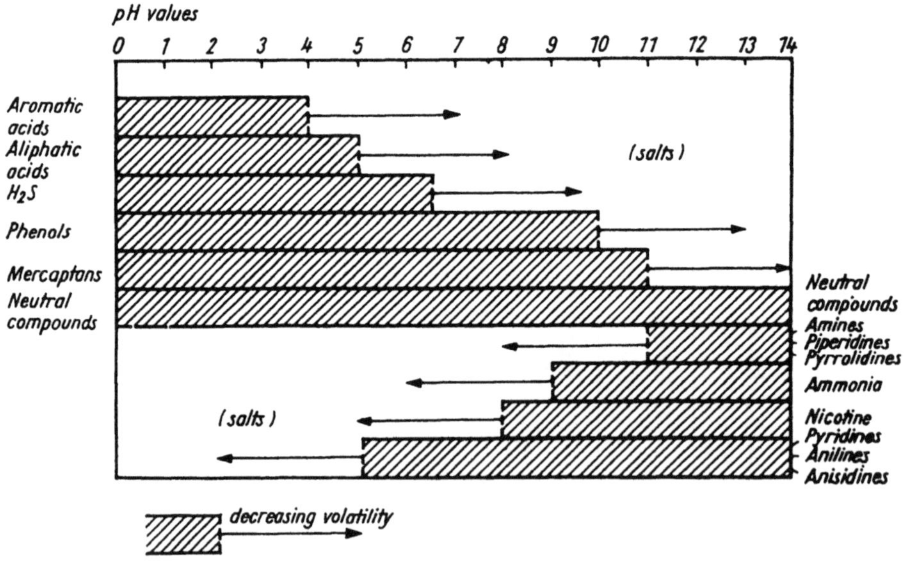

Fig. 2—17
Dissociation of volatile acids and bases as a function of pH according to GROENEN (156). Hatched bars represent the pH ranges over which the ionized fraction of the various groups of compounds increases from negligible to 50%; arrows indicate increases in dissociation up to 99.9%.

Fractionation of odour concentrates into chemical groups can be carried out according to various methods. Of these, fractionation into acidic, neutral and basic classes is probably the easiest one. WEURMAN (155) stressed that the result of such a fractionation depends on the extent to which the compounds are dissociated in a reagent solution. Figure 2-17 shows that in this respect extraction can be made selective by adjusting the pH (156).

Fractionation preceding gas chromatographic separation has been applied in several studies. E.g., KUGLER and KOVÁTS (157) divided tangerine oil into acidic, neutral and phenolic fractions. Extraction with an aqueous solution of sodium carbonate was used to separate the acidic from the non-acidic volatiles of heated pork and beef fat respectively (158, 159). Similar fractionations were applied to the volatiles of coffee (146), molasses (58), peanuts (64, 65) and wood smoke (160, 161). These fractionations are usually carried out while the volatiles are still in dilute solution.

It may be advisable to remove a particular class of compounds from an odorous mixture if they are likely to interfere with the subsequent analysis. HARDY (77) successfully extracted freon solutions with propylene glycol in order to selectively remove alcohols; the lower alcohols were withdrawn almost completely and the higher ones to a great extent. This procedure was used in a study of wine volatiles and allowed further concentration of the compounds of interest than would originally have been possible (163).

Besides fractionation into chemical groups chromatographic procedures are also used to simplify the isolation of individual compounds or to select the important odorous compounds. Liquid chromatography can be applied to fractionate odour concentrates. It was formerly carried out at atmospheric pressure; recently, high-pressure liquid chromatography has successfully been used. Information on the latter procedure is given in Chapter 3.

MURRAY and STANLEY (163) developed a method to separate mixtures of odorous compounds on a silica gel column with a low-boiling solvent ascending in the column as the eluent. After fractionation, the column was cut into sections from which the volatiles were extracted. With this technique, alcohols were isolated from odour concentrates of bananas and peas (164, 165). A modification of this method was described by PALMER (166), who placed the odour concentrate on top of a silica gel column and eluted it with a freon moving down the column, thus allowing the separated fractions to be collected in vials. A silica gel column was also used to separate the volatiles of nutmegs (167), grapes (19), dill oil (168) and roasted turkeys (110).

Several other sorbents can be used to fractionate odour concentrates. Tomato juice volatiles e.g. have been separated with Sephadex G 15 (169), and Sephadex LH 20 was used in a study of the volatiles of molasses (58). WATANABE and SATO (159) applied a silicic acid column for the analysis of the non-acidic volatile fraction of heated beef fat.

Essential oils — like odour concentrates — are complex mixtures, and are often separated in their main fractions prior to gas chromatographic analysis. WALRAVEN (170) used gradient elution to fractionate bergamot oil on silica gel. A similar procedure was followed in a study of the odorous compounds of oakmoss (171), and to separate the hydrocarbons of the essential oil of black pepper (172). Howeer, unspecified changes were reported in the composition of the oil of Origanum vulgare when silica gel was used (173), and SCHEFFER et al. (174) warned that impurities and active sites

in the gel may cause isomerization of several compounds. These isomerization reactions can be avoided by washing the gel with acid and base, and by de-activating the dried gel with 5% water.
In studies of essential oils neutral (173, 175) as well as basic (176) alumina was used. The danger of artifact formation by alumina was shown in a study of lovage root oil (177).
KUBECZKA (178) preferred Kiesel gel to fractionate essential oils and used pentane and benzene as eluents. This sorbent was also used in an investigation of the essential oil from *Elyonurus viridulus* grass (179). MODY et al. (180) used Florosil as a sorbent to fractionate the essential oil of pecan leaves and nuts. The volatiles were eluted with pentane containing gradually increased amounts of ethanol.
Not only physico-chemical separation procedures, but also chemical fractionation of odour concentrates can be used after derivatization of compounds with particular functional groups. The derivatives can either be analysed as such or be regenerated for gas chromatographic investigation. Of the many derivatization reactions (181) only a few will be mentioned here briefly. Carbonyls have often been isolated as dinitrophenylhydrazone derivatives and analysed as such by thin-layer or gas chromatography (cf. Chapter 3). Volatile sulphur compounds have been isolated as heavy metal salts and — after regeneration — have been evaluated sensorially before gas chromatographic analysis (182). This procedure was employed in a study of sulphur containing meat volatiles (183). Derivatization reactions along with several other methods, mentioned above, were applied by WOBBEN et al. (9) in a study of tomato volatiles.

2.1.6. Sensory evaluation of analytical procedures

Whether compounds of interest have indeed been isolated from the product and are present in the concentrates or fractions obtained can only be determined by sensory evaluation. STAHL (184) mentioned already in 1957 that analysis of samples not containing the important odour compounds has only academic value. Some relevant aspects of sensory analysis will be briefly discussed here. More attention to this topic is given in Chapter 5, in which methods for sensory evaluation of the analytical procedures are described in detail.
The loss of volatiles during the isolation and/or concentration procedures can be caused by:

- selectivity of the isolation method
- inadequate extraction efficiency
- volatilization during concentration
- chemical reactions

Despite the fact that most investigators are aware of these losses, surprisingly few reports mention that the various intermediate isolates and/or concentrates were sensorially evaluated. In one of these, TANG and JENNINGS (185), who studied the odour of apricots, reported that the concentrates obtained by molecular distillation had a typical apricot odour; this became apparent only after dilution, since the odour of the concentrate itself was rather overpowering.

While evaluating the odour of isolates, concentrates and fractions, it should be remembered that the concentration ratio of the volatiles in their vapour phase is not the same as in the vapour phase above the food. Furthermore, most extractants used for isolation have a distinct odour themselves, which interferes with the judgement. For this reason freons have sometimes been used as extractant, since they are almost odourless.
From the literature only a few appropriate reports are known dealing with the evaluation of odour distillates and concentrates (80, 186, 187, 188). Likely, elaborate sensory evaluation has not been carried out more often because of the considerable amount of material required. It is time consuming to prepare this, and therefore judgment is usually performed in an informal way by a limited number of persons.
Sensory analysis is also applied after 'gross' fractionation of the odour concentrate before the isolation of individual compounds, because the investigator wants to identify only those compounds that contribute to the odour (166, 169).
Sensory evaluation of an odour concentrate should be applied not only to determine whether the compounds of interest have been sufficiently recovered during isolation and concentration. Odour concentrates that have been stored for some time, should also be sensorially checked immediately before analysis, since they may have changed, as was mentioned in the previous section.
At the end of this chapter, we have to emphasize strongly the necessity of sensorial control of the analytical procedures, so as to ascertain that the fractions under investigation actually contain the relevant compounds. Without sensory evaluation, much of the effort involved in analysis may well prove to have been in vain.

2.2. Isolation and concentration from the vapour phase

J. SCHAEFER

2.2.1. Introduction

The human nose perceives the volatile compounds in the proportions which the air above a product contains. It has been found useful therefore to analyse this air — often referred to as "head space" — when studying the aroma of a product. The choice of the method to be applied depends on the concentration of the volatile compounds in the air as well as on the sentitivity of the available measurement techniques. For gas chromatographic analysis and identification by mass spectrometry quantities of 10^{-8}—10^{-9} g and 10^{-6}—10^{-7} g are required respectively. In table 2—11 the concentrations required for given volumes of air samples are listed.
The concentrations of the main components in food odours are generally in the order of 10^{-4}—10^{-6} g/l of air, while those of the trace components may amount to 10^{-8} g/l or below (155). Since the odour threshold values of many compounds are 10^{-8} g/l or below (189) it is important to analyse the trace components as well because they can contribute considerably to the aroma of a product. In gas chromatography, volumes in the order of 1—10 ml of air can be analyzed directly, which means that only the main compounds can be studied in this way (see Table 2—11). To obtain mass spectral data even the main components require larger samples of air, which is also the case

Table 2—11
Minimum concentration of a substance in a given volume of air required for gas chromatographic analysis or identification by mass spectrometry

	GC	MS
1 ml	10^{-5} — 10^{-6} g/l	10^{-3} — 10^{-4} g/l
10 ml	10^{-6} — 10^{-7} g/l	10^{-4} — 10^{-5} g/l
100 ml	10^{-7} — 10^{-8} g/l	10^{-5} — 10^{-6} g/l
1 l	10^{-8} — 10^{-9} g/l	10^{-6} — 10^{-7} g/l
10 l	10^{-9} — 10^{-10} g/l	10^{-7} — 10^{-8} g/l
100 l	10^{-10} — 10^{-11} g/l	10^{-8} — 10^{-9} g/l
1 m³	10^{-11} — 10^{-12} g/l	10^{-9} — 10^{-10} g/l

if only gas chromatographic analysis of the trace components is to be performed. If volumes larger than 10 ml are to be handled, a cumulation procedure must be inserted. In the course of this procedure five steps are observed which are schematically represented in Figure 2—18; this figure shows that not all steps need always be carried

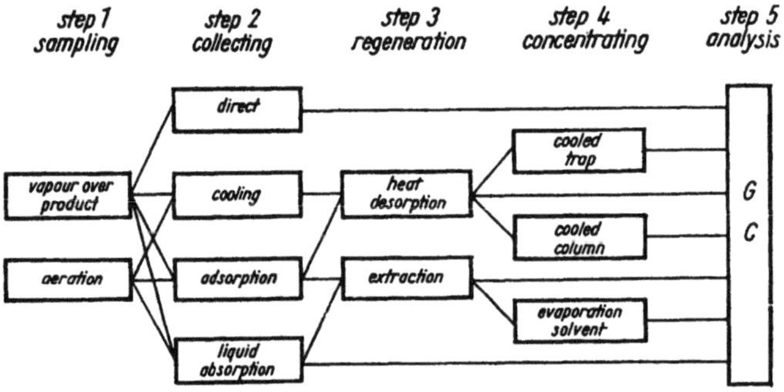

Fig. 2—18
Scheme of cumulation procedure of volatile components from vapour samples.

through, especially no. 4, the concentration step of the collected sample. In the following we will adhere to this scheme up to step 5, the analysis, in many cases being performed by gas chromatography. Chapter 3.2. is devoted to this subject. With each item we shall present a typical example along with some relevant literature references. Since most of the recent developments in cumulation techniques have been reported in the field of air pollution research, some special attention is paid to it. For an excellent review of the cumulation techniques developed up to 1972 the reader is referred to a publication by Weurman (155).
It is not our intention to give preference to any method in particular. The methods in question are equally applicable if some basic rules are observed. The choice of the method will actually depend on the available facilities and the specific requirements in connection with the problem to be solved.

2.2.2. Sampling

In principle vapour samples can be taken from a product in two ways: firstly, directly above the product; secondly, by aerating the product with an inert gas or with air, and collecting the issuing gas which contains the volatile compounds emanated by the product.

2.2.2.1. Sampling above the product

As has been stated in the introduction (2.2.1.), the air above the product (head space) contains the volatile compounds in the proportions as perceived by the nose. Therefore, especially in quantitative analysis samples are taken from the head space, either by the head space technique or by the sweeping technique. The *head space* technique implies that the vapour sample is not replaced (Fig. 2—19a).

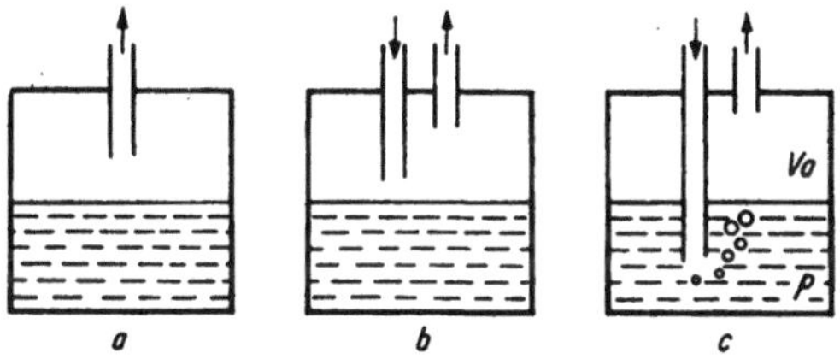

Fig. 2—19
Methods for the collecting of vapour samples. a. Head space sampling; b. Sweeping; c. Aeration.

This is usually performed by taking an air sample of 1—10 ml from a product contained in an infusion bottle with the aid of a syringe (see 3.2). The methods developed by BERTUCCIOLI and MONTEDORO (190) and VON SYDOW et al. (191) for the collection of larger head space samples are schematically represented in Figures 2-20a and 2-20b.

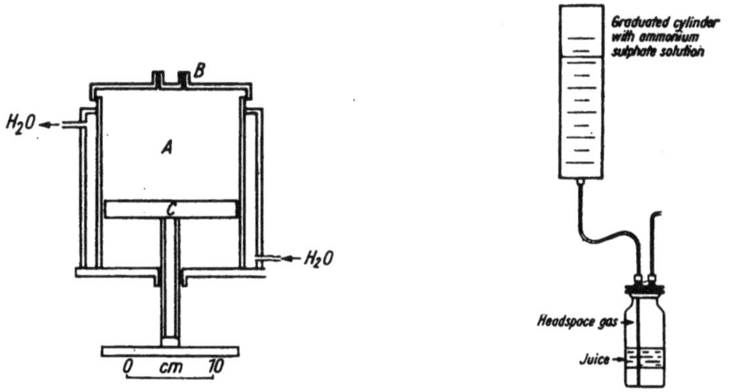

Fig. 2—20
Head space techniques
a. The product, usually solid is placed within a syringe *A*, the piston *C* is pushed upwards, causing the vapour in *A* to escape through *B*, where it can be collected (190)
b. The lipuid product is placed in a bottle, a saturated solution of ammonium sulphate is placed in a higher position, causing it to flow into the bottle with product. The vapour above the product is pushed out and can be collected (191).

Both are based on the principle that the head space is pushed out by a forced decrease of the volume above the product from which 100—400 ml head space is collected in an appropriate trapping system.

The *sweeping technique* (Fig. 2—19b) implies that a vapour sample is taken and replaced by fresh air or N_2. Above the product P, there is an enclosed volume of air V_a, in equilibrium with P. Nitrogen or air is passed for a time interval t at a rate of v ml/min and collected for analysis. Because part of the volatiles is thus carried off, their concentration in the head space changes. If one leaves repletion from P out of consideration and assumes that the incoming gas mixes instantaneously and completely with the head space, the concentration C_t in the outgoing gas at time t obeys the equation:

$$C_t = C_0 \cdot e^{-\frac{vt}{V_a}}$$

In this equation C_0 is the equilibrium concentration. Table 2—12 giving the values for C_t for some volumes $v \times t$ that have been passed, shows that for quantitative analysis

Table 2—12
Sweeping technique. Concentration in the vapour phase (C_t) as a function of total swept volume ($v \times t$)

$v \times t$ fraction of V_a	C_t fraction of C_0
0.1	0.90
0.2	0.82
0.3	0.74
0.5	0.61
1.0	0.37
2.0	0.14
3.0	0.05

a sample size of up to 0.1 V_a is acceptable. If identification is the main purpose, a larger sample, up to 1—2 V_a, may well be taken. Taking still larger samples is not justified in view of the extra time required to recover the comparatively small residues of volatiles remaining in the head space. Mostly, V_a is likely to amount to a few litres; in some cases it may be as much as a few cubic metres, e.g. when sampling in storage rooms for meat, fruit, etc. An example of this technique as applied at our institute is given in Figure 2—21 (192). In this case, the volume V_a of the head space is about one litre so that 100 ml can be collected for quantitative analysis and 1 l for identification.

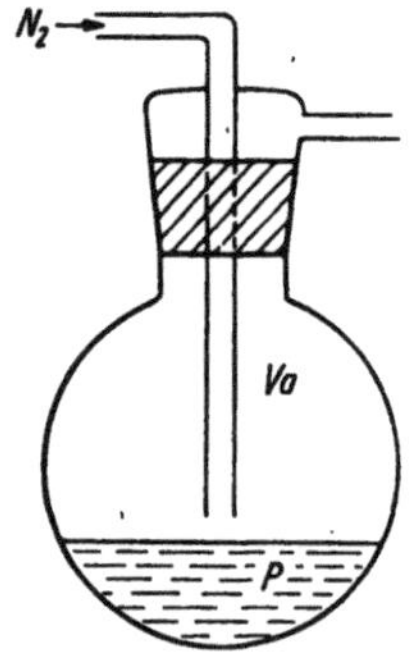

Fig. 2—21
Apparatus for the execution of the sweeping technique.
V_a = vapour volume above the product
P = liquid or solid product.

2.2.2.2. Aeration

Aeration with air or an inert gas (N_2, He) (Figure 2—19c) can be carried out as indicated in Figure 2—22. Note that V_a is comparatively much smaller than in the sweeping technique. This method can only be applied if the product P does not offer much resistance to the free circulation of the gas, which limits the method to liquids and very "porous" solids such as crumbed bread (193).

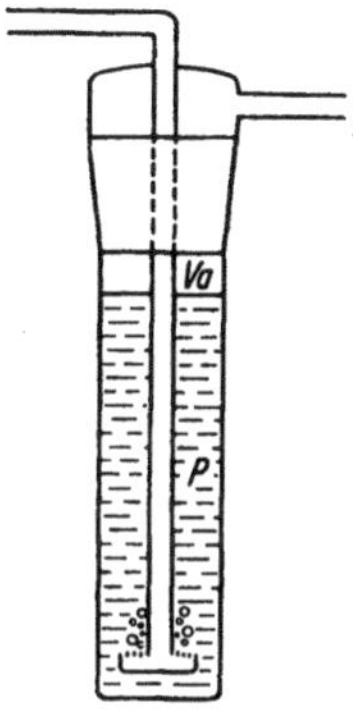

Fig. 2—22
Apparatus for the execution of aeration. V_a, the vapour volume above the product P is small in comparison with that of Fig. 2—21.

Aeration does not render the components in the ratios in which they occur in the air in equilibrium with the product. Virtually, the same takes place as in continuous extraction with a solvent; exhaustive aeration yields the components in the same ratios in which they occur in the product itself. However, in general the aeration will not be pushed so far and the result will lie between the two extremes. Therefore, the method is less suited for application in quantitative analysis, but can quite well be used for identification purposes. One litre of a product will yield 10—20 l of air sample without too strong a lowering in concentration of the volatile compounds. This is due to the high values of the equilibrium constant K (proportion between the concentration in the product and that in air) for the majority of volatile compounds of interest. When handling such quantities, the gases applied should not contain impurities that might interfere in any way wtih the analysis. If necessary, a method for overcoming this problem could be solved by recirculation of the aeration gas, a technique described by NAWAR and FAGERSON (194). A modification of the aeration technique is described by HONKANEN and KARVONEN (195), who added solid carbon dioxide to the product. The volatile components were swept out by the evaporating carbon dioxide and collected.

2.2.3. Collecting

The main volatile components in a vapour sample can be chromatographed directly (see 2.2.1.). A cumulation step has to be inserted for those compounds which are present in too low concentrations. The volatile components are "collected", either by "direct collection" or by "cumulative collection", In direct collection an air sample of 1—10 ml is taken from above the product, usually by means of a hypodermic syringe (for details, see 3.2.) or via sampling loops operated by sampling valves, and is chromatographed. Cumulative collection can be effected by cooling, adsorption and absorption. These three methods are described in detail in this chapter; the examples will be drawn mainly from application in air pollution research; but the methods are equally suitable to be used in aroma research.

2.2.3.1. Cooling

This procedure — often referred to as the cryogenic method — is based on the fact that vapour pressures decrease with temperature and are practically zero at a sufficiently low temperature. The principle is as follows: a vapour sample is passed through a trap of low temperature ("cold trap"), in which all volatile compounds condense, except nitrogen and oxygen and, depending on the temperature, carbon dioxide. The contents of the trap are subjected to regeneration and concentration (steps 3 and 4 of Fig. 2—18) and analysed by gas chromatography. Some refrigerants that can be used for this technique are given in Table 2—13.

Table 2-13
Some refrigerants with their temperatures

ice	0 °C
ice + salt	— 18 °C
solid carbon dioxide (dry ice)	— 78 °C
solid carbon dioxide + acetone	—110 °C
liquid air	—192 °C
liquid nitrogen	—196 °C

The various traps that are in use for the collection of samples in applying the cryogenic method can be divided in three main types:

I A capillary tube of glass or stainless steel, bent or coiled in a suitable shape. A convenient trap consists of a tube with a length of 25 cm and an inner diameter of 0.5 mm.

II A tube of glass or stainless steel (e.g. 20 cm × 3 mm i.d.), filled with glass beads or another inert material. The glass beads with a diameter of 0.1 mm serve as an extended condensation surface.

III A tube of the same material and size as mentioned under II, filled with an active adsorbent. This type is an intermediate between a cooling and an adsorption trap, the principal effect, however, being condensation. If cooling is sufficiently deep, the adsorbent has hardly any effect. Kaiser (196, 197) has tried to gain an extra benefit from the adsorbent by applying a temperature gradient across the trap to effect some preliminary separation, which must be maintained throughout the further analysis procedure. His device is now being marketed and applications are given in the literature (e.g. 198, 199).

Some investigators cool the first part of a gas chromatographic column and condense the volatile compounds in a vapour sample by directly injecting it on the column. The carrier gas is used as the sweeping gas for the sampling procedure. This method was reported by VAN LUNTEREN et al. (200) and is schematically presented in Figure 2—23.

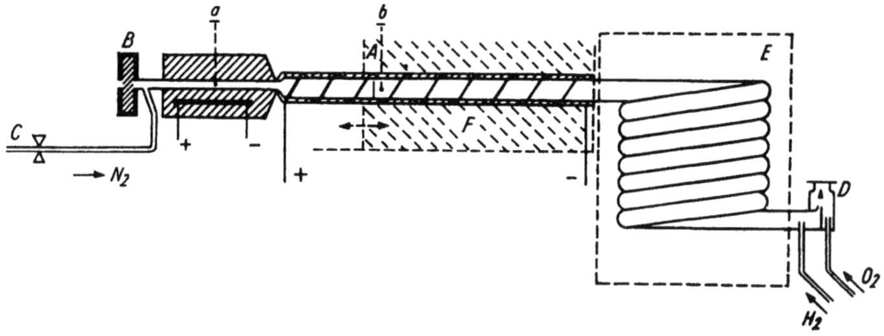

Fig. 2—23
Collection of vapour sample by condensation at the cooled top of a gaschromatographic column (200).
A. Top of column; *B*. injection point with septum; *C*. valve; *D*. detector; *E*. oven; *F*. cooling medium.

GROENEN and TEN NOEVER DE BRAUW (201) injected repeatedly 10—50 ml portions of air on a cooled capillary up to a total volume of 175 ml. In respect of the sorption techniques, the cryogenic method has the advantage of a representative sample being collected that gives a direct indication of the ratios in which the components occur. Also the quantitative composition of the air above the product can thus be determined. A disadvantage is that water, which is likely to be present in relatively high amounts is also collected. Air at 20 °C of 60% relative humidity actually contains approximately 7 mg water per l. The presence of water may cause obstruction of the trap by ice, as a result of which the volume of the sample that can be collected is limited to 2—3 l of air. Drying agents are sometimes employed to remove water from the air sample; in that case there is the risk of other compounds being adsorbed as well. However, it could be a suitable method if the components to be measured are known to pass the drying agent unaltered. FARRINGTON et al. (202) evaluated some drying agents for a number of compounds.

2.2.3.2. Adsorption

Use is made of the property that certain materials adsorb the majority of volatile compounds while nitrogen and oxygen and most of the water from the air are excluded. The adsorbents that are most frequently used can be subdivided in: activated carbon, GC column packings and synthetic polymers.
A large number of compounds, especially less polar ones are strongly adsorbed by *activated carbon*. Adsorption is determined by the adsorption isotherm, which at low concentrations can be represented by the equation:

$$m = k \cdot c_{\text{air}}^{1/n} ,$$

where m = the quantity of substance adsorbed per unit weight of carbon, c_{air} is the concentration in the air, k and n are constants for the substance in question. A typical adsorption isotherm is given in Fig. 2—24.

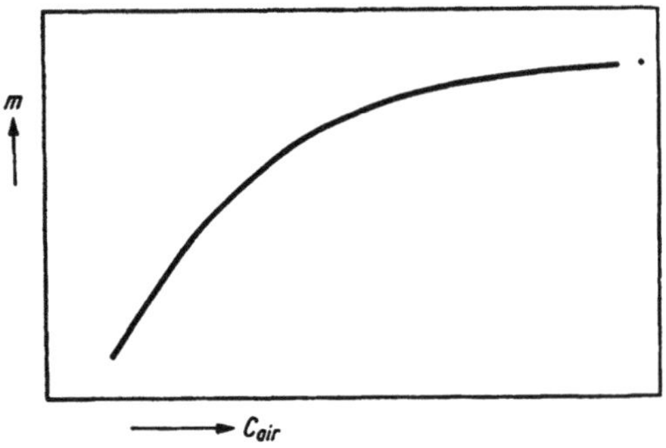

Fig. 2—24
Adsorption isotherm of carbon.
m = quantity of the substance adsorbed per unit weight of carbon
C_{air} = concentration of the substance in air.

It shows that above a certain concentration in the air, saturation occurs and that no more adsorption takes place. It is a rule of thumb that carbon can adsorb a quantity of material equal to its own weight. When it is nearly saturated, displacement phenomena may be observed, in that adsorbed substances are replaced by other ones having greater affinity to carbon. This implies that the volume of air that can be sampled quantitatively on a given active-carbon trap is determined by the total content of volatile compounds in the air. Sample sizes for traps filled with 1—10 mg of activated carbon range from 10 to 100 litres. In the past it was not uncommon to use large amounts of coal, which, however, caused great difficulties with regard to desorption. A classical example is provided by TURK and D'ANGIO (203), who used traps consisting of a steel canister, containing 325 g of activated carbon, to collect some tens of thousands of cubic metres of air. Recently, GROB et al. (204, 205, 206) developed coal traps consisting of a glass tube with an inner diameter of 3 mm, holding 1—2 mg activated carbon between two gold-coated metal filters sealed into the glass. They also were aware of the importance of using the right type of carbon. The properties of commercially available activated carbons are widely different, partly depending on the content of inorganic salts. GROB (204) finally selected one type, which had been developed for use in cigarette filters.

Traps filled with *G.C. column packings* can be considered the classical adsorption traps in aroma research. They consist of a tube with a length of 20 cm and an inner diameter of 3—4 mm, filled with a few hundred mg adsorbent (202, 204, 208, 209).

Recently there has been a trend towards the use of synthetic polymer, such as Tenax and Chromosorb 102 (210—216). This type of trap which was developed in air pollu-

Table 2—14
Some synthetic polymers used in adsorption traps

trade name	monomeric compound
Tenax GC	p-2,6-diphenylphenylene oxide
Chromosorb 101	styrene-divinylbenzene
Chromosorb 102	styrene-divinylbenzene
Chromosorb 105	styrene-divinylbenzene
Chromosorb 106	styrene-divinylbenzene
Porapak P	styrene-ethylvinylbenzene-divinylbenzene
Porapak Q	ethylvinylbenzene-divinylbenzene
Porapak R	N-vinylpyrrolidone
Porapak T	ethyleneglycoldimethylacrylate

tion and medical research, can easily be applied to aroma research, as demonstrated e.g. by LIN (217). Like those filled with G.C. column packings, it consists of a tube filled with a few hundred mg adsorbent. In table 2—14 a number of synthetic polymers is mentioned. For characterization of these polymers and their properties in so far as they are of importance in air sampling, the reader is referred to the literature (218 — 221).

Traps filled with G.C. column packings or synthetic polymers can actually be considered as gas chromatographic columns through which a continuous stream of component-laden air passes. Each component will break through at the moment corresponding to its retention time in normal gas chromatography. This is the frontal chromatography method, which was formerly employed to a certain extent as an analytical method. The detector records the break-through of each component as a corresponding rise of the base line, where V_r is the retention volume and h is a measure for the concentration (Fig. 2—25).

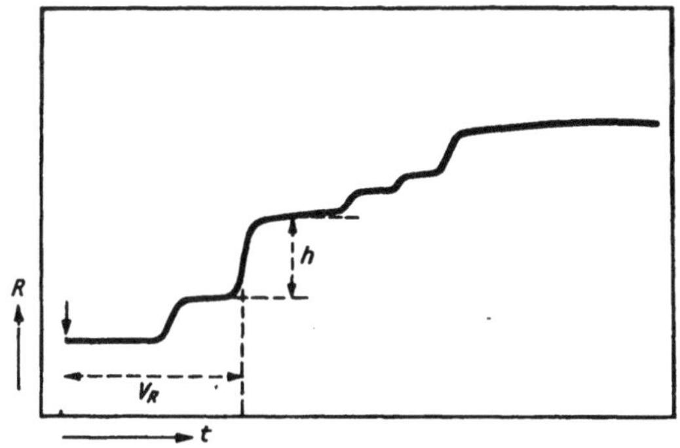

Fig. 2—25
Detector recording of break-through volumes of adsorption traps filled with supports with station ary phases or filled with synthetic polymers.

The break-through volume of a compound, in analogy with the retention time, is independent, within certain limits, of its concentration in the air. For a given component this break-through volume is the maximal sample size that can be trapped quantitatively. It may vary from 1 to 1000 l of air, and can be determined as follows: air, containing the compound in a quantity suitable for direct measurement, is allowed to flow continuously through the trap and a sample loop-and-valve behind it, connected to a gas chromatograph. At regular intervals, e.g. five minutes, the sample valve is switched, so that the contents of one loop are discharged to the gas chromatograph while the other receives a fresh sample from the trap. A peak will not appear in the chromatogram until its break-through volume is reached. Another method follows the principle of frontal chromatography by connecting the trap to the detector either directly or via the gas chromatographic column. Connecting two (or more) traps in series (210) will give an over-all impression of break-through volumes of all the components in a mixture. After having passed a known volume of air sample, e.g. 100 l, the two traps are analysed separately. The ratio of distribution of any compound over the two traps indicates the break-through volume. If for instance 70% of it is found in the first trap and 30% in the second, the break-through volume is 70 l (= 70% of 100 l). Comparison of the performance of the different adsorption traps shows that fairly non-selective adsorption of an air sample is obtained with activated carbon traps. However, they appear not to be suitable for quantitative analysis, owing to the difficulties involved in predicting displacement phenomena at different concentrations.

The traps filled with support covered with stationary phases or with polymer beads are quite suited for quantitative analysis if only the break-through volume has been determined. Components of high volatility, in analogy with their gas chromatographic properties, have low break-through volumes; therefore, the concentration factor is comparatively low.

2.2.3.3. Liquid absorption

The traps used are filled with a liquid; cumulation can be based either on solution in a solvent or on reaction with an aqueous solution of a reagent. Some suitable traps are shown in Figure 2—26 (222). Any solvent, including water, may serve as the absorbing

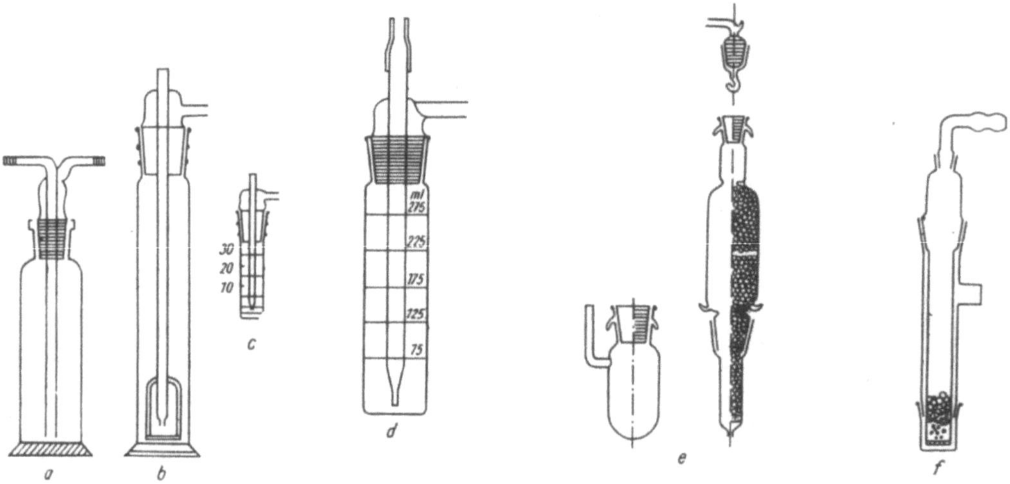

Fig. 2—26
Some liquid absorption traps (222).
a. ordinary washing bottle
b. Greenburg-Smith impinger
c. Midget impinger
d. Standard impinger
e. Pavelka trap
f. Ehmert washing bottle.

liquid. Examples of reactive solutions are (223): 1 N NaOH solution for cumulation of acidic components, 1 N HCl solution for basic components, and a 10% semicarbazide solution for carbonyls. This cumulation method is applicable to large volumes of air (100—1000 l) and is used chiefly in air pollution investigations. In aroma research it seems generally less appropriate, because air samples of the size needed to make this method effective are generally not available in aroma investigations. It might be a suitable method in sampling of store-rooms, the investigation of the release of components during ripening processes, and the study of the processing of food products (e.g. roasting of cocoa and coffee) (223).

2.2.4. Regeneration and concentration

Heat desorption or extraction is applied in regenerating compounds collected in the traps; depending on the concomitant dilution, a concentration step may be necessary.
Heat desorption has the advantage that all the collected components are completely regenerated, so that no changes in the proportions of the compounds occur. A drawback is the fact that at high desorption temperatures artifact formation may occur by decomposition of labile compounds or chemical reactions between compounds in the sample. Since extraction is carried out at ambient temperatures there is no danger of temperature-induced artifact formation. On evaporating the solvent, however, some of the more volatile compounds will also evaporate, causing subsequent losses of these compounds.

2.2.4.1. From cooled traps

Heat desorption under a flow of sweeping gas is the general method for liberating the volatiles collected in a cooled trap. The desorbed volatile compounds are usually analysed on a gas chromatograph. If the volume of sweeping gas is small, it may be applied directly onto the gas chromatograph. The maximum volume in this respect is in the order of 10 ml and 2 ml with packed and capillary columns respectively. The execution of heat desorption from a trap connected directly to a gas chromatogrph is shown in Figure 2—27 (155). The trap is fitted with an injection needle at the front and a septum

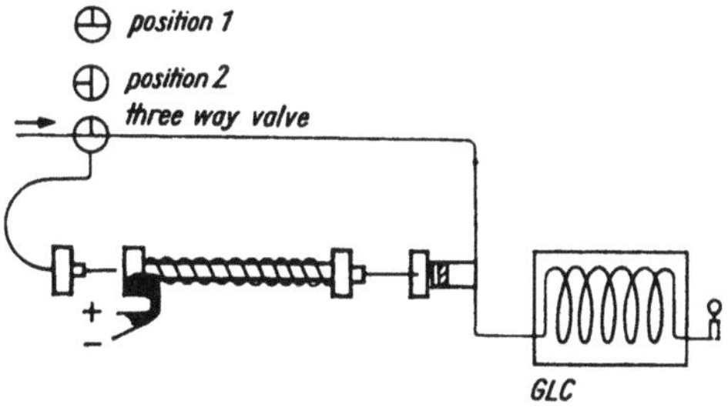

Fig. 2—27
Heat desorption directly to the gas chromatograph (155).

at the end. The carrier gas of the gas chromatograph can be allowed to flow either directly into the gas chromatograph or into the trap, by setting the three-way valve in position 1 or 2. The exit needle of the trap fits into the injection port of the gas chromatograph. The trap is heated with the aid of a heating wire or a heating cartridge to about 200 °C for one minute with the valve in position 1. The valve is then turned to position 2, thus causing the carrier gas to sweep the volatile compounds into the gas chromatograph. This method is feasible if the trap is a capillary tube or a tube packed with glass beads, and occasionally for tubes packed with an adsorbent if highly volatile compounds are concerned. Some applications are described in the literature (224—226). If the volume of the sweeping gas exceeds that which can be applied directly to the gas chromatography before the analysis a concentration step must be inserted. This is generally the case when tubes filled with an adsorbent are used, as

these tubes actually act as a short gas chromatographic column, the compounds especially the high-boiling ones, reaching the end of the tube only after purging with considerable volumes of sweeping gas. The concentration step to be applied in such cases will be described in the next section (2.2.4.2.).
To remove the excessive amounts of water that are collected when the cooling method is used, TYSON and CARLE (237) described an elegant though somewhat laborious method. Here the contents of the trap are volatilized and re-adsorbed on a short sorbitol column; from this column the compounds are eluted to the gas chromatographic column, during an interval in which the water is eluting the effluent of the sorbitol column is vented via a valve system. The system is schematically drawn in Figure 2—28.

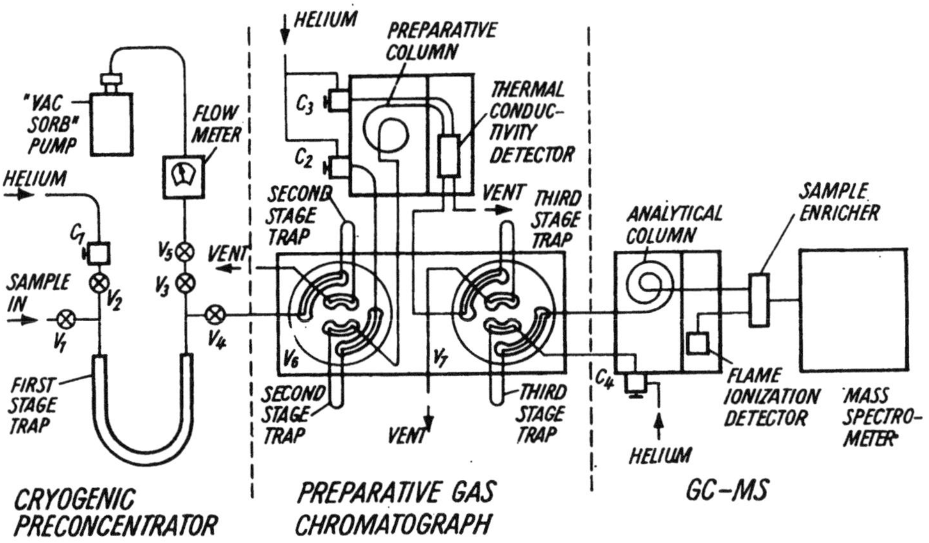

Fig. 2—28
System for removing water from a collected vapour sample. The sample is collected in the first stage trap. After thermal desorption it is condensed in the second stage trap via valve V_4. Again after thermal desorption it is led through a preparative column filled with sorbitol as a stationary phase. The eluted components are collected in the third stage trap with an interval for the time that water is eluted, which is vented. The third stage trap is thermally desorbed to the analytical column (237).

2.2.4.2. From adsorption traps

Regeneration can be achieved either by heat desorption or by extraction with a solvent. When using traps filled with activated carbon, the regeneration step is somewhat different from that in the method with other adsorbents. This is because nowadays 1—10 mg of activated carbon is used in contrast with about 200 mg of adsorbent of other types. For heat desorption the method described in Figure 2—27 can be used. Another method, only feasible for activated-carbon adsorbent, is described by CLARK and CRONIN (228), who placed the contents of the trap into a solid-sample injector of a gas chromatograph. The injector was placed into the injection port and heated to

260 °C, following which the desorbed components were swept off with the carrier gas. After applying solvent extraction the solvent has normally to be evaporated in order to obtain a reasonably concentrated sample for analysis. Using the activated carbon traps described in 2.2.3.2., containing 1–2 mg adsorbent, GROB and ZÜRCHER (206) attained full regeneration by passing three 10 μl portions of carbon disulphide through the trap, thus avoiding the necessity of a concentration step.
If a tube filled with G.C. column packing or with a synthetic polymer is used as an adsorption trap, heat desorption can be carried out as shown in Figure 2-27 for low-boiling compounds, as is demonstrated by LIEBMAN et al. (25) for such compounds as HCN and CO. If components of less volatility are present in the trap, a larger volume of sweeping gas is needed, which would cause severe peak broadening in the chromatograms. Therefore, a concentration step is necessary, which generally consists in collecting the compounds from the sweeping gas on a cooled capillary (see 2.2.3.), followed by heat desorption into the gas chromatograph, as shown in Figure 2–27. This is schematically presented in Figure 2–29 (215).

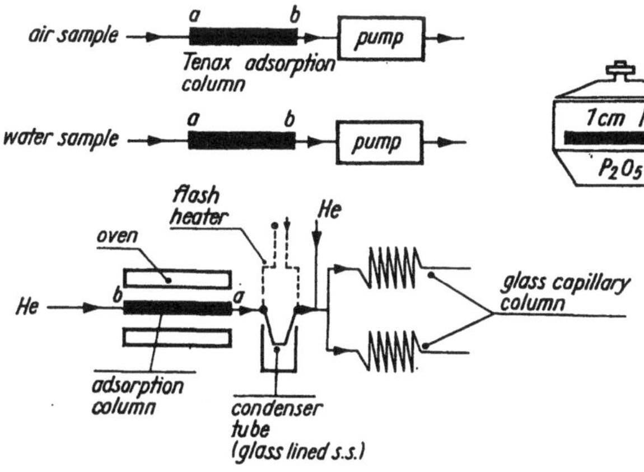

Fig. 2–29
Collection and thermal desorption with solid adsorption traps. After the thermal desorption the components are condensed again on a cooled capillary, and are thermally desorbed to the gas chromatograph (215).

Solvent extraction is effected by simply passing 5–10 ml of a solvent through the adsorbent in the tube and collecting it. This extract can be further concentrated by careful evaporation of the solvent (see 2.1). Details on application and on technical description of the apparatus used can be found in the literature (207, 215, 229).

2.2.4.3. From liquid absorption traps

Regeneration from liquid absorption traps is executed by solvent extraction, followed by careful evaporation of the solvent. If water or ethylene glycol are used as the absorption liquid the extraction solvent could be pentane/ether (1 + 1) or pentane. When using a reagent, the chemically bound components must be regenerated from their derivatives and subsequently extracted with a suitable solvent. For instance, if a 1 N NaOH solution is used as reagent it must be acidified with an HCl solution, after which the liberated acids can be extracted with ether.

References

1. WEURMAN, C., Isolation and concentration of volatiles in food odor research, J. Agric. Food Chem. **17**, 370—384 (1969)
2. FORSS, D. A., Isolation and fractionation of low boiling flavor compounds, Food Prod. Dev. **3**, 77—80, 92, 98 (1969)
3. TERANISHI, R., I. HORNSTEIN, PH. ISSENBERG and E. L. WICK, Flavor Research, Principles and Techniques, Marcel Dekker, Inc., New York 1971, pp. 315
4. WEURMAN, C., Industrial interest in food odour research, Brygmesteren **21**, 179—195 (1964)
5. WEURMAN, C. and G. VAN LUNTEREN, Application of gas chromatography to the study of food odours, Aroma Geschmackstoffe Lebensm., Fortbildungskurs 21—34 (1967)
6. BADINGS, H. T., Cold-storage defects in butter and their relation to the autoxidation of unsaturated fatty acids, Thesis, Agricultural University of Wageningen 1970
7. SOLMS, J., Natürliche, natur-identische und synthetische Geschmackstoffe und deren lebensmittelrechtliche Aspekte, in "Probleme Ernaehr. Lebensm. Wiss." 1 eds. AUERSWALD, W. und B. M. BRANDSTETTER, Wilhelm Maudrich Verlag, Wien 1974, p. 5—10
8. CHANG, S. S., F. M. VALLESE, L. S. HWANG, O. A. L. HSIEH and D. B. S. MIN, Apparatus for the isolation of trace volatile constituents from foods, J. Agric. Food Chem. **25**, 450—455 (1977)
9. WOBBEN, H. J., P. J. DE VALOIS, R. TER HEIDE and R. TIMMER, Investigation into the composition of a tomato flavour, Proc., SOS/74, Int. Congr. Food Sci. Technol., 4th, Vol 1, 22—28 (1974)
10. DRAWERT, F., W. HEIMANN, R. EMBERGER und R. TRESSL, Enzymatische Veränderung des natürlichen Apfelaromas bei der Aufarbeitung, Naturwissenschaften **52**, 304—305 (1965)
11. DRAWERT, F., W. HEIMANN, R. EMBERGER und R. TRESSL, Enzymatische Bildung von Hexen-(2)-al-(1), Hexanal und deren Vorstufen, Justus Liebigs Ann. Chem. **694**, 200—208 (1966)
12. FLEMING, H. P., W. Y. COBB, J. L. ETCHELLS and T. A. BELL, The formation of carbonyl compounds in cucumbers, J. Food Sci. **33**, 572—576 (1968)
13. KAZENIAC, S. J. and R. M. HALL, Flavor chemistry of tomato volatiles, J. Food Sci. **35**, 519—530 (1970)
14. BUTTERY, R. G., R. M. SEIFERT, D. G. GUADAGNI and L. C. LING, Characterization of additional volatile components of tomato, J. Agric. Food Chem. **19**, 524—529 (1971)
15. STONE, E. J., R. M. HALL and S. J. KAZENIAC, Formation of aldehydes and alcohols in tomato fruit from U-14 C-labeled linolenic and linoleic acids, J. Food Sci. **40**, 1138—1141 (1975)
16. MCGLASSON, W. B., Ethylene production by slices of green banana fruit and potato tuber tissue during the development of induced respiration, Aust. J. Biol. Sci. **22**, 489—491 (1969)
17. DRAWERT, F., W. HEIMANN, R. EMBERGER und R. TRESSL, Gas-Chromatographische Untersuchung pflanzlicher Aromen. II. Anreicherung, Trennung und Identifizierung von Apfelaromastoffen, Chromatographia **2**, 57—66 (1969)
18. TRESSL, R., F. DRAWERT und W. HEIMANN, Über die Biogenese von Aromastoffen bei Pflanzen und Früchten. V. Mitteilung: Anreicherung, Trennung und Identifizierung von Bananenaromastoffen, Z. Lebensm. Unters. Forsch. **142**, 249—263 (1970)
19. SCHREIER, P., F. DRAWERT and A. JUNKER, Identification of volatile constituents from grapes, J. Agric. Food Chem. **24**, 331—336 (1976)
20. RIBÉRAU-GAYON, P., J. N. BOIDRON and A. TERRIER, Aroma of muscat grape varieties, J. Agric. Food Chem. **23**, 1042—1047 (1975)

21. JUNK, G. A., J. J. RICHARD, M. D. GRIESER, D. WITIAK, J. L. WITIAK, M. D. ARGUELLO, R. VICK, H. J. SVEC, J. S. FRITZ and G. V. CALDER, Use of macroreticular resins in the analysis of water for trace organic contaminants, J. Chromatogr. **99**, 745—762 (1974)
22. CHANG, S. S., Overcoming problems in flavor component identification, Food Technol. (Chicago) **27**, 27—39 (1973)
23. MAIER, H. G., Bindung flüchtiger Aromatoffe an Lebensmittel, Lebensm. Wiss. Technol. **5**, 1—6 (1972)
24. SOLMS, J., Aromastoffe als Liganden, in "Geruch- und Geschmackstoffe", ed. DRAWERT, F., Verlag Hans Carl, Nürnberg 1975, p. 201—210
25. ARNOLD, R. G., L. M. LIBBEY and E. A. DAY, Identification of components in the stale flavor fraction of sterilized concentrated milk, J. Food Sci. **31**, 566—573 (1966)
26. LIKENS, S. T. and G. B. NICKERSON, Detection of certain hop oil constituents in brewing products, Am. Soc. Brew. Chem. Proc. 5—13 (1964)
27. SCHMIDT, H., Eine spezifische colorimetrische Methode zur Bestimmung der Sorbinsäure, Z. Anal. Chem. **178**, 173—184 (1960)
28. MAARSE, H., J. H. DHONT en M. A. H. RIJK, Regulation on Essence. Methods of analysis (Dutch Community Act) (in Dutch), Report R 4441A, Central Institute for Nutrition and Food Research TNO, Zeist, The Netherlands (1974), pp. 23
29. FALK, A. J., L. BAUER, C. L. BELL and S. J. SMOLENSKI, The constituents of the essential oil from *Achillea millefolium* L., Lloydia **37**, 598—602 (1974)
30. MIQUEL, J. D., H. M. J. RICHARD and F. G. SANDRET, Volatile constituents of Moroccan thyme oil, J. Agric. Food Chem. **24**, 833—835 (1976)
31. KÄMPF, R., Vergleich der Ph. Helv. VI- und der Ph. Eur.-Apparatur zur Bestimmung des ätherischen Oeles in Drogen mittels der Methode der Wasserdampfdestillation, Mitt. Geb. Lebensmittelunters. Hyg. **67**, 192—199 (1976)
32. —, Dosage des huiles essentielles dans les drogues végétales, in "Pharmacopée Européenne", Vol. III, Maisonneuve S.A., 1975, p. 69—72
33. STAHL, E., The influence of the duration of the distillation and of the ion concentration on the yield of azulene and volatile oil from tansy (*Achillea millefolium*), Pharm. Ind. **14**, 262—265 (1952)
34. —, Volatile oil content (in Dutch), in "Nederlandse Farmacopee", 6th ed., Staatsuitgeverij, 's-Gravenhage 1966, p. 72—73
35. TERANISHI, R., E. L. MURPHY and T. R. MON, Steam distillation-solvent extraction recovery of volatiles from fats and oils, J. Agric. Food Chem. **25**, 464—466 (1977)
36. PICKETT, J. A., J. COATES and F. R. SHARPE, Procedure for non-destructive concentration of flavour-active components of beer, J. Inst. Brew. (London) **82**, 228—233 (1976)
37. FORSS, D. A. and G. L. HOLLOWAY, Recovery of volatile compounds from butter oil, J. Am. Oil Chem. Soc. **44**, 572—575 (1967)
38. PERSSON, T. and E. VON SYDOW, The aroma of canned beef: processing and formulation aspects, J. Food Sci. **39**, 406—413 (1974)
39. RÖMER, G. und E. RENNER, Einfache Methoden zur Isolierung und Anreicherung von Aromastoffen aus Lebensmitteln, Z. Lebensm. Unters. Forsch. **156**, 329—335 (1974)
40. GROENEN, P. J., R. J. G. JONK, C. VAN INGEN and M. C. TEN NOEVER DE BRAUW, Determination of eight volatile nitrosamines in thirty cured meat products with capillary gas chromatography-high-resolution mass spectrometry: the presence of nitrosodiethylamine and the absence of nitrosopyrrolidine, IARC Sci. Publ. **14**, 321—331 (1976)
41. WEURMAN, C., An all-glass laboratory apparatus for concentrating volatile compounds from dilute aqueous solutions, J. Food Sci. **26**, 670—672 (1961)
42. BOMBEN, J. L., D. G. GUADAGNI and J. G. HARRIS, Aroma concentration for dehydrated foods, Food Technol. (Chicago) **23**, 83—86 (1969)
43. NAIPAWAR, R. E., R. POTTER, P. VALLON and R. E. ERICKSON, The isolation of flavour components from foods by distillation and adsorption, Flavour Ind. **2**, 465—467 (1971)
44. BADINGS, H. T., personal communication (1971)
45. COBB, W. Y., Quantitation of flavourful food components using isotope dilution, J. Food Sci. **34**, 466—469 (1969)
46. HAYMON, L. W., Characterization of some flavor principles of tabasco peppers, Thesis, North Carolina State University, Raleigh 1970

47. Peterson, R. J., H. J. Izzo, E. Jungermann and S. S. Chang, Changes in volatile flavor compounds during the retorting of canned beef stew, J. Food Sci. **40**, 948—954 (1975)
48. Herz, K. O. and S. S. Chang, An apparatus for the isolation of volatile compounds from foods, J. Food Sci. **31**, 937—940 (1966)
49. Kallio, H., Identification of vacuum steam-distilled aroma compounds in the press juice of arctic bramble, *Rubus arcticus* L., J. Food Sci. **41**, 555—562 (1976)
50. Pickett, J. A., J. Coates and F. R. Sharpe, Distortion of essential oil composition during isolation by steam distillation, Chem. Ind. (London) 571—572 (1975)
51. Azzouz, M. A., G. A. Reineccius and M. G. Moshonas, Comparison between coldpressed and distilled lime oils through the application of gas chromatography and mass spectrometry, J. Food Sci. **41**, 324—328 (1976)
52. Perry, E. S., Technique of organic chemistry, Vol. **4**, Distillation, eds. Perry, E. S. and A. Weissberger, Intersci. Publ., New York 1965, p. 535—598
53. Merritt, Jr., C., S. R. Bresnick, M. L. Bazinet, J. T. Walsh and P. Angelini, Determination of volatile components of foodstuffs. Techniques and their application to studies of irradiated beef, J. Agric. Food. Chem. **7**, 784—787 (1959)
54. Weurman, C., P. J. Groenen and L. J. van Gemert, Experiments on "high-vacuum transfer" in food odor research, Nahrung **14**, 607—616 (1970)
55. Keay, J. N. and A. S. McGill, A method for collection and concentration of fish volatiles, J. Food Technol. **3**, 345—351 (1968)
56. Tyler, J. E. and G. H. Dibdin, Gas chromatography of volatile fatty acids. Method involving separation from biological material by vacuum distillation, J. Chromatogr. **105**, 71—77 (1975)
57. Mulders, E. J., M. C. ten Noever de Brauw and S. van Straten, The odour of white bread. II. Identification of components in pentane-ether extracts, Z. Lebensm. Unters. Forsch. **150**, 305—310 (1973)
58. Ito, H., Isolation and identification of the characteristic sweet-aroma compounds in refinery final molasses, Agric. Biol. Chem. **40**, 827—832 (1976)
59. Wick, E. L., A. I. McCarthey, M. Myers, E. Murray, H. Nursten and Ph. Issenberg, Flavor and biochemistry of volatile banana components, Adv. Chem. Ser. **56**, 241—259 (1966)
60. Badings, H. T., Causes of ribes flavor in cheese, J. Dairy Sci. **50**, 1347—1351 (1967)
61. Brandt, P. E. and B. H. Jensen, Quantitative recovery of volatiles from fats and oils by combined high vacuum degassing and thin film molecular distillation, J. Am. Oil Chem. Soc. **52**, 278—281 (1975)
62. Bills, D. D. and J. L. Sloan, Removal of chlorinated insecticide residues from milk fat by molecular distillation, J. Agric. Food Chem. **15**, 676—678 (1967)
63. Patterson, R. L. S. Boar taint: its chemical nature and estimation, Proc. Symp. Meat Res. Inst. (1969) p. 247—260
64. Johnson, B. R., G. R. Waller and A. L. Burlingame, Volatile components of roasted peanuts: basic fraction, J. Agric. Food Chem. **19**, 1020—1024 (1971)
65. Shu, C. K. and G. R. Waller, Volatile components of roasted peanuts: comparative analyses of the basic fraction, J. Food Sci. **36**, 579—583 (1971)
66. Bruyn, J. de and J. C. M. Schogt, Isolation of volatile constituents from fats and oils by vacuum degassing, J. Am. Oil Chem. Soc. **38**, 40—44 (1961)
67. Lea, C. H. and P. A. T. Swoboda, Simple vacuum distillation procedure for determination of the volatile carbonyl content of autoxidising edible fats, J. Sci. Food Agric. **13**, 148—158 (1962)
68. Nawar, W. W., J. R. Champagne, M. F. Dubravcic and P. R. LeTellier, Recovery and measurement of volatiles from lipids. Hydrocarbons in irradiated fats, J. Agric. Food Chem. **17**, 645—648 (1969)
69. Ferretti, A. and V. P. Flanagan, Characterization of volatile constituents of an Nα-formyl-L-lysine-D-lactose browning system, J. Agric. Food Chem. **21**, 35—37 (1973).
70. Jeon, I. J., G. A. Reineccius, and E. L. Thomas, Artifacts in flavor isolates produced by steam vacuum distillation and solvent extraction of distillate, J. Agric. Food Chem. **24**, 533—534 (1976)

71. Johnson, J. H., W. Gould, A.A. F. Badenhop, and R. M. Johnson, Jr., Quantitative comparison of isoamylol, pentanol, and 3-hexenol-1 in tomato juice, J. Agric. Food Chem. **16**, 255—258 (1968)
72. Edelényi, M., A. Tóth und M. T. Szabo, Gaschromatographische Untersuchung der Geruchs- und Geschmacksstoffe von Sekt, Nahrung **20**, 347—352 (1976)
73. Suomalainen, H. and L. Nykänen, The formation of aroma compounds by yeast in sugar fermentation, Suom. Kemistil. B **37**, 230—232 (1964)
74. Suomalainen, H. and L. Nykänen, The aroma compounds produced by sherry yeast in grape and berry wines, Suom. Kemistil. B **39**, 252—256 (1966)
75. Drawert, F., P. Schreier und W. Scherer, Gaschromatographisch-massenspektrometrische Untersuchung flüchtiger Inhaltstoffe des Weines. III. Säuren des Weinaromas, Z. Lebensm. Unters. Forsch. **155**, 342—347 (1974)
76. Schreier, P., F. Drawert und A. Junker, Identifizierung von 3,7-Dimethyl-1, 5, 7-octatrien-3-ol als flüchtige Komponente des Trauben- und Weinaromas, Z. Lebensm. Unters. Forsch. **155**, 98—99 (1974)
77. Hardy, P. J., Extraction and concentration of volatiles from dilute aqueous and aqueous-alcoholic solution using trichlorofluoromethane, J. Agric. Food Chem. **17**, 656—658 (1969)
78. Williams, A. A. and O. G. Tucknott, The selective extraction of aroma components from alcoholic distillates, J. Sci. Food Agric. **24**, 863—871 (1973)
79. Guymon, J. F. and E. A. Crowell, Gas chromatographic determination of ethyl esters of fatty acids in brandy or wine distillates, Am. J. Enol. Vitic. **20**, 76—85 (1969)
80. Schultz, T. H., R. A. Flath, D. R. Black, D. G. Guadagni, W. G. Schultz and R. Teranishi, Volatiles from Delicious apple essence — extraction methods, J. Food. Sci. **32**, 279—283 (1967)
81. Zenz, H., H. Klaushofer und H. Szep, Erfahrungen mit der CS_2-Extraktionsmethode bei der gaschromatographischen Bestimmung einiger Alkohole, Ester und Säuren des Bieres, Mitt. Versuchsstn. Gärungsgewerbe Wien **27**, 100—108 (1973)
82. Gruss, R., K. Kleber und Ch. Zürcher, Gaschromatographische Bestimmung höherer Alkohole und Ester aus mit Schwefelkohlenstoff extrahierten Bierdestillaten, Monatsschr. Brau. **28**, 293—299 (1975)
83. Wong, N. P. and O. W. Parks, Simple technique for extracting flavor compounds from fatty foods, J. Dairy Sci. **51**, 1768—1769 (1968)
84. Jolly, R. C. and F. V. Kosikowski, Quantification of lactones in ripening pasteurized milk blue cheese containing added microbial lipases, J. Agric. Food Chem. **23**, 1175—1176 (1975)
85. Schultz, W. G. and J. M. Randall, Liquid carbon dioxide for selective aroma extraction, Food Technol. (Chicago) **24**, 94—98 (1970)
86. Vitzthum, O. G., P. Werkhoff and P. Hubert, Volatile components of roasted cocoa: basic fraction, J. Food Sci. **40**, 911—916 (1975)
87. Blakesley, C. N. and J. Loots, A convenient method for multiple extraction of volatile flavor components from food slurries and pulps using a two-chambered glass bomb extractor and dichlorodifluoromethane (Freon 12) solvent, J. Agric. Food Chem. **25**, 961—963 (1977)
88. de Mets, M. and M. Verzele, The aroma of hops. I. Origin and identification of hop aroma substances, J. Inst. Brew. (London) **74**, 74—81 (1968)
89. Nelson, P. E. and J. E. Hoff, Paraffin oil for extraction and collection of volatile microconstituents in food, Food Technol. (Chicago) **22**, 61—64 (1968)
90. Nelson, P. E. and J. E. Hoff, Food volatiles: gas chromatographic determination of partition coefficients in water-liquid systems, J. Food. Sci. **33**, 479—482 (1968)
91. Martin, D., Extrahieren von festen und flüssigen Stoffen im Laboratorium, GIT Fachz. Lab. **14**, 1404—1410 (1970)
92. Martin, D., Extrahieren von festen und flüssigen Stoffen im Laboratorium, GIT Fachz. Lab. **15**, 33—36 (1971)
93. Stage, H. und L. Gemmeker, Verfahren und Apparaturen zur flüssig-flüssig-Extraktion im Laboratorium, Chem. Ztg., Chem. Appar. **88**, 517—532 (1964)
94. Ahnoff, M. and B. Josefsson, Simple apparatus for on-site continuous liquid-liquid extraction of organic compounds from natural waters, Anal. Chem. **46**, 658—663 (1974)
95. Schwarze, J. W., D. A. M. Geddes, and M. N. Gilmour, Quantitative ether extraction of acidic fermentation products, Biotechnol. Bioeng. **13**, 151—155 (1971)

96. KENYHERCZ, T. M. and P. T. KISSINGER, A new approach to the phenolic components in beer. Application to the determination of sinapic, ferulic, and p-coumaric acids, J. Agric. Food Chem. **25**, 959—961 (1977)
97. STAFFORD, M., M. G. HORNING and A. ZLATKIS, Profiles of volatile metabolites in body fluids, J. Chromatogr. **126**, 495—502 (1976)
98. SCHULTZ, W. G., T. H. SCHULTZ, R. A. CARLSON and J. S. HUDSON, Pilotplant extraction with liquid CO_2, Food Technol. (Chicago) **28**, 32—36 + 88 (1974)
99. KARWOWSKA, K., Eine neue Methode zur Extraktion von natürlichen Aromen, Schweiz. Z. Obst Weinbau **108**, 625—627 (1972)
100. LIBBEY, L. M. and J. P. WALRADT, 3,5-di-tert-butyl-4-hydroxytoluene (BHT) as an artifact from diethyl ether, Lipids **3**, 561 (1968)
101. NICKERSON, G. B. and S. T. LIKENS, Gas chromatographic evidence for the occurrence of hop oil components in beer, J. Chromatogr. **21**, 1—5 (1966)
102. BUTTERY, R. G., R. M. SEIFERT, D. G. GUADAGNI, D. R. BLACK and L. C. LING, Characterization of some volatile constituents of carrots, J. Agric. Food Chem. **16**, 1009—1015 (1968)
103. BUTTERY, R. G., R. M. SEIFERT, D. G. GUADAGNI and L. C. LING, Characterization of some volatile constituents of bell peppers, J. Agric. Food Chem. **17**, 1322—1327 (1969)
104. BUTTERY, R. G., R. M. SEIFERT and L. C. LING, Volatile tomato components: characterisation of 6,10,14-trimethylpentadec-5,9,13-trien-2-one, Chem. Ind. (London) 238 (1969)
105. FLATH, R. A., R. R. FORREY and D. G. GUADAGNI, Aroma components of olive oil, J. Agric. Food Chem. **21**, 948—952 (1973)
106. BUTTERY, R. G., R. M. SEIFERT, D. G. GUADAGNI and L. C. LING, Characterization of volatile pyrazine and pyridine components of potato chips, J. Agric. Food Chem. **19**, 969—971 (1971)
107. MILUTINOVIĆ, L., R. E. BARGMANN, K. Y. CHANG, M. CHASTAIN and J. J. POWERS, Comparison of flavor and volatiles of tomato products and of peanuts, J. Food Sci. **35**, 224—228 (1970)
108. CRONIN, D. A. and M. K. WARD, The characterisation of some mushroom volatiles, J. Sci. Food Agric. **22**, 477 —479 (1971)
109. HORVAT, R. J., Identification of some volatile compounds in cooked chicken, J. Agric. Food Chem. **24**, 953—958 (1976)
110. CRAWFORD, L. and M. J. KRETSCH, GC-MS identification of the volatile compounds extracted from roasted turkeys fed a basal diet supplemented with tuna oil: some comments on fishy flavor, J. Food Sci. **41**, 1470—1478 (1976)
111. CLARK, R. G. and H. E. NURSTEN, Volatile flavour components of brazil nuts *Bertholletia excelsa* (Humb. and Bonpl.), J. Sci. Food Agric. **27**, 713—720 (1976).
112. MAARSE, H. and R. E. KEPNER, Changes in composition of volatile terpenes in Douglas fir needles during maturation, J. Agric. Food Chem. **18**, 1095—1101 (1970)
113. SCHAEFER, J. and R. TIMMER, Flavor components in cognac, J. Food Sci. **35**, 10—12 (1970)
114. MAARSE, H., Samenstelling van de vluchtige olie van *Origanum vulgare* L. ssp. *vulgare* gedurende de ontwikkeling van de plant, Thesis, University of Groningen 1971
115. BEMELMANS, J. M. H. and M. C. TEN NOEVER DE BRAUW, Chloroanisoles as off-flavor components in eggs and broilers, J. Agric. Food Chem. **22**, 1137—1138 (1974)
116. MACLEOD, A. J. and S. J. CAVE, Volatile flavour components of eggs, J. Sci. Food Agric. **26**, 351—360 (1975)
117. MACLEOD, G. and B. M. COPPOCK, Volatile flavor components of beef boiled conventionally and by microwave radiation, J. Agric. Food Chem. **24**, 835—843 (1976)
118. MACLEOD, G. and B. M. COPPOCK, A comparison of the chemical composition of boiled and roasted aromas of heated beef, J. Agric. Food Chem. **25**, 113—117 (1977)
119. FARMER, J. W., A. HUME and J. R. BURT, Efficiency of Likens and Nickerson distillation method for carbon compounds up to C_{11}, Chem. Ind. (London) 279—280 (1973)
120. MCGILL, A. S. and R. HARDY, Artefact production in the Likens-Nickerson apparatus when used to extract the volatile flavorous components of cod, J. Sci. Food Agric. **28**, 89—92 (1977)
121. BUTTERY, R. G., D. G. GUADAGNI, L. C. LING, R. M. SEIFERT and W. LIPTON, Additional volatile components of cabbage, broccoli, and cauliflower, J. Agric. Food Chem. **24**, 829—832 (1976)
122. FLATH, R. A. and R. R. FORREY, Volatile components of papaya (*Carica papaya* L., Solo variety), J. Agric. Food Chem. **25**, 103—108 (1977)

123. WILLIAMS, A. A., An apparatus for continuous steam distillation of materials completely miscible with water, Chem. Ind. (London) 1510—1511 (1969)
124. FORSS, D. A., V. M. JACOBSON and E. H. RAMSHAW, Concentration of volatile compounds from dilute aqueous solutions, J. Agric. Food Chem. **15**, 1104—1107 (1967)
125. SHIPTON, J. and F. B. WHITFIELD, A technique for the recovery of higher-boiling flavour volatiles from aqueous ethanolic solutions, Chem. Ind. (London) 2124—2125 (1966)
126. POWERS, J. E., Recent advances in crystallization. New developments in crystallization indicate it can compete with the more conventional separation processes, Hydrocarbon Process. Pet. **45**, 97—102 (1966)
127. SWOBODA, P. A. T., Freeze concentration of dilute aqueous solutions of volatile flavour compounds, in "Int. Symp. Gas Chrom., 6th.", ed. Littlewood, A.B. Institute of Petroleum (London) 1966, paper 22, p. 1—8
128. GOUW, T. H., Normal freezing, in "Progress in separation and purification". Vol. 1, ed. PERRY, E. S. Interscience Publishers, New York 1968, p. 57—82
129. BAKER, R. A., Trace organic contaminant concentration by freezing — I. Low inorganic aqueous solutions, Water Res. **1**, 61—77 (1967)
130. BAKER, R. A., Trace organic contaminant concentration by freezing — II. Inorganic aqueous solutions, Water Res. **1**, 97—113 (1967)
131. SHAPIRO, J., Freezing-out, a safe technique for concentration of dilute solutions, Science **133**, 2063—2064 (1961)
132. KOBAYASHI, S. and G. F. LEE, Freeze-concentration of dilute aqueous solutions, Anal. Chem. **36**, 2197—2198 (1964)
133. SMITH, G. H. and M. P. TASKER, Freeze concentration of dilute solutions, Anal. Chim. Acta **33**, 559—560 (1965)
134. WILSON, T. E., D. J. EVANS JR. and M. L. THERIOT, Freezing-out technique applied to the concentration of biologically active materials, App. Microbiol. **12**, 96—99 (1964)
135. SHAPIRO, J., Concentration of volatile substances in aqueous solution and production of water free of organics by freezing out, Anal. Chem. **39**, 280 (1967)
136. KEPNER, R. E., S. VAN STRATEN and C. WEURMAN, Freeze concentration of volatile components in dilute aqueous solutions, J. Agric. Food Chem. **17**, 1123—1127 (1969)
137. SCHIDLKNECHT, H., Zoneschmelzen, Verlag Chemie, Weinheim 1964, pp. 226
138. HUCKLE, M. T., A low temperature zone melting technique for the concentration of flavours, Chem. Ind. (London) 1490—1492 (1966)
139. SCHILDKNECHT, H., Zone melting and column crystallization as analytical tools, Anal. Chim. Acta **38**, 261—273 (1967)
140. SCOTT, R. P. W., Recent techniques in flavour analysis, Chem. Ind. (London) 797—804 (1969)
141. ESSEN, M. J. VAN, P. F. J. VAN DER MOST and W. M. SMIT, Rapid purification of organic substances, in "Fractional solidification". Vol. 2, eds. ZIEF, M. and W. R. WILCOX, Marcel Dekker, New York 1969, p. 201—208
142. ARKENBOUT, G. J., Concentration of minimal quantities considerably improved (in Dutch), Innovatie (6) 1972, no page numbering
143. STEINKE, R. D. and M. C. PAULSON, Phenols from grain. The production of steam-volatile phenols during the cooking and alcoholic fermentation of grain, Agric. Food Chem. **12**, 381—387 (1964)
144. JUNK, G. A., H. J. SVEC, R. D. VICK and M. J. AVERY, Contamination of water by synthetic polymer tubes, Environm. Sci. Technol. **8**, 1100—1106 (1974)
145. LEONI, V., G. PUCCETTI, R. J. COLOMBO and A. M. D'OVIDIO, The use of Tenax for the extraction of pesticides and poly-chlorinated biphenyls from water. II. Tests with artifically polluted and natural waters, J. Chromatogr. **125**, 399—407 (1976)
146. VITZTHUM, O. G. and P. WERKHOFF, Steam volatile aroma constituents of roasted coffee: neutral fraction, Z. Lebensm. Unters. Forsch. **160**, 277—291 (1976)
147. NURSTEN, H. E. and A. A. WILLIAMS, The drying of organic solutions containing food volatiles, Chem. Ind. (London) 2188 (1966)
148. SWAIN, J. W., Volatile flavor constituents of pork cured with and without nitrite, Thesis, University of Missouri, Columbia 1972

149. FREHSE, H. und H. NIESSEN, Die Analyse von Pflanzenschutzmittelrückständen, Z. Anal. Chem. **192**, 94—137 (1963)
150. NURSTEN, H. E. and M. R. SHEEN, Volatile flavour components of cooked potato, J. Sci. Food Agric. **25**, 643—663 (1974)
151. GOLDBERG, M. C. and L. DELONG, Extraction and concentration of organic solutes from water, Anal. Chem. **45**, 89—93 (1973)
152. MOINAS, M., Arôme volatil de produits laitiers — Extraction et identification, Mitt. Geb. Lebensmittelunters. Hyg. **64**, 60—65 (1973)
153. BEMELMANS, J. M. H. and N. L. A. TE LOO, unpublished data, 1976
154. BADINGS, H. T., personal communication, 1976
155. WEURMAN, C., Sampling in airborne odorant analysis, in "Human responses to environmental odors", eds. TURK, A., J. W. JOHNSTON and D. G. MOULTON, Academic Press, Inc., New York 1974, p. 263—328
156. GROENEN, P. J., personal communication, 1970
157. KUGLER, E. und E. sz. KOVÁTS, Zur Kenntnis des Mandarinenschalen-Öls (*Citrus reticulata* Blanco bzw. *Citrus nobilis* var. *deliciosa* Swingle "Mandarin"), Helv. Chim. Acta **46**, 1480 to 1490 (1963)
158. WATANABE, K. and Y. SATO, Lactones in the flavor of heated pork fat, Agric. Biol. Chem. **33**, 242 —249 (1969)
159. WATANABE, K. and Y. SATO, Gas chromatographic and mass spectral analyses of heated flavor compounds of beef fats, Agric. Biol. Chem. **35**, 756—763 (1971)
160. LUSTRE, A. O. and PH. ISSENBERG, Volatile components of hardwood sawdust smoke. Components of phenolic fraction, Agric. Food Chem. **17**, 1387—1393 (1969)
161. ISSENBERG, PH., M. R. KORNREICH and A. O. LUSTRE, Recovery of phenolic wood smoke components from smoked foods and model systems, J. Food Sci. **36**, 107—109 (1971)
162. HARDY, P. J. and E. H. RAMSHAW, Analysis of minor volatile constituents of wine, J. Sci. Food Agric. **21**, 39—41 (1970)
163. MURRAY, K. E. and G. STABLEY, Class separation of flavour volatiles by liquid chromatography on silica gel at 1°, J. Chromatogr. **34**, 174—179 (1968)
164. MURRAY, K. E., J. SHIPTON, F. B. WHITFIELD, B. H. KENNETT and G. STANLEY, Volatile flavor components from green peas (*Pisum sativum*). 1. Alcohols in unblanched frozen peas, J. Food Sci. **33**, 290—294 (1968)
165. MURRAY, K. E., J. K. PALMER, F. B. WHITFIELD, B. H. KENNETT and G. STANLEY, The volatile alcohols of ripe bananas, J. Food Sci. **33**, 632—634 (1968)
166. PALMER, J. K., Separation of compo nents of aroma concentrates on the basis of functional group and aroma quality, J. Agric. Food Chem. **21**, 923—925 (1973)
167. BALDRY, J., W. S. MATTHEWS, F. V. ROBINSON and J. NABNEY, Chemical composition and flavour of nutmegs of different geographical origins, Proc., SOS/74, Int. Congr. Food Sci. Technol., 4th ,Vol. 1, 99—104 (1974)
168. SCHEFFER, J. J. C., A. KOEDAM, and A. BAERHEIM SVENDSEN, Analysis of essential oils by combined liquid-solid and gas-liquid chromatography. Part II. Monoterpenes in the essential seed oil of *Anethum graveolens* L., Medd. Nor. Farm. Selsk. **39**, 161—188 (1977)
169. NOBLE, A. C., Gel chromatographic fractionation of flavor components for their individual sensory evaluation, J. Agric. Food Chem. **24**, 321—323 (1976)
170. WALRAVEN, J. J., Identification in complex volatile organic mixtures, Thesis, University of Eindhoven 1968
171. HEIDE, R. TER, N. PROVATOROFF, P. C. TRAAS, P. J. DE VALOIS, N. VAN DER PLASSE, H. J. WOBBEN and R. TIMMER, Qualitative analysis of the odoriferous fraction of oakmoss (*Evernia prunastri* (L.) Ach.), J. Agric. Food Chem. **23**, 950—966 (1975)
172. DEBRAUWERE, J. and M. VERZELE, Constituents of peppers: IV. The hydrocarbons of pepper essential oil, J. Chromatogr. Sci. **14**, 296—298 (1976)
173. MAARSE, H. and F. H. L. VAN OS, Volatile oil of *Origanum vulgare* L. ssp. vulgare. I. Qualitative composition of the oil, Flavour Ind. **4**, 477—481 (1973)
174. SCHEFFER, J. J. C., A. KOEDAM and A. BAERHEIM SVENDSEN, Occurrence and prevention of isomerization of some monoterpene hydrocarbons from essential oils during liquid-solid chromatography on Silica gel, Chromatographia **9**, 425—432 (1976)

175. BUTTKUS, H. A., R. J. BOSE, and D. A. SHEARER, Terpenes in the essential oil of sagebrush (*Artemisia tridentata*), J. Agric. Food Chem. **25**, 288—291 (1977)
176. HUNTER, G. L. K. and W. B. BROGDEN JR., Analysis of the terpene and sesquiterpene hydrocarbons in some citrus oils, J. Food Sci. **30**, 283—287 (1965)
177. LAWRENCE, B. M., J. W. HOGG and S. J. TERHUNE, Artefacts found during alumina chromatography, J. Chromatogr. **42**, 261—262 (1969)
178. KUBECZKA, K.-H., Vortrennung ätherischer Öle und ähnlich komplexer Stoffgemische für die GC-Analyse durch modifizierte Trockensäulen-Chromatographie, Chromatographia **6**, 106—108 (1973)
179. HEFENDEHL, F. W. und L. ROMERO F ONSECA, Analyse des ätherischen Öls von *Elyonurus viridulus*, Planta Med. **30**, 135—140 (1976)
180. MODY, N. V., P. A. HEDIN, and W. W. NEEL, Volatile components of pecan leaves and nuts, *Carya illinoensis* Koch, J. Agric. Food Chem. **24**, 175—177 (1976)
181. SMITH JR., W. T. and J. M. PATTERSON, Functional group analysis, Anal. Chem. **48**, 83R—86R (1976)
182. GOLOVNYA, R. V., YU. N. ARSENYEV and N. I. SVETLOVA, Use of heavy metal salts in the analysis of organic sulphur compounds, J. Chromatogr. **69**, 79—86 (1972)
183. GARBUSOV, V., G. REHFELD, G. WÖLM, R. V. GOLOVNYA and M. ROTHE, Volatile sulfur compounds contributing to meat flavour. Part I. Components identified in boiled meat, Nahrung **20**, 235—241 (1976)
184. STAHL, W. H., Gas chromatography and mass spectrometry in the study of flavor, in "Chemistry of natural food flavors, a symposium", eds. MITCHELL, J. H. JR., N. J. LEINEN, E. M. MRAK and S. D. BAILEY, National Academy of Sciences, Washington 1957, p. 58—75
185. TANG, C. S. and W. G. JENNINGS, Volatile components of apricot, J. Agric. Food Chem. **15**, 24—28 (1967)
186. PYNE, A. W. and E. L. WICK, Volatile components of tomatoes, J. Food Sci. **30**, 192—200 (1965)
187. SOEDARMO MOELJOHARDJO, D., On the volatile flavour compounds of cooked trassi, a cured shrimp paste condiment of the Far East, Thesis, Agricultural University of Wageningen 1972
188. DIJKSTRA, F. Y. and T. O. WIKÉN, Studies on mushroom flavours. 1. Organoleptic significance of constituents of the cultivated mushroom, *Agaricus bisporus*, Z. Lebensm. Unters. Forsch. **160**, 255—262 (1976)
189. GEMERT, L. J. VAN and A. H. NETTENBREIJER, Compilation of odour threshold values in air and water, National Institute for Water Supply, Voorburg, Netherlands, and Central Institute for Nutrition and Food Research TNO, Zeist, Netherlands 1977, pp. 79
190. BERTUCCIOLI, M. and G. MONTEDORO, Concentration of the headspace volatiles above wine for direct chromatographic analysis, J. Sci. Food Agric. **25**, 675—687 (1974)
191. SYDOW, E. VON, J. ANDERSSON, K. ANJOU, G. KARLSSON, D. LAND and N. GRIFFITHS, The aroma of bilberries (*Vaccinium myrtillus* L.) II. Evaluation of the press juice by sensory methods and by gas chromatography and mass spectrometry, Lebensm. Wiss. Technol. **3**, 11—17 (1970)
192. SCHAEFER, J., unpublished results
193. MULDERS, E. J., H. MAARSE and C. WEURMAN, The odour of white bread I: Analysis of volatile constituents in the vapour and aqueous extracts, Z. Lebensm. Unters. Forsch. **150**, 68—74 (1972)
194. NAWAR, W. W. and I. S. FAGERSON, Technique for collection of food volatiles for gas chromatographic analysis, Anal. Chem. **32**, 1534—1535 (1960)
195. HONKANEN, E. and P. KARVONEN, Isolation of volatile flavour compounds from fats and oils by vacuum carbon dioxide distillation, Acta Chem. Scand. **20**, 2626—2627 (1966)
196. KAISER, R. E., Enriching volatile compounds by a temperature gradient tube, Anal. Chem. **45**, 965—967 (1973)
197. KAISER, R. E., Trenn-Casetten zur Gas-Chromatographie, ein neues technisches System zur Erleichterung normaler und extrem schwieriger Analysen, besonders in der Umwelt-Analytik, Chromatographia **7**, 688—695 (1974)
198. ANGERER, J. und A. HAAG, Zur Luftanalyse im Ultraspurenbereich; Untersuchungen zur praktischen Verwendung der Kombination Thermogradientrohr-Gaschromatograph-Massenspektrometer, Z. Klin. Chem. Klin. Biochem. **12**, 321—326 (1974)

199. BERGERT, K. H., V. BETZ und D. PRUGGMAYER, Untersuchung flüchtiger organischer Mikroverunreinigungen in Luft und Wasser mit Hilfe von Tieftemperatur-Kapillar-GC/Massenspektrometrie, Chromatographia **7**, 115—121 (1974)
200. LUNTEREN, G. VAN, S. VAN STRATEN and C. WEURMAN, Irregularities in gas chromatographic direct vapour analysis, Troisiéme Colloque International sur la Chimie des Cafés, Trieste 1967, p. 191—196
201. GROENEN, P. J. and M. C. TEN NOEVER DE BRAUW, Determination of volatile N-nitrosamines in the vapour phase of the smoke from various tobacco products, Beitr. Tabaksforsch. **8**, 113—123 (1975)
202. FARRINGTON, P. S., R. L. PECSOK, R. L. MEEKER and T. J. OLSON, Detection of trace constituents by gas chromatography; analysis of polluted atmosphere, Anal. Chem. **31**, 1512 — 1516 (1959)
203. TURK, A. and C. J. D'ANGIO, Composition of natural fresh air, J. Air Pollut. Control Assoc. **12**, 29—33 (1962)
204. GROB, K. and G. GROB, Gas-liquid chromatographic-mass spectrometric investigation of C_6-C_{20} organic compounds in an urban atmosphere; an application of ultra trace analysis on cappillary columns, J. Chromatogr. **62**, 1—13 (1971)
205. GROB, K., Organic substances in potable water and in its precursor I: methods for their determination by gas-liquid chromatography. J. Chromatogr. **84**, 255—273 (1973)
206. GROB, K. and F. ZÜRCHER, Stripping of trace organic substances from water, equipment and procedure, J. Chromatogr. **117**, 285—294 (1976)
207. AUE, W. A. and P. M. TELI, Sampling of air pollutants with support-bonded chromatographic phases, J. Chromatogr. **62**, 15—27 (1971)
208. DRAVNIEKS, A. and B. K. KROTOSZYNSKI, Collection and processing of airborne chemical information, J. Gas Chromatogr. **4**, 367—370 (1966)
209. NOVAK, J., V. VASAK and J. JANAK, Chromatographic method for the concentration of trace impurities in the atmosphere and other gases, Anal. Chem. **37**, 660—666 (1965)
210. BERTSCH, W., E. ANDERSON, and G. HOLZER, Trace analysis of organic volatiles in water by GC-MS with glass capillary columns, J. Chromatogr. **112**, 701—718 (1975)
211. DRAVNIEKS, A. and A. O'DONNELL, Principles and some techniques of high-resolution headspace analysis, J. Agric. Food Chem. **19**, 1049—1056 (1971)
212. LIEBMAN, S. A., D. H. AHLSTROM and C. I. SANDERS, Automatic concentrator, G.C. system for trace analysis, Intern. Laboratory 1975 Nov./Dec., p. 43—48
213. PELLIZZARI, E. D., J. E. BUNCH, R. E. BERKLEY and J. MCRAE, Determination of trace hazardous organic vapor pollutants in ambient atmospheres by gas chromatography/mass spectrometry/computer, Anal. Chem. **48**, 803—807 (1976)
214. TASSAN, C. G. and G. F. RUSSELL, Sensory and gas chromatographic profiles of coffee beverage headspace volatiles entrained on porous polymers, J. Food Sci. **39**, 64—68 (1974)
215. VERSINO, B., H. KNÖPPEL, M. DE GROOT, A. PEIL, J. POELMAN, H. SCHAUENBURG, H. VISSERS and F. GEISS, Organic pollutants in air and water: sampling, gas chromatographic-mass spectrometric analysis and computer identification, J. Chromatogr. **122**, 373—388 (1976)
216. ZLATKIS, A., H. A. LICHTENSTEIN and A. TISHBEE, Concentration and analysis of trace volatile organics in gases and biological fluids with a new solid adsorbent, Chromatographia **6**, 67—70 (1973)
217. LIN, S. S., Alkylpyrazines in processed American cheese, J. Agric. Food Chem. **24**, 1252—1254 (1976)
218. BUTLER, L. D. and M. F. BURKE, Chromatographic characterization of porous polymers for use as adsorbents in sampling columns, J. Chromatogr. Sci **14**, 117—122 (1976)
219. MURRAY, K. E., The concentration of head space, airborne and aqueous volatiles on Chromosorb 105 for examination by gas chromatography and gas chromatography-mass spectrometry, J. Chromatogr. **135**, 49—60 (1977)
220. NOVOTNY, M., M. L. LEE and K. D. BARTLE, Some analytical aspects of the chromatographic head space concentration method using a porous polymer, Chromatographia **7**, 333—338 (1974)
221. SAKODYNSKII, K., L. PALINA and N. KLINSKAYA, A study of some properties of Tenax, a porous polymer sorbent, Chromatographia **7**, 339—344 (1974)

222. LEITHE, W., Die Analyse der Luft und ihrer Verunreinigungen, Wissenschaftliche Verlagsgesellschaft GmbH, Stuttgart 1968, pp. 288
223. MAARSE, H. and J. SCHAEFER, Quantitative headspace analysis: total and specific group analysis, in "Analysis of foods and beverages". Head space techniques, ed. CHARALAMBOUS, G., Academic Press, New York, San Francisco, London 1978, p. 17—35
224. ELLIN, R. I., R. L. FARRAND, F. W. OBERST, C. L. CRAUSE, N. B. BILLUPS, W. S. KROON, N. P. MUSSELMAN and F. R. SIDELL, An apparatus for the detection and quantitation of volatile human effluents, J. Chromatogr. **100**, 137—152 (1974)
225. HURST, R. E., A method of collecting and concentrating head space volatiles for gas chromatographic analysis, Analyst (London) **99**, 302—305 (1974)
226. NARAIN, C., P. J. MARRON and J. H. GLOVER, The analysis of gaseous pollutants, in "Gas Chromatography 1972", eds. PERRY, S. G. and E. R. ADLARD, Applied Science Publishers Ltd., Barking 1973, p. 1—7
227. TYSON, B. J. and G. C. CARLE, Gas chromatographic separation of water from cryogenically collected air samples — enhanced concentration of trace organic volatiles prior to GC-MS analyses, Anal. Chem. **46**, 610—613 (1974)
228. CLARK, R. G. and D. A. CRONIN, The use of activated charcoal for the concentration and analysis of headspace vapours containing food aroma volatiles, J. Sci. Food Agric. **26**, 1615—1624 (1975)
229 MIEURE, J. P. and M. W. DIETRICH, Determination of trace organics in air and water, J. Chromatogr. Sci. **11**, 559—570 (1973)

3. Separation

3.1. Principles of chromatography

S. VAN STRATEN

The term chromatography applies to all separation processes of components, based on their distribution between a stationary and a mobile phase. The mass transfer between the two phases is effected either by distribution of the molecules of a mixture between the two phases, or in adsorption-desorption on the surface of the stationary phase. Partition chromatography involves the partition of a solute between the mobile phase — being gaseous or liquid — and the stationary phase, which can be liquid or solid. Gas-liquid chromatography (GLC) ist most often used in aroma research. Liquid-liquid chromatography and liquid-solid chromatography under high pressure are also important, e.g. for aroma compounds with high boiling points.
In thin-layer chromatography, both adsorption and partition processes are encountered the latter being typical for impregnated plates.
Depending on the technique, a chromatogram is developed for a certain time (e.g. paper and thin-layer chromatography) or the solutes are allowed to emerge at the end of the column (elution chromatography).
The following scheme (Fig. 3—1) is based on the various possibilities.
The term gas chromatography (GC) comprises GLC and GSC.
The chromatographic behaviour of a solute can be described by the retention volume (V_R), the retention time (t_R) or the retardation factor (R value) of the compound.

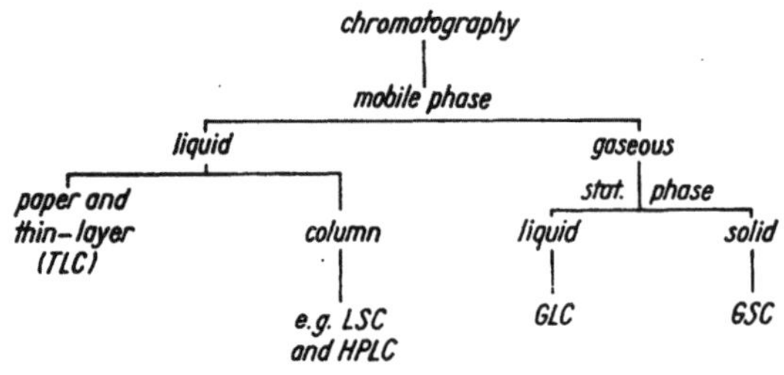

Fig. 3—1
Scheme of chromatographic techniques

Both R value and retention volume or time give an indication of the fraction of time that a molecule of the solute spends in the mobile phase.
The R value is defined as the quotient of the velocities of the solute and the mobile phase. It follows that

$$R = \frac{t_M}{t_M + t_S} \qquad \text{eq. 3.1.}$$

in which t_M and t_S are the respective times spent by a molecule of a solute in the mobile and stationary phases. The R value varies between zero and one; $R = 0$ means that a solute does not move at all and $R = 1$ that a solute moves at the same speed as the mobile phase. In paper and thin-layer chromatography the R value is exqressed by R_F

$$R_F = \frac{\text{distance travelled by a solute}}{\text{distance travelled by the mobile phase}} \qquad \text{eq. 3.2.}$$

In elution chromatography the volume of the mobile phase having left the column exit at the moment that a solute emerges at its maximum concentration, is called retention volume.
The time between the introduction of a solute into the column and the moment that the solute emerges at maximum is the retention time. If the rate of flow of the mobile phase is constant, the retention time is simply proportional to the retention volume, the proportionality factor being the flow rate.
As stated earlier, the separation of solutes in a chromatographic system is based on the distribution of substances between the mobile and stationary phase. In partition chromatography it is indicated by the partition coefficient (K) for a solute in the system

$$K = \frac{C_S}{C_M} \qquad \text{eq. 3.3.}$$

in which C_S is the concentration of the solute in the stationary phase and C_M the concentration in the mobile phase.
The partition isotherm is said to be linear if the partition coefficient is independent of the concentration. This is generally true at low concentrations. It is clear that the following equation is applicable to the retention volume V_R

$$V_R = V_M + K \cdot V_S \qquad \text{eq. 3.4.}$$

in which V_M and V_S are the respective volumes of the mobile and stationary phases in the column. If the partition ratio (k) is defined as

$$k = \frac{C_S}{C_M} \cdot \frac{V_S}{V_M} = K \cdot \frac{V_S}{V_M}$$ eq. 3.5.

while the retention times are proportional to retention volumes, equation 3.4. can be transformed to

$$t_R = t_M + k \cdot t_M$$ eq. 3.6.

in which t_R and t_M are the respective retention times of the given substance and an unretained solute. In addition, t_M — the retention time of a solute not retained by the stationary phase — is defined as the retention time of the air peak and the corresponding volume V_M as the dead volume of that column. The partition ratio k is alternatively called the capacity factor and is equal to the ratio of the amounts of a solute in the two phases within the column.

3.2. Gas chromatography

S. VAN STRATEN

3.2.1. Theory

Although detailed information on theory is to be found in many textbooks (e.g. 1—4) and publications, some of the most frequently used expressions are described here.

3.2.1.1. Retention time

As pointed out in section 3.1., the following formula is used to describe the chromatographic process

$$t_R = t_M + k \cdot t_M$$ eq. 3.7. (= 3.6.)

in which t_R and t_M are the respective retention times of a solute and an unretained solute and k is the partition ratio. In the form

$$k = \frac{t_R - t_M}{t_M} = \frac{t'_R}{t_M}$$ eq. 3.8.

the adjusted retention time t'_R is introduced. This is the retention time as corrected for the retention time of unretained solute, or dead volume of a column, and is used for the calculation of retention indices.
Relative retention of two peaks in a chromatogram is the quotient of their respective adjusted retention times under actual gas chromatographic conditions. As follows from equations 3.5. and 3.8., this is equal to the ratio of the partition coefficients

$$\frac{t'_{R(2)}}{t'_{R(1)}} = \frac{K_2}{K_1}$$ eq. 3.9.

in which $t'_{R(1)}$ and $t'_{R(2)}$ are the adjusted retention times of peaks no 1 and no 2. This equation is often used in qualitative analysis to relate the position of the peaks in a chromatogram to one specific standard peak. KOVATS developed a well known and useful identification method based on this system, which will be discussed in 3.2.6.

3.2.1.2. Separation efficiency

From eq. 3.8. one can draw the conclusion that two compounds having different K values in a chromatographic system at a given temperature will have different retention times. This does not mean, however, that these compounds injected as a mixture will always be observed as two distinct peaks in the chromatogram. They might partially overlap. The separation of two substances in a column is determined by the difference of the K values and the shape of the peaks. Peak shape is related to the quality of a column for a specific problem. If two peaks are overlapping, there are three possibilities for improving the separation. Firstly, to choose another system, e.g. a stationary phase of different polarity in order to increase the difference of the K values; secondly, to improve the quality of the column to give narrower peaks; and thirdly, to increase the length of the column.
In analogy to a distillation column the separation efficiency of a chromatographic column can be expressed as the number of theoretical plates, for which the formula reads

$$n = 16\left(\frac{t_R}{W_b}\right)^2 \qquad \text{eq. 3.10.}$$

in which W_b is the peak width at base. An alternative formula is

$$n = 5.54\left(\frac{t_R}{W_h}\right)^2 \qquad \text{eq. 3.11.}$$

in which W_h is the peak width at half height.
The "height equivalent to a theoretical plate" (HETP) is defined as

$$H = \frac{L}{n} \qquad \text{eq. 3.12.}$$

in which L is the total length of the column.
The "plate theory" is very useful in explaining the band broadening that occurs when a solute, brought into a column as a small band or spot, is transported through the column. The plate concept can be used to calculate parameters from an actual chromatogram.
An entirely different approach has been followed by some investigators. From a kinetic point of view, band broadening is caused by non-ideal behaviour of the chromatographic process. Conditions for ideal chromatography in a packed column would be:

- constant proportions of both phases at all points of the column,
- uniform flow of the mobile phase,
- no molecular diffusions of a solute in longitudinal direction in either of the two phases,
- instantaneous equilibration between the two phases.

In practice the molecules of a solute move along a variety of paths of different lengths between the solid particles of a packed column. This is similar to a diffusion phenomenon known as "Eddy diffusion", which depends on particle size and regularity of the packing material. Eddy diffusion and the usual molecular diffusion cause some molecules of a solute to quickly reach the end of a column while others are retained for a very long time. VAN DEEMTER devised a formula for the plate height, being in a simplified form

$$H = A + \frac{B}{\bar{u}} + C \cdot \bar{u} \qquad \text{eq. 3.13.}$$

in which A, B and C are constants respectively representing the contributions of Eddy diffusion, molecular diffusion and non-equilibrium effects; $\bar{u}$ = the average linear velocity of the carrier gas.

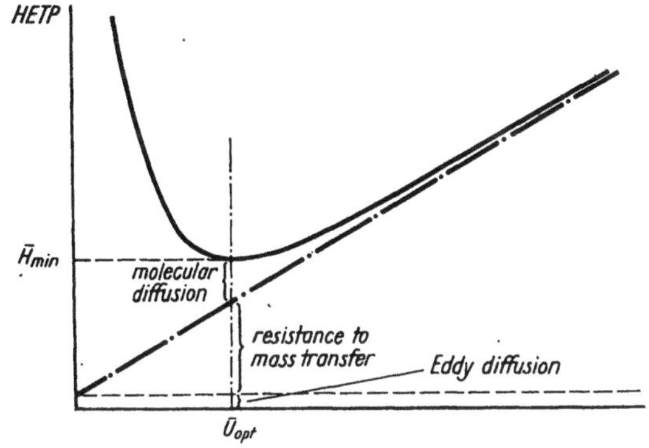

Fig. 3—2
HETP as a function of the mean linear velocity of the carrier gas in the column (2).

Figure 3—2 gives a curve of H vs $\bar{u}$, showing that a packed column has a minimum H at a distinct $\bar{u}$ value; both slower and faster flows of the mobile phase correspond to increasing plate heights. In practice, somewhat higher gas-flows than the optimum are chosen so as to keep retention times within practical limits. As can be seen from the figure, a slightly high gas velocity results in only a small rise of the plate height. With open tubular columns (3) coated on the inner wall with a thin film of stationary phase, the so-called Golay equation for the plate height reads

$$H = \frac{B}{\bar{u}} + C_G\bar{u} + C_L\bar{u} \qquad \text{eq. 3.14.}$$

in which B depends on the longitudinal diffusion in the mobile phase; C_G and C_L are factors for the resistance to mass transfer in the gas and stationary phase respectively. It can be calculated from the Golay equation that the minimum plate height is proportional to the column radius. The minimum plate height is a theoretical value which can never be reached in practice: columns with 50% of the obtainable maximum plate number are very useful. Coating efficiency can be defined as

$$\frac{\text{HETP}_{\text{minimum}}}{\text{HETP}_{\text{measured}}} \cdot 100\% \qquad \text{eq. 3.15.}$$

$\text{HETP}_{\text{measured}}$ can be calculated from the chromatogram (eq. 3.12.).

3.2.1.3. Resolution

The degree of separation of two compounds leaving a column is expressed by the resolution, R. This quantity can be calculated from the chromatogram; it is defined as

$$R = \frac{t_{R(2)} - t_{R(1)}}{\frac{1}{2}\{W_{b(1)} + W_{b(2)}\}} = \frac{2\Delta t}{W_{b(1)} + W_{b(2)}} \qquad \text{eq. 3.16.}$$

in which $t_{R(1)} < t_{R(2)}$ and $W_{b(1)}$ is the width at the base of peak no 1 and $W_{b(2)}$ of peak no 2, expressed in time units.
Resolution is a convenient expression for the separation of two adjacent peaks in a chromatogram. Resolutions $R > 1.5$ mean that two compounds are completely separated. In practice, quantitative analysis can be applied with success when two peaks have a resolution $R = 1$. Peaks overlapping to a larger extent — resolution $R < 1$ — can only be quantitatively determined by special calculations based on peak shapes.
Kaiser (5) introduced the calculation of the separation number (in German "Trennzahl") in gas chromatography

$$SN \text{ (or } TZ) = \frac{t_{R(n+1)} - t_{R(n)}}{W_{h(n)} + W_{h(n+1)}} - 1 \qquad \text{eq. 3.17.}$$

in which $t_{R(n+1)}$ and $t_{R(n)}$ are the retention times, $W_{h(n)}$ and $W_{h(n+1)}$ are the peak widths at half peak height for two n-alkanes with n and $n + 1$ carbon atoms respectively. SN is used as a parameter for the separating power of a column, especially capillary ones and when temperature programming is applied. It indicates the number of peaks that can be separated in between the two peaks of the n-alkanes, assuming that two peaks have a retention time difference of at least one peak width.

3.2.2. Packed columns

Metal or glass packed columns, with a length of 4 m and an i.d. of 2 to 4 mm, are still of great value to fractionate mixtures of aroma components. Some of the characteristics of columns and their possible applications in aroma research are presented in table 3—1.
Metal is usually preferred as material for the column tubing, because of its excellent mechanical properties. In our laboratory we make use of aluminium tubing, which can be easily cleaned and bent to fit in gas chromatographic ovens.
In aroma research one has to be aware of the instability of some groups of aroma compounds, e.g. sulphur or unsaturated compounds. Glass tubing being less reactive is used in such cases.
Among the different types of support for the stationary phase, diatomaceous earth (diatomite) (6, 7, 8) or porous polymers are usually employed. To meet as much as possible the requirements of an ideal support, treatments such as acid-washing and silylation with hexamethyldisilazane (HMDS) or dimethyldichlorosilane (DMCS) are applied. The particle size of a support is usually expressed in British Standard mesh size. Depending on their i.d., analytical columns are packed with supports of mesh sizes 60—80, 80—100 or 100—120; preparative columns with mesh size 40—60.

Table 3—1
Columns in aroma research

Tubing	i.d. (mm)	length (m)	type of column	capacity mg/component	applications
metal	4	4	packed	1	fractionation; separation of simple mixtures; useful for stable compounds only
glass	4	4	packed	1	as above, also useful for unstable compounds
metal	0.8	50—150	WCOT	0.25	separation of rather complex mixtures; fractionation of very complex mixtures; not for use with very unstable compounds
glass	0.8	50—150	WCOT	0.1	as above, also for unstable compounds
glass	0.8	50—150	PLOT	0.25	as above; higher loading capacity
metal	0.25	25— 50	WCOT	0.005	separation of very complex mixtures or fractions; metal columns for stable compounds only
glass	0.25	25— 50	WCOT	< 0.005	
glass	0.25	25— 50	PLOT	0.01	

A short compilation of some of the commercially available supports is given in Table 3—2. It must be emphasized that the choice of support depends to a great extent on the type of compounds to be analysed.

The following is a general classification of supports:

— "pink" supports: large surface area, loading capacity to 30% or more, polar components are easily adsorbed

Table 3—2
Solid supports and porous polymers for coating with stationary phase

Support	Type	Supplier[1]	Density[2] g/ml
Anakrom C 22	Diatomite	A	0.37
Anakrom U	Diatomite	A	0.27
Chromosorb P	Diatomite	J.M.	0.32—0.38
Chromosorb W	Diatomite	J.M.	0.21—0.27
Chromosorb G	Diatomite	J.M.	0.49
Gas Chrom S	Diatomite	A.S.	0.19
Anaport Kel F	Polymer	A.	0.4
Chromosorb T	Polymer	J.M.	0.49
Fluoropak 80	Polymer	A.	0.4
Porasil A	Polymer	W.	—
Anaport	Glass	A.	1.5

[1] A = Analabs, J.M. = Johns-Manville, A.S. = Applied Science, W = Waters Associate
[2] free-fall density; of Chromosorb W coated with 25% of stationary phase, 4 g is packed in a 1 m × 4 mm i.d. column.

— "white" supports: small surface area, loading capacity to 25%, less adsorptive than the pink ones, poor mechanical strength
— high-density supports: very small surface area, recommended loading capacity to 10%, highly inert
— polymers: surface area small to large, depending on type; used with or without stationary phase, very inert.

Furthermore, glass beads — which can only be loaded with a low percentage of stationary phase — are sometimes used for the separation of high-boiling components (9).
The selection of a *stationary phase* can be made from more than 200 commercially available phases. A practical classification of phases is based on polarity. ROHRSCHNEIDER (10) and McREYNOLDS (11) introduced a polarity scale that might be helpful in choosing the best stationary phase for a given separation problem. Apolar phases are mostly employed in separating non-polar compounds, and polar phases to separate polar compounds. The separation effected with apolar phases follows more or less the differences in boiling points of the compounds. In polar columns, on the other hand, compounds are retarded corresponding to their boiling points as well as to their polarities. For the analysis of aroma compounds a choice can be made from the limited number of apolar and polar phases given in Table 3—3.

Table 3—3
Stationary phases

	solvent	max. temp. (°C)[1]	polarity[2]
Apiezon L or M	toluene	250	15
SF 96	toluene	250	42
SE 30	chloroform	300	44
OV 101	chloroform	350	45
Ucon LB-550	methanol	200	158
OV 17	chloroform	375	162
Carbowax 20 M	chloroform	225	368
FFAP	chloroform	250	397
SP 2300	chloroform	275	446
DEGA	acetone	200	460
DEGS	acetone	200	593
TCEP	chloroform	175	759
Carbowax 400	chloroform	100	—

[1] from a catalogue of Supelco company.
[2] ΔI of 2-pentanone (see 3.2.6.2); from McREYNOLDS (11).

Mention has already been made of a number of polymers that are useful as gas chromatographic supports; some have been developed for special purposes, e.g. to overcome adsorption problems with very polar aroma compounds, and are used without coating. Generally, polymers have a good thermal stability up to 250 °C, little bleeding and higher capacities than conventional column packings have. Polymers with chemically bonded stationary phases (12, 13, 14) can be used at temperatures at least 50 °C higher than a conventionally packed column with the same phase. Some of the polymers are listed in Table 3—4.

Table 3—4
Porous polymers for use without coating

	Supplier[1]	Max. temp. °C	Applications
Chromosorb 101[2]	J.M.	275	e.g. separation of free fatty acids
Chromosorb 103[2]	J.M.	275	e.g. separation of amines
Porapak P[3]	W	250	wide applicability
Porasil A[3]	W	600	wide applicability
Durapak:			
OPN/Porasil C	W	150	separation of low-boiling compounds
Carb. 400/Porasil C	W	175	separation of low-boiling polar compounds
Octane/Porasil C	W	175	separation of low-boiling non-polar compounds

[1] J.M. = JOHNS-MANVILLE, W = Waters Associates Inc.
[2] Types indicated by numbers 101 to 108.
[3] Types indicated by capitals.

Micro-packed columns (15—18) with an i.d. of 0.5—1.0 mm are conventionally packed columns, filled with small particles. Column lengths of up to 15 m have been reported, a disadvantage of which is the large pressure drop across the column. Although micro-packed columns appear to be less important for aroma research, they can be useful when both high performance and large capacity are necessary, e.g. for the separation and collection of components from complex mixtures for I.R. and N.M.R. analyses.

3.2.3. Capillary columns

3.2.3.1. Introduction

After the introduction of capillary columns by GOLAY in 1957 (19) and 1958 (20), the number of applications of this type of columns increased considerably. Although some preliminary work was done with nylon and glass tubing, metal tubing was favoured in the sixties. To prevent artifact formation on copper tubing, stainless steel columns are preferred by aroma research workers.

During the last decade many aroma workers changed to glass capillary columns, which are generally less reactive than metal ones; nevertheless, metal columns — especially the stainless steel ones — are still a valuable device for aroma research. Nickel capillary columns as developed by BERTSCH et al. (21) are reported to be also useful for the analysis of instable compounds (22).

Only a few of the many publications on the theory, preparation and application of these columns will be mentioned here. For more information the reader is referred to the handbooks on the subject (e.g. 3, 4).

Actually, the correct name for a capillary column is "open tubular column", which denomination is still adhered to with regard to some types: wall-coated open tubular (WCOT), porous layer open tubular (PLOT), surface-coated open tubular (SCOT), and stabilized porous layer open tubular (SPOT) columns. The more common name "capillary" will be used in this book.

The internal diameter of a capillary column is usually between 0.25 and 1.0 mm (0.01 in. and 0.04 in.). With columns described as narrow-bored ones (i.d. approx. 0.25 mm), having a high efficiency and a low loading capacity, an injection splitter or a specific splitless injection technique (see 3.2.5.) is a necessity. These columns are preferred when a high separation efficiency is needed, e.g. for identification by G.C.-M.S. Large-bored columns (23, 24) (i.d. 0.5—0.75 mm) with splitless injection are chosen when, for example, fractions or single compounds emerging at the end of the column have to be collected. Some of the characteristics of capillary columns and possible applications in aroma research are presented in table 3-1.

3.2.3.2. Metal capillary columns

Much attention has been given to the cleaning of metal capillary columns, which is essential in order to get rid of any impurities present in new tubing and in previously coated columns. Rinsing the metal tubing with organic solvents will not always remove all contaminants, whereas more drastic cleaning reagents, such as inorganic acids or ammonia, may cause chemical changes in the metal surface layer. The cleaning, testing for adsorption, and coating of a stainless steel column will be described here.

Depending on the column volume — 5 ml for a 100 m column of 0.25 mm i.d., and 45 ml for a column of 0.75 mm i.d. —, portions of 25 ml for the narrow-bored column and 100 ml for the large-bored one, of each of the following sequence of solvents: chloroform, acetone, water, concentrated nitric acid, water, ammonia, water, acetone and chloroform are pushed through, the column by nitrogen. A stainless steel reservoir of 50 ml capacity for the narrow-bored and of 200 ml capacity for the large-bored columns is used.

If a used column is being cleaned, continuous washing with larger amounts of solvents, especially nitric acid, may be necessary. To remove very persistent material, washings with a hot 10% methanolic potassium hydroxide solution (25, 26) or 20% ammonia in methanol (27) might help. After cleaning with ammonia, the water washing should be continued till neutral.

Warning: one should be sure that the type of stainless steel chosen for the reservoir etc., can withstand nitric acid.

After drying a cleaned column, it is tested for adsorption properties. The necessity of consecutive cleanings to remove adsorptive materials and testing after the cleaning was described by Ryba (28). In general, new not cleaned tubing material very easily adsorbs polar and high-boiling apolar compounds. The following is a simple test procedure: the column is heated in a gas chromatographic oven at 100 °C and small vapour samples of methanol and petroleum ether are injected successively. Both samples must give single, rather sharp peaks with the same retention time. If too much "tailing" is noticed, the cleaning procedure has to be repeated. Slight tailing at the base of a peak can usually be eliminated by treating the tubing with a surface-active compound, such as Gas-Quat L (29); see also 3.2.3.4. Alternatively, a small amount of surface-active material can be added to the coating solution.

Coating is effected by either the dynamic or the static method. With the dynamic method, an amount of a coating solution is forced through the column by nitrogen.

The amount of solution will fill the column completely, or partly as a "plug" (4). The velocity of the plug should be kept constant, e.g. by using a dummy length of tubing attached to the end of the column. When all the coating solution is in the dummy column, the latter is removed and the actual column dried in a flow of nitrogen. With the static method, the column is filled completely with the solution and one end is closed; the solvent is evaporated by drawing the column through an oven (see also "Static coating of glass capillary columns"). With the dynamic method of coating, a solution of 3—10% of liquid phase in a suitable solvent is used; with the static method, approx. 1% concentrations.
In spite of the fact that film thickness is easier controlled and predicted in static-coated columns, the dynamic method is generally employed because of its advantage that columns can be shaped before coating. A honey-comb spool of 150 m columns was described by VERBEEK and MAARSE (30).

3.2.3.3. Porous layer open tubular columns

From the theoretical point of view, PLOT columns are intended to provide an increased amount of stationary phase without increasing the film thickness, implying that a larger k value for a solute is reached without loss of efficiency. This means in practice that such columns have a larger sampling capacity than WCOT columns with the same diameter. This might be an important advantage in aroma research when trace components have to be determined and/or collected at the column exit.
A review on PLOT columns is given by ETTRE and PURCELL (31).
Porous layers in glass capillary columns are mostly intended to improve the wettability of the glass wall and the uniformity of the film of the stationary phase. This technique and other methods for improving the wettability of glass capillaries are described in the following.

3.2.3.4. Glass capillary columns

Pre-treatments:
During the last decennium, extensive studies have been made of the preparation of glass capillaries (32). The lack of stability of the liquid film of stationary phase on the column wall was soon recognized to be one of the main problems. In general, the wettability of an untreated glass wall is poor and consequently many stationary phases tend to form droplets instead of a homogeneous film.
Two main methods are followed in improving the wettability of the glass wall. The first one, chemically changes the glass wall, while the second is based on adhering an inert layer to the glass. The pre-treatment of the glass wall depends on the type of glass: soft soda-lime glass or hard borosilicate glass. Glass capillaries can be drawn from both types of glass tubing. Two glass-drawing machines — Hupe-Busch and Shimadzu — are commercially available. With both machines, coiled capillaries of variable inner diameter, depending on the size of the original glass tubing and the drawing speed, can be produced. The usual inner diameters range from 0.25 to 0.75 mm, and the usual lengths from 25 to 100 m. Preferably the glass tubing is cleaned with chromic acid before being drawn, followed by successive washings with water and acetone (33) or potassium permanganate in sodium hydroxide (34).

Soft glass can be etched with gaseous hydrogen chloride (35, 36). After drying the column with dry nitrogen at temperatures above 100 °C (36, 37) it is etched, either dynamically in a constant stream of hydrogen chloride (37) or statically as a sealed column filled with hydrogen chloride (38—41) by heating at 350 °C for several hours. Crystals of sodium chloride will form a uniform layer on the wall if time and temperature are well chosen (37). Both methods are non-laborious and can be executed with commonly available laboratory equipment (Fig. 3—3).

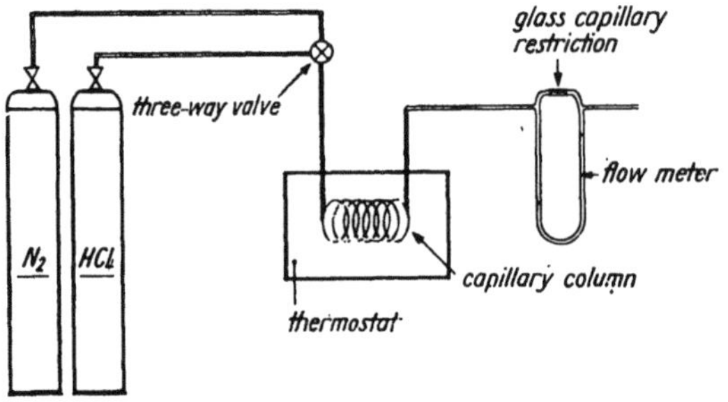

Fig. 3—3
Schematic set-up for the HCl etching of glass capillary columns (37).

Hard glass can be etched to some extent by gaseous hydrogen chloride although fluorine-containing compounds are preferred. TESAŘIK and NOVOTNY (42) recommended gaseous methyltrifluorochloromethyl ether which liberates hydrogen fluoride upon heating, in a sealed capillary column. Under the circumstances described by SCHIEKE et al. (43—48) and SANDRA and VERZELE (49), the etching is even more rigorous and the resulting "whiskers" on the glass give an ideal and enlarged surface. ONUSKA and COMBIA (50) conditioned columns with gaseous hydrogen fluoride.
The second method for solving the problem of poor wettability of the glass wall consists of covering the glass, prior to the actual coating, with a support material. These types of columns and their preparation and properties have been reviewed by ETTRE and PURCELL (31). Carbon black (51—53), Silanox 101 (54—58), Aerosil 380 (40) and recently barium carbonate (59, 60) and Carb-O-Sil (61, 62) and colloidal silicic acid (63) are being used. The advantages of these columns include a longer lifetime, possibility for thicker films, and higher loading capacity. A specific disadvantage of barium carbonate columns is their absorption of acidic substances. The porous layer of the Carb-O-Sil columns is very stable because of a network being formed, and consequently the authors designated these as SPOT (stabilized porous open tubular) columns.
To obtain a satisfactory separation of substances — especially polar ones — on some types of columns, de-activation is necessary (32). Apart from silylating agents, which silylate the active sites of the glass wall, surface-active compounds are employed to de-activate the surface of untreated, etched or porous-layer columns. Benzyltriphenylphosphonium chloride (BTPPC) (34, 40, 64, 65) is one of them, and is used either to de-activate the column prior to coating or is added to the coating solution. Polar stationary phases (59, 66) or triethanolamine (49) are also used. Trioctadecylmethylammonium bromide (Gas-Quat-L) (29) thermally decomposes at temperatures above 200 °C, and is therefore less effective than BTPPC, which can withstand temperatures up to 250 °C. It must be mentioned that BTPPC, when used in considerable amounts — 5 to 10% by weight of the stationary phase — will affect the ROHRSCHNEIDER constants of the phase; consequently, the order of elution of compounds might be altered.

The necessity for de-activating a column is less stringent with polar phases, since these themselves have a de-activating effect. However, surface-active compounds have been found to increase the lifetime of such columns. An efficient de-activation of glass capillary columns and of all connecting tubing material, including the liner in the injection port, the connection to the detector and that to a mass spectrometer, is obtained with Carbowax 20 M. This method stems from AUE et al. (67) who de-activated Chromosorb with Carbowax 20 M by heating the coated support to 280 °C. CRONIN (68) heated a sealed capillary, coated with a 2% w/v solution of Carbowax 20 M in dichloromethane at 280 °C for 16 h. The method of GROB et al. (59, 60) — "coating" with an 0.05% solution and heating at 280 °C for 20 minutes in a stream of nitrogen — gives good results in our laboratory. With some types of glass, Emulphor de-activation (59) gives better results. As pointed out by GROB (69), even noble metals can be very active.

Coating

Uniformity of the layer of stationary phase on the column wall is of great importance for the performance of a column. Two coating procedures are in general use — the dynamic (32, 70) and the static (71) one.

With the first, a plug of a solution of a stationary phase is forced through the capillary by pressure (Fig. 3—4). After evaporation of the solvent from the liquid layer adhering to the wall, a film of stationary phase is left behind. Film thickness depends on the concentration of the stationary phase in the solution, the diameter of the capillary, the coating velocity (i.e. rate of travel of the plug) and the viscosity and surface tension of the solvent (72, 73). Recommended coating velocities are approximately 2 cm/sec, resulting in a layer of approximately 1 μm of stationary phase on the wall of the column. Porous-layer columns are usually coated with a thicker layer of stationary phase.

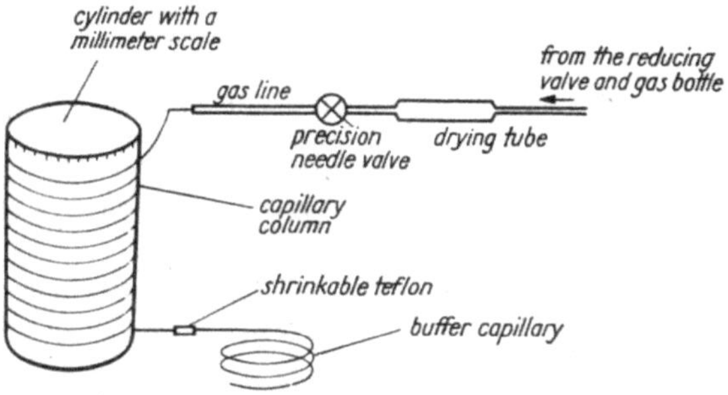

Fig. 3—4
Equipment for dynamic coating of a glass capillary column (32).

To improve the uniformity of the film in wall-coated columns, SCHOMBURG and HUSMANN (74) coated capillaries with rather concentrated solutions of stationary phase, pushing them through the column with a plug of mercury. VENEMA and VAN DER VEN (75) investigated the influence of the mercury plug length, the concentration of the coating solution and the coating velocity on the film thickness.

The static coating procedure according to BOUCHE and VERZELE (71) (Fig. 3—6) has the advantage that a uniform film is guaranteed and that the film thickness can be calculated from the volume of the capillary and the concentration of the coating solution. In this method, the capillary is filled completely with the coating solution, closed at one end and the solvent evaporated by a vacuum at the other end. Although generally better results are reached with this procedure — especially for polar phases — the dynamic method is preferred being less time-consuming.

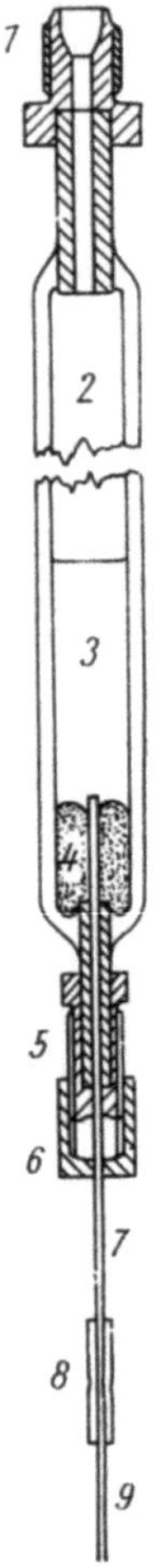

Fig. 3—5
Coating apparatus for glass capillaries using the mercury drop method (74).

1. Connection for pressured N_2 (1—20 bar)
2. Glass-tube with fused metal connections
3. Solution of stationary liquid
4. Mercury
5. Connection part to glass capillary
6. Nut with silicon rubber sealing
7. Stainless steel capillary (moveable)
8. Teflon shrinking tube
9. Glass capillary (column).

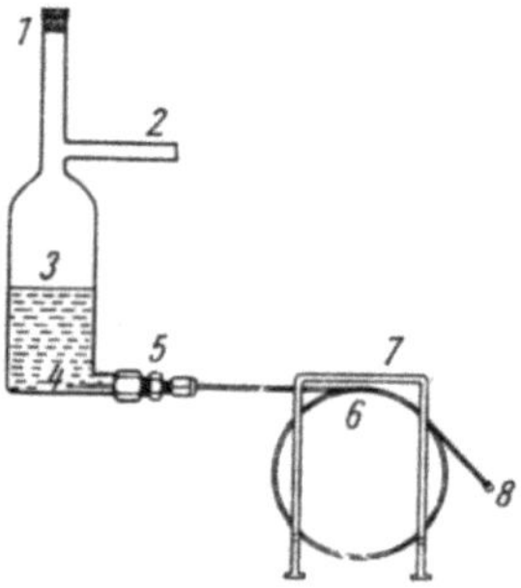

Fig. 3—6
Filling a glass capillary column for static coating (71).

1. Filling side
2. To pressure regulator
3. Coating solution
4. Column head
5. Metal connecting nut
6. Glass capillary column
7. Metal support
8. Column end with silicate bead.

The static method according to Ilkova and Mistryukov (76) and used by Jennings et al. (34) with a modified drying apparatus, is a rather laborious method. As stated by Jennings (77), the actual process during this coating procedure is still unknown, and even small alterations in the procedure may have strong effects on the performance of the columns.

Column connections

After a coated column has been dried, arrangements have to be made to facilitate its connection to the injector and detector, and mounting the column in the thermostatted oven. The use of commercially available cassette systems (78) is feasible in many types of apparatus, and solves mounting problems.

Connections are easily made to glass-lined injector and detector systems with shrinkable Teflon tubing (40, 79). Gas-tight seals can also be made with metallic connections using Vespel or graphite ferrules or silicone rubber disks or using silicone gum (80). Narrow-bored capillary columns need special attention to avoid dead volumes. As demonstrated by Hartigan and Ettre (81), specially constructed fittings, with a T-piece for the addition of make-up gas on the detector side, are required.
More practical information on manipulation and operation of glass capillary columns is given by Grob and Jaeggi (82).

Column performance

Capillary columns can either be checked for adsorption behaviour towards specific compounds, or tested for their general chromatographic behaviour. A mixture of compounds of different polarity — e.g. dibutylketone, 2-propylcyclohexanol, naphthalene and undecane (65, 83) — can be used to this end after column coating or even after each pretreatment. As reported by Grob and Grob (59, 84), columns are tested for their basic or acidic character with a mixture of 2,6-dimethylphenol and 2,6-dimethylaniline. It is recommended to further test the performance of a column with a standard test mixture containing the compounds to be analyzed.
General information on column performance can be gathered on the basis of the number of theoretical plates, separation number, or coating efficiency. In practice, with capillary columns the separation number is favoured as a direct indication of the number of peaks that can be seen separately in a chromatogram.

3.2.4. Detectors

A detector is a device placed at the exit of a GC column to monitor the eluting components. Detectors can be classified into general and selective ones. General detectors give a response for almost all organic compounds. Selective detectors (see 4.6.) on the other hand, respond to a select group of substances with a minimal response to others. In aroma research, two types of general detector are extensively used, viz.

(1) the thermal conductivity detector (TCD) or katharometer;
(2) the flame ionization detector (FID).

Details of construction and principles of these detectors can be found in a number of books (85—87).
A newly designed type of photo-ionization detector (PID) could be of interest in aroma research.
The principle of the *TCD* is based on the measurement of electrical resistivity changes of filaments caused by thermal conductivity changes of the mobile phase as substances are being eluted. The TCD being a non-destructive detector, eluted compounds or fractions can be collected at the outlet for further investigation.
Although many factors affect the output signal of a TCD, users of commercially bought apparatus cannot influence the signal other than by selection of the carrier gas, the temperature and the Wheatstone-bridge current. The choice of carrier gas is very important. As table 3—5 shows, hydrogen and helium have large thermal conductivities compared to nitrogen and organic compounds.

Table 3—5
Thermal conductivity of carrier gases and organic compounds (88)

	Thermal conductivity at 100 °C (cal cm^{-1} s^{-1} $°C^{-1}$ · 10^{-5})
N_2	7.5
H_2	53.4
He	41.6
Nonane	4.5
Benzene	4.1
Acetone	4.0
Ethanol	5.3
Ethyl acetate	4.1

Consequently, the largest electrical resistivity changes are obtained when using hydrogen or helium, which are therefore the carrier gases of choice. To avoid the danger of explosions by leakage of hydrogen, the more expensive helium is preferred. With hydrogen or helium, the thermal conductivity decreases when an organic compound elutes; with nitrogen, the changes in thermal conductivity are small and positive, but occasionally negative and W-shaped peaks can occur under certain conditions.

As the output signal is proportional to the temperature difference between the filaments and the detector block, filaments should be as hot as possible and the detector block as cold as possible. It should be taken into account that too hot filaments can lead to damage and noise, and a too cold detector block, to condensation. The limitations of TCD's in aroma research are their moderate sensitivity, the presence of metal surfaces which might give rise to alterations of some aroma compounds, and the large volume — approx. 2 ml for a cell of standard size. For applications of a TCD with capillary columns, special micro-katharometers with cell volumes as low as 100 µl or less are commercially available (the cell of a Micro-Katharometer of Servomex Controls Ltd. measures 8.5 × 1.2 × 0.25 mm).

A *flame ionization detector* (FID) is nowadays one of the most popular detectors, because of its high sensitivity towards almost all organic compounds. No signal, or at least only a small one, is provoked by such compounds as formic acid, carbon disulphide, carbon monoxide, carbon dioxide and water. An FID is a destructive detector; organic compounds eluting from the column are burnt in a hydrogen flame and generate ions which are recorded as an electric current. The carrier gas — nitrogen or helium — is mixed at the column exit with the necessary amount of hydrogen. With capillary columns, make-up gas has to be supplied — usually by mixing extra gas with the hydrogen — to avoid heating the burner tip too much, causing noise. Air or oxygen is added to the flame in excess. The amounts of hydrogen and air are generally recommended by the supplier. With oxygen, the same or sometimes better sensitivity can be reached than with air; depending on the detector construction the noise level, however, is often higher. In contrast to a TCD, an FID can be arranged so as to avoid any contact of compounds eluting from the column with metal parts, the burner tip being made of glass or quartz and the column brought just at the bottom of the burner tip. Capillary columns can be brought within the tip.

As pointed out above, an FID is a general detector for almost all organic compounds; it responds to the C—H configurations in a molecule and its signal is influenced negatively by the presence of oxygen in a molecule. Carbondisulphide and formic acid give only a small signal; the first compound is often used as extractant and the second one is frequently added to the carrier gas to eliminate adsorption effects when analyzing free fatty acids. The insensitivity to water is very convenient e.g. in the analysis of vapour samples.

The *PID*, designed by HNU Systems Inc. (89, 90), which became available recently, could be of much interest in investigations with high-resolution columns. This detector consists of a small ionization chamber of less than 0.2 ml volume, and sensitivity is claimed to be better than that of the FID. Furthermore the detector is nondestructive and peaks emerging from it can be trapped or directly subjected to aroma evaluation.

The principle of the system is ionization — by absorption of ultraviolet radiation — of molecules passing through the ionization chamber, and measuring the resulting current between two electrodes, one of which is positive. The UV-source emits the 10.2 eV Lyman α-line, and therefore the system will detect all compounds whose ionization potential is less than 10.2 eV (see table 3—6)

This includes all aliphatic hydrocarbons with 5 or more C atoms, and aromatics, carbonyls, esters, heterocyclics, amines, organic sulphur compounds and many in-

Table 3—6
Ionization potentials (I.P.) for selected molecules (89)

compound	I.P. (eV)
Diethyl sulphide	8.43
m-Xylene	8.56
Dimethyl sulphide	8.68
Toluene	8.82
Cyclohexene	8.94
1,3-Butadiene	9.07
Benzene	9.25
Pyridine	9.32
Trichloroethylene	9.45
Allyl alcohol	9.67
Acetone	9.69
Methyl ethyl ketone	9.53
Tetrahydrofuran	9.54
Cyclohexane	9.88
Vinyl chloride	9.95
Carbon disulphide	10.08
Acrolein	10.10
Hexane	10.17
Ethyl alcohol	10.48
Ethylene	10.51
Oxygen	12.07
Water	12.59
Carbon dioxide	12.80
Methane	12.98
Carbon monoxide	14.01
Nitrogen	15.55

organic compounds. Oxygen, nitrogen, helium, water and some other compounds are not detected.
The linear dynamic ranges and sensitivities of the TCD, FID and PID are shown in figure 3-7.

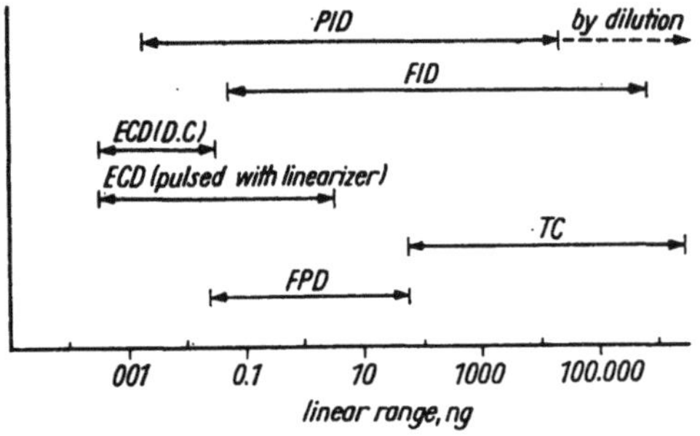

Fig. 3—7
Comparison of linear dynamic ranges of various detectors.

3.2.5. Sampling and derivatization

3.2.5.1. Introduction

Liquid and vapour samples are usually introduced in a gas chromatographic column by means of a syringe. Only a limited group of food products can be analyzed without pretreatment, i.e. liquid products with a low amount of non-volatile components, and vapours of products if the concentration of the volatile components is sufficiently high. Isolation and concentration procedures often have to be carried out to isolate sufficient material for qualitative and quantitative analyses, since aroma compounds generally occur in very low concentrations. As pointed out in chapter 2, artifacts might be produced during isolation and concentration of aroma compounds by chemical and biochemical reactions. During GC analysis formation of artifacts is also possible, viz. in the injection port, in the column or in the detector. Using different types of injection systems and at different temperatures Bruhn (91) noticed decomposition for terpenic compounds ranging from 0% to 100%. Testing a GC system with a mixture of heat-labile or otherwise labile compounds is recommended.
Injection of the vapour phase of a product is one of the most ideal sampling methods.
The advantages are evident: the analyzed vapours are the same as those observed by smelling, artifact formation arising during pretreatments is eliminated, and quantitative ratios of constituents of a vapour remain unchanged (92). Unfortunately, this method is limited to aroma compounds present in the vapour in concentrations above the detection limits, while the number of components occurring in sufficient amount to be identified by means of spectroscopic methods is even smaller. Part of these limitations can be overcome by cumulative sampling (see 2.2.).

3.2.5.2. Vapour sampling and injection

Direct vapour analysis is based on Henry's law

$$K = \frac{C_g}{C_l} \qquad \text{eq. 3.18.}$$

where C_g is the concentration of a component in the gas phase and C_l its concentration in the liquid.

HENRY's law holds for dilute solutions at a given temperature. Only a very limited group of food products, such as diluted fruit juices, can be expected to strictly obey this law. However, some correlation still exists even in non-ideal solutions. Up to now quantitative analysis was mostly based on the comparison of a sample with standard solutions imitating the analyzed sample as closely as possible. DROZD and NOVAK (93) proposed the use of the standard additions method. For qualitative analysis the non-ideal behaviour of a solution is of minor importance, and the method can even be used with partially or completely solid products retaining the major advantage: prevention of artifacts.

With direct vapour analysis a product is contained in an infusion flask closed with a rubber septum secured by a screw cap. The flask is filled to one quarter with the sample and is placed in a thermostatted water bath for 30 to 45 minutes. During this time an equilibration between product and air is established. Vapour samples are taken with a syringe by piercing the septum with the syringe needle (94—96). Injection into packed columns is possible with sizes up to 20 ml. On capillary columns, however, sample sizes of 0.5 to 2 ml, depending on the i.d. of the column, are the maximum (97). Inorganic salts are added to aqueous products in order to enlarge the concentration of compounds in the vapour phase. KEPNER et al. (94) reported the use of sodium chloride and ammonium sulphate, which are inexpensive, even in a very pure form; these salts are used by the present author also.

Losses by absorption of volatiles in the rubber septum can be prevented by placing a piece of aluminium foil under the septum to avoid contact of vapour with the septum. Also the use of special flasks, which can be closed with a glass stopcock during the equilibration period, is reported (98, 99). Absorption losses play a role only if solid products are investigated. Volatile components of liquid samples generally evaporate sufficiently fast to make up for absorption losses. Applications of the method have been described for many food products, such as beer (94, 95, 100—103), raspberries (96), rum (100), pears and carrots (104), apple juice (105), onion (106), cereal grain (107), milk (108, 109), horseradish (110) and apple (111). Application is also reported in the field of canned or bottled foods (112, 113), not only for analysis of aroma compounds but also of permanent gases.

JENTZSCH et al. (114) described a dispensing system to be used for automatic sampling of vapour samples of a series of 30 bottles. With non-standardized absolute peak-height evaluation, a relative standard deviation of less than 0.7% was obtained. MIETHKE (115) applied the method in routine analysis, e.g. of alcoholic beverages. He also pointed out that the volatility of the higher alcohols and esters is diminished by the presence of a large concentration of ethanol. In the present author's laboratory alcoholic beverages with more than 15 volume % ethanol are diluted to this percentage before vapour analysis.

Specific studies have been made of the volatility of aroma components of complex mixtures (116), irregularities that might occur during the injection of large head space samples (117) and volatilities of compounds in aqueous solutions (118). The sampling of vapour samples of beer directly from the bottles as described by DREWS et al. (119) and BÄRWALD (120), excludes any possible influence of oxygen.

3.2.5.3. Injection of liquids

Injection on packed columns

Compared with direct vapour sampling, introduction of liquid samples into a gas chromatographic column sometimes requires the additional step of evaporation.
The ideal procedure is to bring a sample on top of a column in such a way that it is flash-evaporated and carried along by the carrier gas as a narrow band. The "on-top-of-column injection", which means that a sample is brought directly into the column packing at the top of the column, gives satisfactory results in most cases.
Mostly, however, the sample is injected in the carrier gas stream in a *liner* (Fig. 3-8). A liner is an easily interchangeable, 10—20 cm length of tubing with an i.d. of 1 to 3 mm, either of glass or stainless steel, and is located in the injection port.
A restriction upstream in the carrier gas lead impedes back-flushing, which otherwise could lead to condensation, the carrier gas lead being less heated than the injection block (117, 121).
Septum bleeding and adsorption-desorption effects, resulting in an unstable base line or ghost peaks, have been studied by many investigators (123—130). This problem became more evident with the increasing sensitivities of GC detectors and the high-temperature applications of GC especially in programmed-temperature analysis. A septum, which is usually made of silicone rubber, contains low-molecular components. Upon heating the septum in a gas stream — i.e. the actual circumstances in an injection port — these compounds evaporate and enter the column. This steady stream of contaminants is measured by the detector as part of the total bleeding of the system

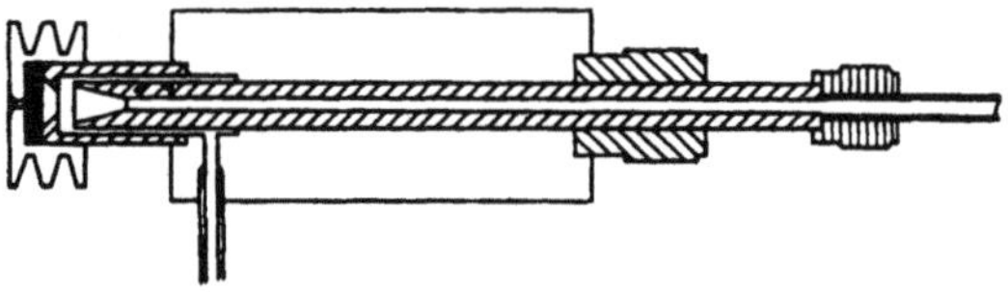

Fig. 3—8
Injection port for use with packed columns (122).

consisting of injection port, column and detector. Under isothermal conditions of injection port and column, bleeding may be expected to be constant and can be compensated electrically, so that even at high sensitivity a straight base line is observed. However, by heating a newly-installed septum or when raising the temperature, bleeding is enlarged and may sometimes result in extra peaks. Tamsma et al. (124) eliminated the bleeding of septa to a large extent by preconditioning the septa in a vacuum oven at 250 °C overnight. Nevertheless, compounds formed by degradation of the silicone polymers give rise to bleeding when septa are heated too high. Most commercially available septa can be used at temperatures as high as 250 °C without too much bleeding. At still higher temperatures, precautions must be taken so as to equilibrate the septum at the high temperature. Septum bleed can also be prevented by cooling the septum (131, 132) or by continuously washing it with a side stream of the carrier gas (133).
Another effect which might be misinterpreted as septum bleed, is caused by adsorption of compounds to the septum or other parts of the injection port, and release of these products during the following injection, resulting in so-called memory peaks.

DRESSMAN (122) observed memory peaks when successive aqueous solutions were injected. He eliminated these peaks by changing the construction of the injection port in such a way that the "dead volume" was decreased.

Injection on capillary columns

With capillary columns, on-column injection cannot be used. A sample must be evaporated in the injection system and the vapour swept into the column in as short a plug as possible. Just as with packed columns, here also a sample can be injected into a liner. An injection port with a glass liner for use with glass capillary columns is given in figure 3—9.

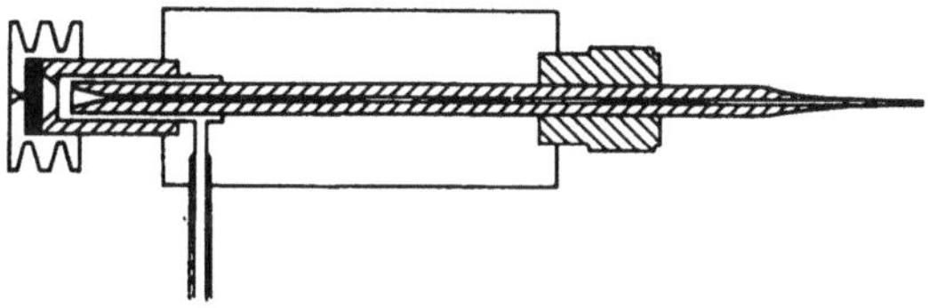

Fig. 3—9
Injection port for use with glass capillary columns (22).

On large-bored capillary columns (i.d. 0.5—1.0 mm), injection sizes of 0.5 μl of sample can be applied without column deterioration or peak broadening of the sample constituents.
Since the loading capacity of most capillary columns is low, especially of narrow-bored columns, sample-splitting devices have often to be used. This technique, however, has a number of disadvantages:

— errors in quantitative analyses are hard to prevent;
— additional equipment is needed;
— at least 90% of the sample is lost.

Therefore, splitless injection is favoured in aroma research. Many efforts have been made to develop satisfactory injection systems, e.g. an on-column micro-trap by BARTEL and VAN DER WAL (134), a splitless system especially for high-boiling components in a low-boiling solvent by EVRARD and GUIOCHON (135), a glass splitter by JENNINGS and ADAM (136, 137), and a direct sample system by CRAMERS and KESSEL (138).
With the injection system as described by GROB and GROB (133, 139) (Fig. 3—10), sample sizes of 1 μl of very diluted samples, can be brought into narrow-bored capillary columns. The system consists of an evaporation chamber and regulated restrictions for two by-pass streams of carrier gas. Under working conditions, valves A en B are open and consequently the incoming gas is divided in three parts, one of which streams along the septum to divert peak memory from the septum, a second through the splitter outlet, and one as carrier gas through the column. During injection of large volumes — e.g. 1 μl of a dilute extract — the valves A and B are both closed and the complete sample is introduced on the column which for the moment is kept below working temperature. Under these circumstances, the compounds dissolved in the low-boiling solvent are concentrated at the top of the column, while the solvent is carried through the column. After opening valves A and B, the flows are established for work-

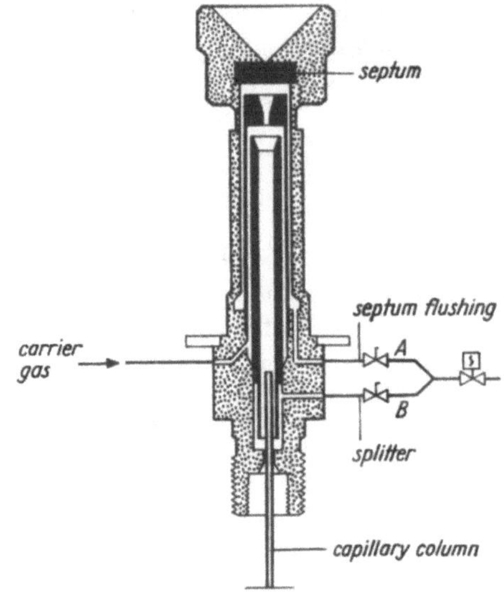

Fig. 3—10
Inlet port with built-in septum flushing (133).

ing condition, and the column temperature is increased, either fast to working temperature, or programmed.
It should be emphasized that splitless injection with a rather simple injection system is only feasible with columns that can withstand a large amount of sample. Large-bored capillary columns with a carrier gas flow of approximately 5 ml/min form an ideal combination with a splitless injection system. For identification purposes this system can be directly connected with a mass spectrometer.
An all-glass injection system (79) that can be used for solvent-free injection of solids or high-boiling compounds is shown in figure 3—11. The sample is brought onto a glass needle and the solvent blown off. In the injection mode the needle is brought in a position close to the top of the column; the sample evaporates by the heat of the column oven. This system is certainly not suitable for quantitative analyses of low-boiling compounds, nor even for qualitative analyses of very low-boiling ones. Loss of volatile compounds during evaporation of the excess solvent from an extract is described in chapter 2. Using the above-described injection system and blowing off the solvent, losses of volatile compounds are much greater. Generally speaking, compounds of b.p. below 200 °C are lost partially if not completely.
Krupčík et al. (41) reported the application of this system in the analysis of methyl esters of fatty acids of more than 10 carbon atoms.
Analysing complex mixtures of aroma compounds, many important ones of which have low boiling points, necessarily restricts the choice of injection technique. Moreover, the choice depends on other characteristics of the sample, the available amount and, last but not least, the wanted information. Generally, splitless injection systems are to be preferred to splitting systems when only small amounts are available and/or low boiling compounds are to be analyzed. The use of high-performance narrow-bored columns is generally limited to the final separation of prefractionated aroma mixtures; the capacity of this type of column is usually too low to collect enough sample for further analysis. For such purposes, packed columns or large-bored capillaries, both with simple injection systems, are preferred. Further analysis, e.g. combined GC-MS

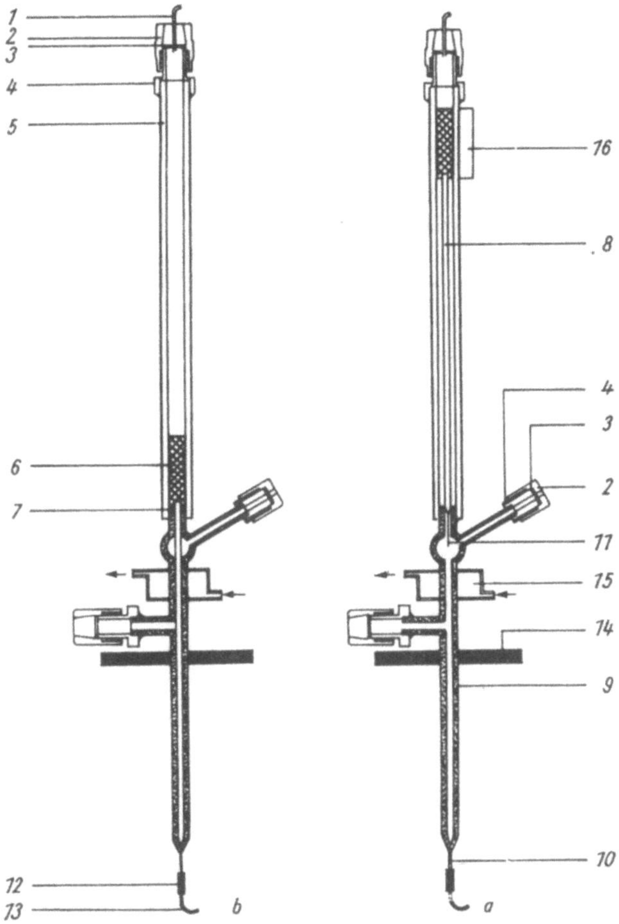

Fig. 3—11
Gas chromatographic injection system for solids (79).

1. Capillary restriction, i.d. 0.25 mm
2. Screw cap
3. Septum
4. Glued connection to glass with thread
5. Glass supporting tube 160 mm, (i. d. 6 mm)
6. Iron plunger
7. Polytetrafluoroethylene washer
8. Glass or quartz needle, o.d. 1.8 mm
9. Glass supporting tube, i.d. 2.0 mm
10. Drawn-out supporting tube, i.d. 0.3 mm
11. Drawn-out needle, o.d. 0.2 mm
12. Polytetrafluroethylene shrinking connection
13. Glass capillary column, o.d. 1 mm, i.d. 0.25 mm
14. Oven wall
15. Cooler
16. Magnet

a. indicates the loading position
b. indicates the injection position.

analysis, is executed with either narrow- or large-bored capillaries and the special injection systems described. To prevent any loss of substance by adsorption to active sites of the injection system, deactivation is recommended (see 3.2.3.).

3.2.5.4. Injection of solids and GC fractions

The injection of volatiles from a *solid product* into a column — wrongly called "solid injection" — is carried out with a modified injection port. Basically the device consists of two parts: a sample holder that can be loaded with the sample when the holder is disconnected from the injection port which enables introduction of the sample holder without disturbing the carrier gas flow. In figure 3—12 a solid sample injector, designed by KARLSEN and BAERHEIM SVENDSEN (140) and modified by MAARSE (141) for the analysis of plant material, is illustrated.

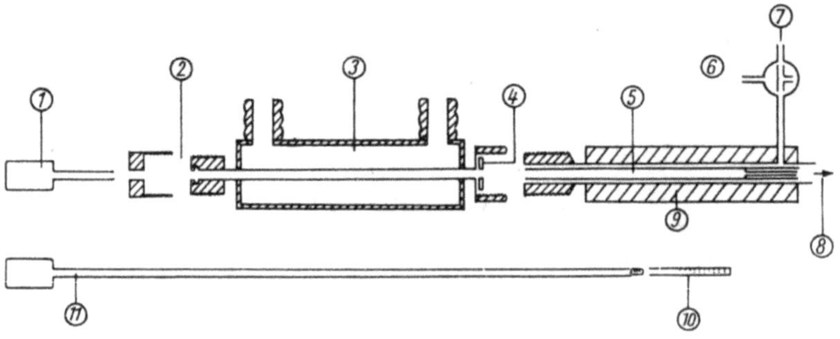

Fig 3—12
Solid sample injection (141).

1. Plug
2. O-ring
3. Cooling mantle
4. Septum
5. Liner
6. Three-way valve
7. Carrier gas inlet
8. To column
9. Injection port heater
10. Sample holder
11. Injector handle.

The construction of an injection system by SCHOMBURG (142) also allows small amounts of solid materials or samples of high-boiling compounds to be brought into the evaporation chamber (Fig. 3—13).

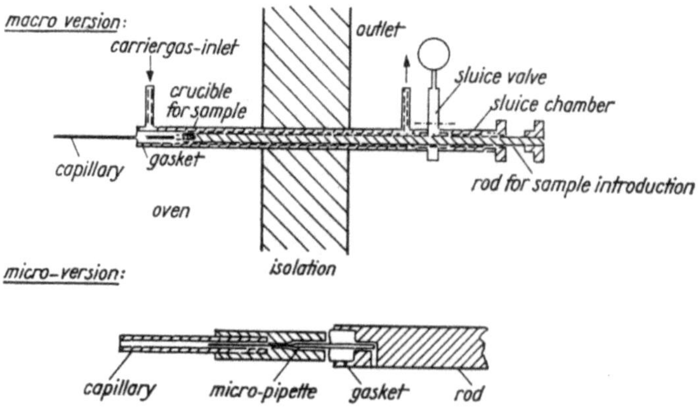

Fig. 3—13
Direct on-column injection of liquid samples with capillary columns (142).

Some other types of injection system for solid or high-boiling materials were reported by MEYER (143) and WONG and SCHWARTZ (144).

Fractions collected in traps at the end of a column can be injected either by use of a crushing device for capillary traps, or by connection of a trap in a loop circuit, or after the addition of a solvent to the contents of the trap. This last possibility is the simplest; it is obligatory with large traps and facultative with small ones. A GC fraction in a straight capillary tube of approximately 5 to 10 cm length and 0.5—1.0 mm i.d. can be dissolved in 1 to 2 µl of a suitable solvent. While holding the capillary tube, opened at both ends, in a horizontal position, the solvent is brought in with a syringe, moved carefully through the tube, and sucked into the syringe from the other end. Volumes up to 1 µl can be injected into packed and even capillary columns with splitless injection systems.
A crushing device, used by HARDING et al. (145), is illustrated in figure 3—14. Fractions must be presented in capillary traps.

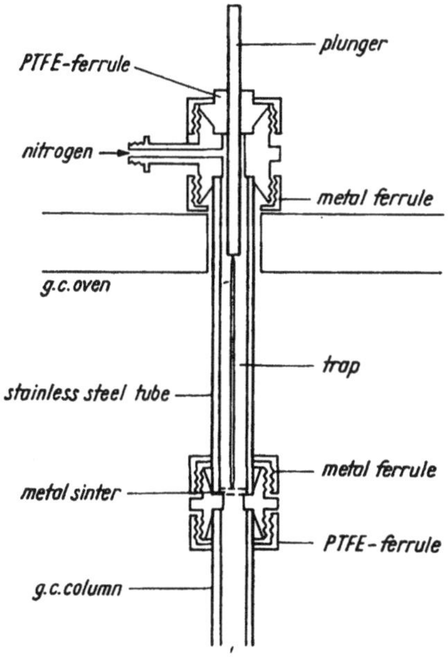

Fig.3—14
Injection port for rechromatography of trapped fractions.

With a loop system, both large-bored and narrow-bored traps can be connected to the injection system. As BADINGS et al. (40, 146) have shown, fractions collected in capillary zig-zag traps are easily brought on capillary columns (Fig. 3—15).
A method for bringing a straight capillary trap into a modified injection port has been developed by KLIMES et al. (147). They claim that small amounts of substances trapped in a short trap can be "injected" on another column without loss (Fig. 3—16).

3.2.5.5. Derivatization

Derivatization in gas chromatography is usually carried out either to increase the volatility of compounds of too high boiling points to be analyzed as such, or to reduce the adsorption effect of polar compounds to the column packing or column wall.

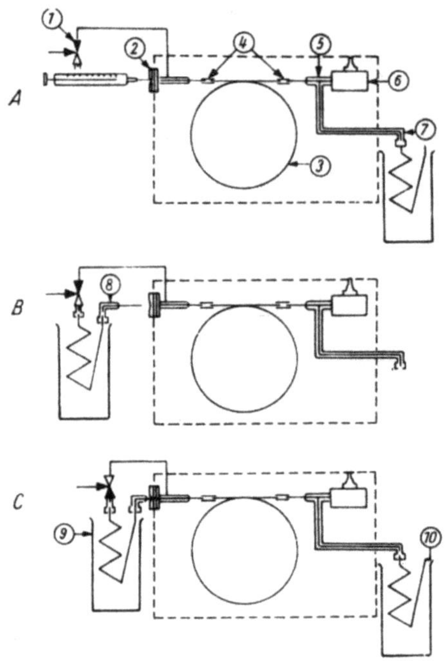

Fig. 3—15
Large-bored capillary gas chromatography system for separation, micro-collection and re-injection of fractions (146).
A. Injection of sample and micro-collection of separated fractions
B. Mounting of capillary trap for re-injection
C. Re-injection from micro-trap and collection of subfractions
1. Three-way valve in carrier gas line
2. Injector
3. Large-bored capillary column
4. Teflon shrinking tubing
5. Splitter (1:10)
6. Flame ionization detector
7. Mounting of glass capillary trap in cooling bath
8. Mounting of capillary trap in cooling bath for re-injection
9. Micro-trap in position for re-injection (carrier gas flow switched on trap in heating bath)
10. Capillary trap for collection of sub-fractions.

Although derivatization is applicable to many types of compounds in gas chromatographic analysis (148), its use in GC aroma research has only been reported for some types of compounds. This technique introduces an extra source of errors due to possible impurities in the reagents and solvents used and incomplete conversion to derivatives.

Derivatives of aldehydes and ketones

Isolation of carbonyl compounds with 2,4-dinitrophenylhydrazine (2,4-DNPH) is frequently applied. Usually the derivatives — 2,4-DNPH's — are analyzed by TLC (see 3.4). Direct gas chromatographic analysis of 2,4-DNPH's, on the other hand, is seemingly in contradiction with the purposes of derivatization in GC; carbonyl compounds are neither very polar nor do they have extremely high boiling points. In this

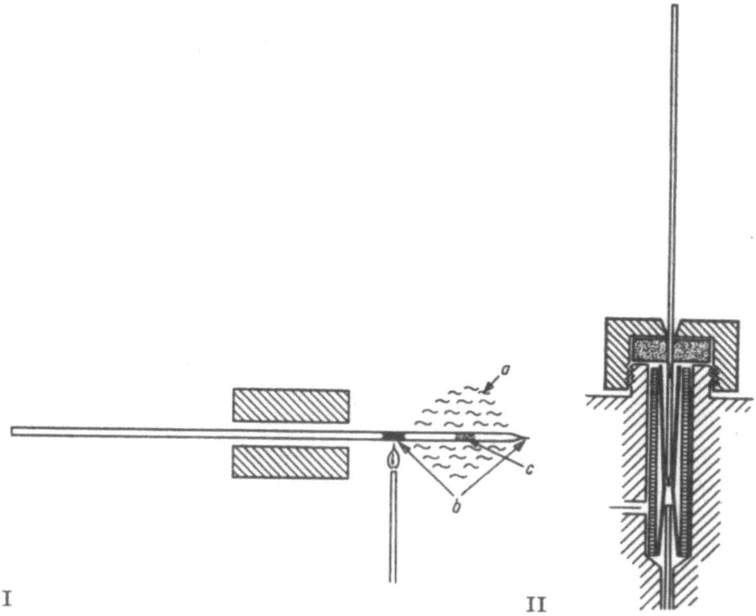

I II

Fig. 3—16 I

Preparation of a capillary trap for re-injection.
a. Dry ice b. Sealing points c. Trapped material (147).

Fig. 3—16II

Re-injection of the trapped material (147).

case the isolation method has an advantage, since 2,4-DNPH is a selective reagent for carbonyl compounds. Releasing the carbonyl compounds from the 2,4-DNPH's, after derivatization, has often been employed and reported for determination of carbonyls in food products, but HASHIMOTO (149) reports that the regeneration of carbonyl compounds from the 2,4-DNPH's by α-ketoglutaric acid, which is commonly used, gives regeneration recoveries of 70%, which is generally unacceptable for quantitative analysis. Direct gas chromatographic analysis of 2,4-DNPH's is reported (150—154) and is used by KALLIO (155) for the quantitative analysis of aldehydes from Arctic bramble. Contents as low as 20 μg/kg could be detected with an electron capture detector (155). The occurrence of double peaks for the 2,4-DNPH derivatives of most aldehydes and some of the ketones is explained by KOROLCZUK et al. (156) to be caused by the existence of the *syn* and the *anti* form.

Derivatization with Girard T reagent and subsequent regeneration of the carbonyls with formaldehyde or hydrochloric acid gives unsatisfactory results in quantitative analysis (157). In qualitative analysis, however, the method is used as a specific isolation method for carbonyl compounds. In an aqueous or ethanolic solution the carbonyls are converted to their Girard T derivatives and, after evaporation or extraction of the other volatile aroma compounds, the carbonyls are released again from their derivatives.

A more successful method for quantitative as well as qualitative analysis is derivatization with semicarbazide and subsequent regeneration of the carbonyl compounds with an inorganic acid. In our institute the method is applied with success in quantitative analysis of low-boiling saturated aldehydes in aqueous (158) and/or ethanolic solutions. After a reaction time of 15 minutes of the carbonyl compounds with the reagent in a sodium acetate solution at room temperature, the excess of water and/or ethanol and other volatile compounds can be distilled off in vacuum. Addition of, e.g., sulphuric acid releases the carbonyl compounds, which can then be determined accurately by direct vapour analysis.

Derivatives of alcohols

Although gas chromatographic analysis of alcohols does usually not present any problem, peak tailing especially of aromatic alcohols on some columns hampers their quantitative analysis. Derivatization may be expected to bring improvement. Another example where derivatization is of specific advantage, is the regeneration of the alcoholic moiety of acetals; in fact, GALETTO et al. (159) have used 3,5-dinitrobenzoyl-chloride in the isolation of the alcoholic part of acetals and subsequent analysis of the 3,5-dinitrobenzoates. This reagent has, however, the disadvantage that it is sensitive to water which might be present in an aroma concentrate. Applications of 4-nitro-phenyl-azobenzoyl chloride as reagent are reported for raspberry (160) and tomato (161) aromas. The reagent is preferred because of its resistance to humidity and the possibility of regenerating the alcohols from the isolated esters under rather mild conditions.

Trimethylsilyl ethers of alcohols are less important in aroma research. SEIDENSTÜCKER (162) analyzed the trimethylsilyl derivatives of some high-boiling terpene alcohols to prevent the destruction of tertiary alcohols by catalytic effects of metal parts of the gas chromatographic apparatus.

As shown by TULLBERG et al. (163), who investigated trimethylsilyl ethers of both alcohols and phenols, the ratio between the retention times of the free substance and the derivative depends on the chemical structure. In this way, extra information on qualitative analysis is gathered.

Derivatives of amines

Analysis of free amines on special types of column packings (e.g. Chromosorb 103) is applied with excellent results. On most column packings it is advisable to use less polar derivatives in order to avoid difficulties inherent to the character of the amino group.

CRATHORNE and SAXBY (164) reported the gas chromatographic separation of tri-fluoroacetyl derivatives of aliphatic amines. Derivatives of amines with a dinitro-phenyl group, used in the study of biochemical problems, can be detected with an electron capture detector; quantitative analysis of cyclohexylamine in a concentration of 1 mg/kg has been reported (165). GASCO and BARRERA (166) converted primary and secondary amines into N-alkylbenzamides: a mixture of isolated amine hydro-chlorides was benzoylated according to the SCHOTTEN-BAUMANN method. HOSHIKA reported the conversion of primary amines into their corresponding SCHIFF bases with benzaldehyde (167, 168) or thiophene aldehyde (169) and the subsequent analyses.

Derivatives of carboxylic acids

Free carboxylic acids can be directly analyzed by GC. The carboxyl group in the molecule necessitates, however, the use of special techniques to prevent the loss of compounds by adsorption in the column. Furthermore, the volatilities of high-molecular acids are generally too low for chro atography with reasonable retention times and column temperatures.
Usually the acids, especially the higher ones, are converted to methyl esters before chromatography. Methylation of free acids is carried out with e.g. methanolic boron trifluoride (170, 171) or diazomethane (172).
The separation of methyl esters is executed with good results on different types of polar stationary phases, such as diethylene glycol succinate or SP2300. Both packed and capillary columns are used, although the capillary ones (173) are favoured because of the complex nature of mixtures of fatty acids isolated from food products.

3.2.6. Qualitative analysis

3.2.6.1. Introduction

The chromatographic behaviour is one of the characteristics to be used in the qualitative analysis of unknown compounds. The most important parameters measured in qualitative GC analysis are the retention times of a compound. Other characteristics of an unknown compound are usually measured following a GC analysis, and include spectroscopic and organoleptic properties (see Chapters 4. and 5.). In many cases, an emerging component has to be *trapped* for such further analysis. We will also devote a section to that subject.

3.2.6.2. Retention data

Even on a column with a large separating power, an aroma concentrate will not be fully separated into its constituents. Two or more components of a mixture will emerge from the column, at the same time resulting in one peak. Additional information on the possible complexity of such a peak can be obtained by three GC methods, i.e. analysis of the same sample on another column of different polarity, re-analysis of a trapped compound or fraction on another type of column (see 3.2.6.3.), or specific detection (see 4.6.).
As mentioned already, identification with GC methods is based on the measurement of the retention times of the components (174).
Under constant GC conditions, the retention time of a compound also remains constant. In practice, however, these conditions vary for which reason the absolute retention time is unreliable. Improvement of the retention time method is reached by calculating relative retention times, obtained by relating the retention time of a compound to that of a standard compound, chromatographed under identical conditions. Such a compound can be analyzed either in a separate run, or added to the actual sample. Although this last possibility — the internal standard method — is preferred, the first one — the external standard method — is still in use.

In 1958, KOVATS (175) introduced more than one standard; his system, the so-called KOVATS Index System, is now in general use. The standards are the n-paraffins, to which by definition an index is assigned according to the number of C atoms multiplied by 100. The retention index or KOVATS index of any compound is calculated from the following equation

$$I = 100 \frac{\log (t'_R)_x - \log (t'_R)_z}{\log (t'_R)_{z+1} - \log (t'_R)_z} + 100z \qquad \text{eq. 3.19}$$

in which $(t'_R)_x$ is the adjusted retention time of a solute x and $(t'_R)_z$ and $(t'_R)_{z+1}$ are those of two n-paraffins with z and $z+1$ carbon atoms.
In determining KOVATS indices of a component of a mixture, the reference paraffins, chosen to emerge just before and after the peak of the unknown component, are preferably added directly to the mixture.
KOVATS indices are determined either numerically or graphically (176, 177). In the first case, the index of a component of a mixture is calculated using equation 3.19.; in the second case, adjusted retention times of the n-paraffins are plotted on the logarithmic scale of half-logarithmic paper and the KOVATS indices on the linear scale, resulting in a straight line. It does indeed follow from thermodynamic considerations that in any homologous series the logarithm of adjusted retention time should depend almost linearly on chain length, at least under isothermal conditions. This linearity is one of the advantages of the KOVATS Index System. Using the resulting graph, the KOVATS index of any compound is found by intersecting its adjusted retention time with this graph.
In some cases the retention indices of n-paraffins are not sufficiently reproducible, especially on polar columns. Improvement can then be effected by using other standards than n-paraffins. ANDERSON et al. (178) and MAARSE and VAN OS (179) selected some sesquiterpene hydrocarbons to calculate the KOVATS indices of other sesquiterpenes; VAN DEN DOOL and KRATZ (180) used ethyl esters as secondary references.
Retention indices can be found in many papers, while the books of MCREYNOLDS (181) and SCHUPP and LEWIS (182) list many of them on different stationary phases.
In linear programmed-temperature gas chromatography (PTGC) retention indices may be determined in a similar fashion as in isothermal GC. Assuming that a linear programme starts at the moment of injection and is not stopped until the last peak has emerged, one can plot the retention times or calculate the indices as shown by GUIOCHON (183), HARGOOD and HARRIS (184) and VAN DEN DOOL and KRATZ (180). The difference between isothermal and programmed-temperature analysis is that in the first case one relates the logarithms of the retention times to the retention index, whereas in the latter one uses the retention times themselves; the formula for linear programming conditions reads as follows:

$$I_{pr} = 100 \frac{(t'_R)_x - (t'_R)_z}{(t'_R)_{z+1} - (t'_R)_z} + 100z \qquad \text{eq. 3.20.}$$

Note: due to the linear relationship, the adjusted retention times in this formula may be substituted by the simple retention times or by the retention temperatures.

The extra operation parameter — the programme rate — makes the utilization and the comparison of published data very difficult. Detailed information on the influence of temperature programming on retention indices is given in the book of HARRIS and HABGOOD (185).
Qualitative analysis by retention indices is based on calculating or measuring the indices and comparing them with indices of known compounds under identical conditions. Refinements include the use of ΔI values and the determination of indices on more than one column. The ΔI value of a substance at a certain temperature is the difference between the retention indices on a polar and an apolar column (186):

$$\Delta I = I_{pol} - I_{apol} \qquad \text{eq. 3.21.}$$

and reflects the polarity of the analyzed substance (see table 3—7).

Table 3—7
I and ΔI-values of C_6-compounds (186)

name	I_{100}^{S}	I_{100}^{PPG}	ΔI
Hexane	600.0*	600.0*	—
2-Methylpentane	571.3	568.4	2.9
Hexene-1	586.9	610.5	23.6
Cyclohexane	675.7	699.2	23.5
1-Chlorohexane	838.3	922.6	84.3
Benzene	660.7	767.9	107.2
Hexanal	758.3	899.0	140.7
2-Hexanone	745.6	891.0	145.4
Cyclohexanone	855.1	1038.2	183.1
1-Hexanol	831.2	1103.8	272.6

* by definition
I^{S} = Index in Squalane 100 m, i.d. 0,25 mm, glass column 100 °C
I^{PPG} = Index in Polypropylenglycole 100 m, i.d. 0.25 mm, glass column 100 °C
ΔI = $I^{PPG} - I^{S}$

Relations between molecular structure and retention indices of substances have been extensively studied (187—189). Much work in this field has been done by GOLOVNYA and co-workers (190—197) in the systematic identification of aliphatic components in mixtures of aroma compounds.
Analyses of fruit aromas carried out on more than one column and identifications by comparison of retention indices have been reported by DRAWERT and co-workers (e.g. 198, 199).
It must be emphasized that identification by means of retention data alone, without any further information, can never lead to unambiguous identification (see 4.1).

3.2.6.3. Trapping

The collection of single compounds or fractions at a column exit, in order to re-use the trapped components for further chromatographic analysis or spectrometric investigation, has been extensively studied. A prerequisite for trapping is a nondestructive detector or a splitting device at the column exit.

Trapping has for many years been executed at the outlet of systems consisting of an analytical column with a katharometer. Application of capillary columns, and the development of more sensitive — mostly destructive — detectors, have made it necessary to adapt trapping techniques. The collection devices are divided into two main types:

— open tubular traps
— traps filled with a solid material.

Collection of eluted compounds in a short column, filled with e.g. a porous material, a column packing material or carbon black, has been described in section 2.2. The collection in open tubular traps will be discussed here. The kind of open tubular trap to be used depends on the type of column and detector, the amount and nature of the compounds to be trapped and, last but not least, the subsequent analysis of the trapped material.
Although metal, U-shaped, traps have been used in the past, glass traps in various shapes and sizes are very common nowadays in order to exclude formation of artifacts as much as possible. U-shaped glass traps with an inner diameter of 3 mm, and zig-zag traps (figure 3-17), as described by BADINGS et al. (200) have been used with success for trapping at the outlet of a katharometer detector connected to an analytical column. With these types of traps, efficiencies of 80% and higher can be attained. Cooling is effected by an ice-water mixture, solid carbon dioxide, or liquid nitrogen. To prevent premature condensation of components in the connecting tubing, the katharometer outlet must be heated. Inefficient trapping, which is often noticed when smelling at the outlet of the trap, is caused either by inefficient cooling or too strong cooling, so that the droplets of the condensed vapour are carried along by the gaseous phase. In practice, cooling temperatures of approx. 200 °C below the boiling points of the compounds to be trapped are low enough. Condensation of water and other materials entering the trap upstream must be prevented, e.g. by connecting a tube filled with molecular sieve to the outlet of the trap.
The large volume and inner surface of the above-described traps is a disadvantage, because large quantities of solvent are needed to rinse the condensed compound(s) from the trap. A more suitable kind of trapping in this respect, especially for small-

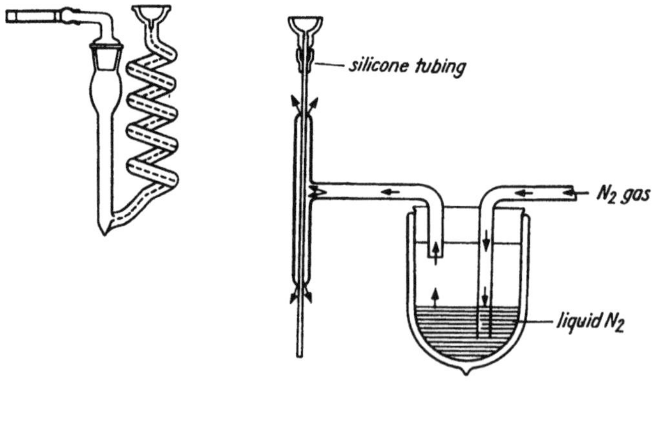

Fig. 3—17
Trap according to BADINGS and WASSINK (200) and capillary trapping device according to BURTON and KENNER (201).

sized samples, is best effected with capillary tubes, either straight or bent. The inner diameter of such capillaries is 0.5—1.0 mm.
A trapping device according to BURSON and KENNER (201) (Fig. 3—17) is being used in our institute with success (e.g. 202). Trapping efficiencies up to 90% have been attained by collection of compounds at the outlet of a katharometer or microkatharometer.
When using a destructive detector, a splitting device has to be connected at the outlet of the column to direct part of the effluent to the trapping system, by-passing the detector. Glass capillary tubes, straight (145, 203, 204) or bent zig-zag (40), have been used as traps (Fig. 3—18).

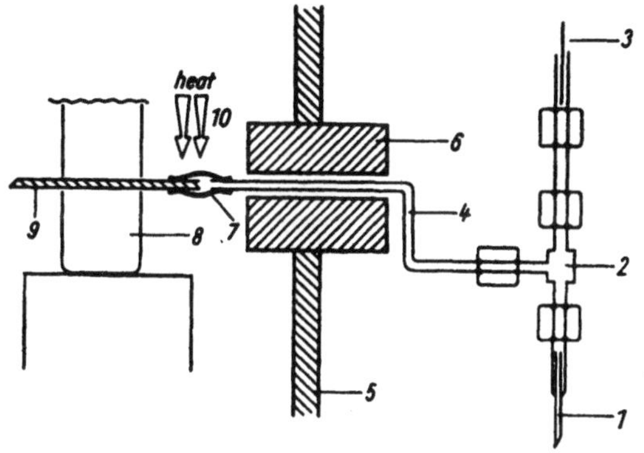

Fig. 3—18
Schematic representation of a peak trapping system (203).
1. Hypodermic connection to column
2. o.d. 1.55 mm Simplifix three-way connector
3. Restricting wire to FID
4. Stainless-steel capillary transfer line
5. Oven wall
6. Outlet heater
7. Silicone rubber sleeve (i.d. 1 mm)
8. 25 ml plastic beaker with dry-ice column
9. Trap
10. Heat from hair dryer.

The advantages of these glass traps are evident:

— easy sealing by melting both ends with a micro-burner
— small volume.

As described in 3.2.5., trapped compounds are re-injected after addition of a small amount of solvent, or by crushing the trap or incorporating it in the flow system (Fig. 3—15).

3.2.7. Quantitative analysis

3.2.7.1. Introduction

Modern gas chromatographic and spectroscopic equipment, e.g. a gas chromatograph-mass spectrometer-computer system (see 4.2.), enables the aroma worker to identify

aroma compounds, some of which may be hardly or not at all discernible as separate peaks in a chromatogram of a complex mixture. In some food products, more than 300 volatile compounds have been identified by means of such techniques (205).
Quantitative analysis of the most interesting aroma components of a food product is the logical follow-up of identification. The interest in quantitation of quality differences, relation of organoleptic characteristics to analytical composition and control of manufacturing processes and products, has given aroma research an impetus to deal with quantitative analysis.

3.2.7.2. Sampling

Depending on the concentration and volatility of the compounds, they are determined either by direct vapour analysis (see 3.2.4.) or by direct liquid analysis, or after isolation and concentration.
In the case of isolation and concentration, fewer precautions need be taken than in qualitative analysis, since possible alterations in the sample do not necessarily interfere with the analysis of the components under study (206—208). Thus, steam distillation at normal pressure and extractions for several hours which are favourable circumstances for the formation of artifacts can be applied as long as the compounds to be determined are not produced or decomposed.

3.2.7.3. Recovery determinations

When isolation methods are used, *recoveries* must be determined. Generally recoveries must be at least 80%, because low recoveries tend to be irreproducible. Three different methods for the determination of recoveries will be described:

(1) Known amounts of the compound(s) under investigation are added to the food product, the recoveries of which are determined after isolation and concentration.
The method has some obvious disadvantages. Because of the compound(s) under investigation may be expected to show natural variations, preferably two or more test samples should be prepared with added amounts of this (these) compound(s). However, in the case of solid or semi-solid products it is difficult to obtain an even distribution without partial volatilization. The added quantity should in any case be of the same order of magnitude as that already present.
(2) A known amount of a component similar to those under investigation is added and its recovery is determined. With this method only one test sample needs to be prepared, because fluctuations in sample composition do not come into play, and a test substance can be chosen of somewhat lower volatility than the compound(s) to be determined and can be added in a relatively small amount. The ideal choice would be one yielding a free-lying peak, but in fact some mixtures are so complex that a "free place" in the chromatogram can hardly be found. A disadvantage of this method is that the recovery can only be determined on the basis of one compound. In fact, even within a homologous series of compounds the recoveries of two consecutive members can be different.

(3) A "reference food product" containing known amounts of the components of interest is made and the recoveries are determined after isolation and concentration. For complex food products, including solid, semi-solid and fatty products, the method is seldom used, whereas it is recommended for liquid products not containing nonvolatile material. The other methods can be applied alternatively. For liquid products containing non-volatiles such as sugars and salts, the first or second method is generally chosen. For solid and partially solid products in most cases the first method is used when only a limited number, and the second when a large number of components has to be determined.

3.2.7.4. Calculations

The quantities of the compounds are basically determined by measuring the peak areas in a chromatogram. Peak areas are proportional to the amounts of the compounds eluted from the column. This means that quantitative gas chromatography can be divided into two steps: determination of peak areas and conversion of areas to amounts or concentrations.
The peak area can be determined by different methods of which we mention the following in decreasing order of sophistication: automatic integration, planimetry, multiplying the peak height by the peak width at half peak height, multiplying peak height by retention time, even quantitative methods based on peak height only are used. For overlapping peaks approximations must be made. Minor peaks on the trailing flank (the "tail") of a major one are commonly measured by drawing the interpolated flank of the main peak and calculating the area above. Automatic integrators calculate such minor peaks by a tangent method. Two overlapping peaks of comparable size are commonly divided by drawing a straight line at valley-point as a perpendicular to the base line.
Two methods are generally used in aroma research to convert peak areas to amounts or concentrations.
In the first or "external standard" method, one employs a synthetic mixture containing the compounds to be investigated in known concentrations approximating those in the actual sample. The solvent must be the same for both samples. If equal volumes of both samples are analyzed under identical GC conditions, the concentration of a compound (C) in the actual sample is:

$$C = \frac{A}{A'} \cdot C' \qquad \text{eq. 3.22.}$$

in which A and A' are the peak areas of the compound in the sample and the reference respectively, and C' is the concentration of the compound in the reference sample.
The method can be improved by plotting concentrations and peak areas from several reference samples with different concentrations. If non-linear behaviour of the detection system is expected, e.g. with electron capture detection, this method is obligatory.
In the second or "internal standard" method, a reference sample is also used, but an additional compound, not occurring in the original sample, is added to both samples. This internal standard compound should yield a free-lying peak in the chromatogram

in the vicinity of the compounds under investigation and preferably have a peak area comparable in size to those compounds. With the help of the internal standard compound, corrections can be made for varying injection sizes. The formula for calculation of the concentration C of a compound reads

$$C = \frac{A}{A'} \cdot \frac{A'_{st}}{A_{st}} \cdot C' \qquad \text{eq. 3.23.}$$

in which A_{st} and A'_{st} respectively, are the peak areas of the internal standard compound in the actual and the reference sample injection.
Plotting A/A_{st} against concentration is also possible (see External Standard Method).

Generally, the internal standard method is recommended, but in some cases — e.g. no free place being available in the chromatogram for an internal standard compound or no compound with desired characteristics — the external standard methods has to be applied.

3.3. Liquid chromatography

J. Schaefer

3.3.1. Introduction

Since the same theoretical considerations as those dealt with in 3.1. and 3.2. on gas chromatography are applicable to liquid chromatography, only those characteristics specific to the latter method will be discussed here.
For a more theoretical approach to the relevant mechanisms the reader is referred to the books written on the subject by Snyder (209) and Kirkland (210).

An essential parameter in chromatography is the partition coefficient of a component between the stationary and the mobile phase. In gas chromatography the main determining factor is the vapour pressure of the component, while its polarity plays a secondary role. To be at all amenable to gas chromatography a substance should have a certain vapour pressure, which means that gas chromatography can only be applied to the analysis of volatile compounds. Since in liquid chromatography the partition takes place between liquid and solid or between liquid and liquid, no vapour pressure is involved. Therefore liquid chromatography is applicable to volatile as well as to non-volatile compounds. Another important difference with regard to gas chromatography is the fact that in liquid chromatography the separation is also determined by the properties of the mobile phase.
The parameter governing the rate of transfer of matter between the stationary and the mobile phase is the rate of diffusion. In order that partition equilibrium between the two phases will be established as quickly as possible, it is important that the distances across which diffusion has to take place are as short as possible. In practice this means that the stationary phase in the column has to consist of small particles. Moreover, it is important that the particles are of uniform size, so that the profile of flow of the mobile phase will be homogeneous. In modern liquid chromatography, therefore,

accurately sieved fractions are used with a uniform particle size of about 10 μm diameter. A column filled with such small particles presents strong resistance to the flow of the mobile phase. The only way to make it pass at such a rate that the analysis will not take so long as to induce undue peak broadening, is the application of high pressures, i.e. in the order of a few hundreds of atmospheres. Liquid delivery pumps not only meeting this requirement but also working pulsefree are available since a few years and have given the impetus to a rapid development of high-pressure liquid chromatography (HPLC).

3.3.2. Equipment

Figure 3-19, schematically shows the main parts of which a liquid chromatograph consists. Reservoir *A* containts the mobile phase, (usually an ordinary solvent),

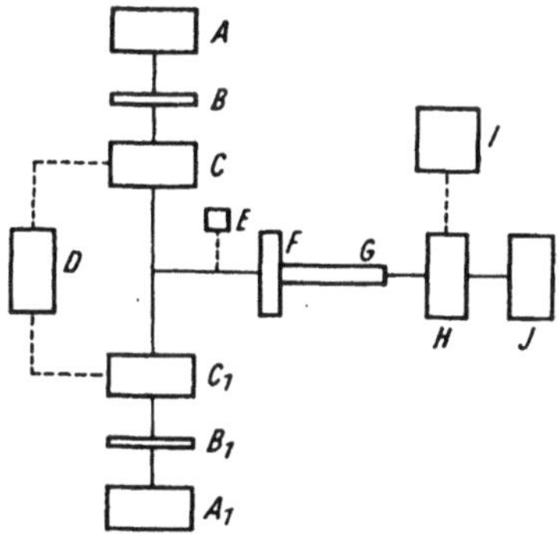

Fig. 3—19
Schematic diagram of a high performance liquid chromatograph.
A. solvent reservoir; B. filter; C. liquid pump; D. solvent programmer; E. pressure gauge; F. Injection system; G. separation column; H. detector; I. recorder; J. fraction collector.
The solid line represents the solvent path. The dotted line represents electronic connections.

which passes through filter *B* to pump *C*. If necessary, another solvent is delivered by a second pump (C_1); the proportion of the two solvents is controlled by solvent programmer *D*. The solvent then reaches separation column *G* via sample introduction system F. After having passed the column the eluted compounds then pass detector H the signal of which is displayed by recorder *I*. Finally, the eluted compounds plus the mobile phase leave the liquid chromatograph and can be collected in fraction collector *J* for further analysis. The applied pressure is recorded on pressure gauge *E*. If required, column *G* may be placed in a thermostatted oven. As a rule this is not necessary since the equilibrium is not much affected by temperature, in contrast to the situation in gas chromatography.

The main component parts as used in present-day equipment will be very briefly dealt with below. For more detailed descriptions the reader is referred to the specialized literature by, for instance, BAKER et al. (211), BERRY and KARGER (212), CHANDLER (213), KIRKLAND (214), LAWRENCE and FREI (215), LEITCH and DE STEFANO (216), LOCKE (217), MCNAIR and CHANDLER (218) and MCNAIR (219).

Pump

The pump must meet very stringent requirements, because under high working-pressure a constant and pulse-free flow rate of the mobile phase should be maintained.

In present-day equipment, two main types of pumps are in use: the continuous delivery systems and the piston pumps. Continuous delivery systems are equipped with a plunger which is forced downwards in a cylinder by an accurately adjusted motor; the mobile phase is delivered at a very constant flow, entirely free of pulses. This system is rather costly but important in such cases that require the greatest possible sensitivity in the analytical procedure. In piston pumps, the mobile phase is delivered by means of a dual piston system: the forward stroke pushes liquid to the chromatographic system, while the backward stroke sucks liquid from the reservoir. Here pulses do occur which are, however, minimized in modern pumps.
If it is necessary to vary the composition of the mobile phase during a chromatographic run (= solvent programming), a second pump plus a solvent programmer are usually required. Nevertheless, systems do exist in which solvent programming can be effected with only one pump.

Column

There are four basic types of liquid chromatography, each of them requiring their own specific column packings:

1. Adsorption chromatography; the packing consists of solid particles, e.g. silica, with a diameter of approximately 10 μm or glass beads coated with a layer of silica (pellicular packings), are used, the diameter in this case being a few tens of μm.
2. Liquid-liquid chromatography; the packing consists of a solid support with a layer of liquid stationary phase as in gas chromatography. The mobile phase being a solvent, the danger exists of the stationary phase dissolving in the mobile phase. Therefore, nowadays one mostly employs stationary phases chemically bound to the support (217). Otherwise a small saturation column has to be inserted in front of the separation column, in order to saturate the mobile phase with the stationary phase.
3. Ion-exchange chromatography; anion or cation exchangers, often chemically bound to the support, are employed for the separation of ionogenic compounds.
4. Exclusion chromatography;packings are applied that effect separations according to molecular size, which means separation according to molecular weight for compounds of similar molecular shape.

In general, for analytical purposes columns with a length of 10—60 cm and a diameter of 0.2—1.0 cm are used, with plate numbers of 1,000—10,000/m.

Injection system and detector

The most common injection systems are not essentially different from those in gas chromatography: septum injection and injection by means of a sampling loop. At present systems are available permitting any desired sample volume to be injected pressureless outside the flow system and brought into the column with the aid of a valve.
Two classes of detectors are in general use: universal detectors, responding to an effect that any compound in the effluent will manifest to a greater or lesser degree, and

specific detectors, responding to an effect caused by a physical property shown only by a limited number of compounds.
Of the *universal* detectors, the one most often employed is the refractive index (RI) detector. It is based on continuous measurement of the refractive index of the eluate against that of the pure mobile phase. A difference of 10^{-7} RI units is still detectable; this generally corresponds to a quantity of substance in the order of 1 μg, which is not particularly sensitive. Another type of universal detector is the moving wire detector. Here a small part of the effluent is, with the aid of a wire, transported to a solvent evaporation chamber. The components are then transferred to a reaction chamber where they are catalytically transformed into methane, which is finally detected with a flame ionization detector as in gas chromatography.
Among the class of *specific* detectors, the ultra-violet (UV) detector is the one most often employed. It measures the UV absorption of the eluate against that of the pure mobile phase or air. The simplest UV detectors operate at a fixed wavelength, usually 254 nm; the more sophisticated ones have a wavelength adjustable from 200 to 700 nm. The sensitivity is high towards UV-absorbing compounds, for instance of a compound with $\varepsilon = 10^4$ a quantity of 1 ng is detectable. Another type of specific detector is the fluorescence detector, which is very sensitive towards some special compounds. Several other detectors, which are more or less in the experimental stage, include electrochemical ones (220, 221), monitors of dielectric constant (215), monitors of radio-activity (215), and infra-red detectors (222). An interesting development is the coupling of a liquid chromatograph to a mass spectrometer as described by McLafferty (223).

3.3.3. Applications

3.3.3.1. Food, general aspects

Liquid chromatography is often applied in the analysis of amino acids, fats, acids, and pesticide residues in foods. It also is a suitable method for the analysis of sugars, steroids and various food additives.
One example will be discussed here, namely the determination of benzo(a)pyrene in condensates of wood smoke, that are used in the curing of foods (224). An important feature of liquid chromatographic analysis is the necessity of pre-treatment and 'clean-up' of the sample by methods of extraction, classic column chromatography, thin-layer chromatography and the like, in order to ensure that the final sample to be analyzed contains as few compounds as possible. This is necessary because of the limited separation power of liquid chromatography. While with gas chromatography, especially with capillary columns, one can obtain chromatograms of two hours' duration with up to 200 well defined peaks, a liquid chromatogram should preferably be completed within 10—15 minutes because beyond such a period the peaks would become very broad as a result of which especially quantitative analysis would become rather hard to perform. This comparatively limited separation power necessitates previous removal of all substances that would give interfering peaks in the chromatogram.
Returning now to the above mentioned study (224), first of all an aqueous solution of the smoke condensate was extracted with isooctane, the extract was passed through a

'Florisil' column that held back the polyaromatic hydrocarbons, then eluted from the column with benzene and chloroform, leaving many of the accompanying compounds behind. After distilling off the solvent a separation was effected by thin-layer chromatography on acetylated cellulose plates. The spots corresponding to aromatic hydrocarbons were scraped out and finally the determination of benzo(a)pyrene was carried out by high-pressure liquid chromatography (HPLC) (Fig. 3—20). Although this rather

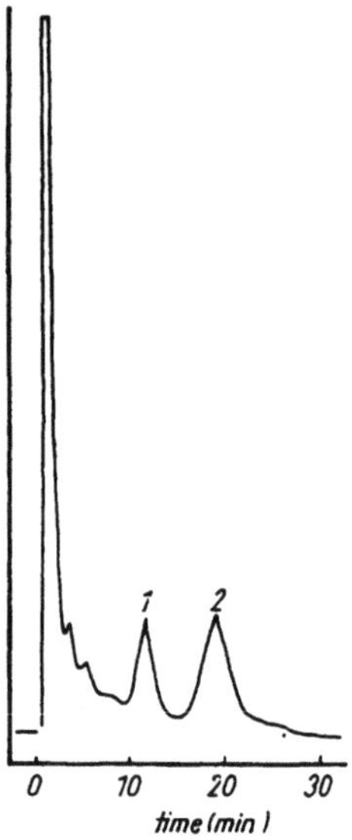

Fig. 3—20
HPLC separation of polycyclic aromatic hydrocarbons in aqueous smoke.
column: VYDAC Reverse Phase
solvent: ethanol-toluene-water 17:5:2 flow rate 0,75 ml/min.
1. benzo(a)pyrene; 2. 1,2,5,6-dibenzanthracene.

elaborate sample preparation may seem somewhat exceptional, each analysis of any component of such complex products as most foods are, does require at least one of these steps of pretreatment and/or clean-up.

3.3.3.2. Aroma research

Application of liquid chromatography in aroma research is still in a stage of development. In the relevant literature emphasis is put on the potentials of this method for the separation of aroma concentrates. Lee and Chang (225) have tried these possibilities, using a synthetic mixture of eight aroma compounds. By a series of separation procedures on different columns, they succeeded in isolating each one of the eight compounds, though they admitted the procedure to be long and tedious. Komae and Hayashi (226) have pointed to the feasibility of separating essential oils by means of HPLC. The low temperatures at which separation takes place are an advantage over gas chromatography; this is especially important when dealing with thermo-labile compounds as may occur in such oils. Figure 3—21 shows the method for some oils.
Schmit et al. (227) give an example of the analysis of rum oil and orange oil by HPLC. The oils are first separated into fractions of different molecular weight by exclusion chromatography at atmospheric pressure. The resulting fractions are then analyzed by HPLC. Another application is presented by Ross (228), who was able to draw distinctions between three oils by maens of HPLC: cinnamon bark oil, cinnamon leaf

oil and cassia oil. The first oil is by far the most valuable and attempts have been made to adulterate this by the addition of cinnamaldehyde to the other two oils. This can be detected by measuring the ratio cinnamaldehyde/eugenol with the aid of HPLC.

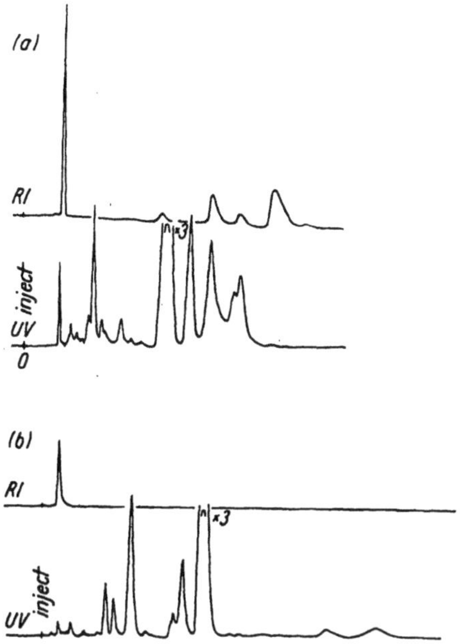

Fig. 3—21
HLPC separation of essential oils.
column: μ-Bondapak C 18
solvent: methanol-water 1:1 flow rate 1 ml/min.
UV: recorded with ultra-violet detector
RI: recorded with refractometric detector
a. *L. umbellata* THUMB
b. *L. sericea* var. *glabrata* BLUME.

HPLC is also employed for the analysis of components that contribute to the flavour without being appreciably volatile. This is the case for hop components in beer, which have been analyzed by CHARALAMBOUS et al. (229, 230), who obtained a good separation of polyphenols, phenolic acids and nucleosides, which for lack of volatility, cannot be well separated by gas chromatography.
Liquid chromatography can be applied to subdivide an aroma concentrate into fractions in order to judge these sensorially. Further investigation can then be concentrated on those fractions that are of interest from the aroma viewpoint. This type of separation has been effected with an extract from wood (231), containing compounds that are important to the aroma of brandy and that enter the brandy during aging in wooden barrels. The relevant chromatogram is presented in Figure 3-22; ten fractions were collected and sensorially judged. One fraction proved to carry the aroma of wood; non-aged brandy acquired the aroma typical of aged brandy when this fraction was added to it.

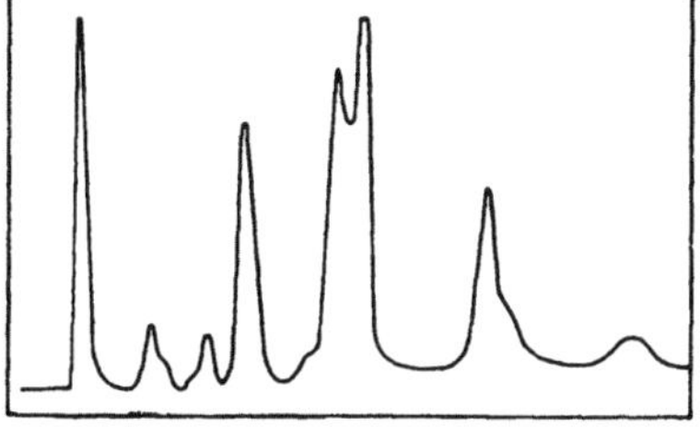

Fig. 3—22
HPLC separation of extract of oak wood.
column: OPN/Porasil C
solvent: hexane (10 min.), hexane-chloroform 1:1 (10 min.), chloroform (10 min.), flow rate 2 ml/min.
detector: UV. — 254 nm.

3.4. Thin layer chromatography

J. H. DHONT

3.4.1. Introduction

In their extensive article on the history of TLC, PELICK et al. (232) state that "The belief is often expressed today that a significant scientific contribution requires the effort of an interdisciplinary team and the extensive utilization of electronic equipment. However, all the most useful chromatographic methods currently employed in chemical and biochemical research were discovered by individuals working under rather adverse conditions and whose funds for equipment and supplies were extremely limited".

Among the chromatographic methods developed since the time TSWETT added column chromatography to the arsenal of analytical tools in the first years of this century, only paper chromatography, first described by CONSDEN et al. in 1944 (233), and TLC, invented by the Russian workers IZMAILOV and SHRAIBER (234) in 1938, have remained simple and inexpensive methods without sophisticated electronic equipment and without a real theory.

A rapid expansion of publications concerning the application of TLC has started since 1956 when STAHL invented an applicator for coating glass plates together with other auxiliary equipment (235). Several books on TLC are now available (e.g. 236—238).

Although TLC has a low plate height as compared to gas- and liquid-liquid column chromatography, resulting in a slight resolution, if any, of straight chain compounds from their branched chain isomers having the same carbon number, the plate height suffices to separate series of homologues differing from each other by one carbon atom.

The usefulness of TLC is mainly due to the outstanding selectivity of the method. Classes of compounds and some forms of isomers are easily separated.

In this chapter the TLC of volatile aroma compounds will be described. Since the thin layers are exposed to the atmosphere in the developing tank, generally compounds having a boiling point of less than 180° largely evaporate from the layer during development. It is therefore necessary to convert these compounds to non-volatile derivatives before spotting them onto the thin-layer plate.

Although a large number of different kinds of precoated plates is now commerically available the presence of an applicator for manufacturing home-made plates is a must for every worker who will make a rational use of TLC, since impregnated plates are not commercially available. Moreover, mixing the adsorbent with the stationary phase before applying the sorbent onto the glass plates is still the only proper way to obtain satisfactory layers for partition chromatography.

Practical and theoretical aspects of TLC are described in several handbooks to which the reader may be referred for details (236—238). In this chapter details are given for the practical application of TLC to problems in which volatile aroma compounds are involved.

Part of the material presented in this chapter has been collected from the relevant literature. In accordance with normalization prescriptions (239) the composition of solvent mixtures is given in volume percentages. The compositions of the solvent mixtures taken from the literature were not normalized but expressed as in the original publication. Much of the material, however, is the result of work done in the author's laboratory, with the assistance of miss J. C. DE BEAUVESER and mr. J. RUS. Dr. R. GOLOVNYA[1]) contributed much of the material presented in the part on TLC on acetylcellulose.

3.4.2. Definitions of R_F, ^nR_F and R_M; standardization of R_F

The most important single parameter in TLC is the R_F-value of a compound. It is defined as the ratio between the distances the compound and the solvent front have travelled from the starting line. In a well-defined chromatographic system, R_F ought to be a characteristic of the compound in question.

Usually the length of a TLC chromatogram is 10—15 cm. In order to enlarge the effective distance travelled by the solvent front, and to ensure a better separation of the substances involved without using larger plates, "multiple development" can be used. In this technique, the plate is first developed in the normal way. After the solvent front has reached a predetermined line (10—15 cm from the start) the plate is removed from the tank and dried. Development is then repeated with the same solvent. This operation can be repeated n times. Denoting the measured "R_F-value" after n developments as ^nR_F, the following relations hold:

$$^nR_F = 1 - (1 - R_F)^n \quad \text{and} \quad R_F = 1 - \sqrt[n]{1 - {}^nR_F}. \qquad \text{eq. 3.24.}$$

For the theoretical aspects concerning multiple development, the reader is referred to the literature (240—243).

As is the case in paper chromatography, the use of R_M-instead of R_F-values is of advantage in the study of structural interrelationships and in the study of the behaviour of homologous series, R_M being defined as:

$$R_M = \log\left(\frac{1}{R_F} - 1\right) \qquad \text{eq. 3.25.}$$

As stated above, the R_F-value ought, in principle, to be a constant in a given chromatographic system. The determination of this constant, however, suffers from the difficulty that many of the conditions in the determination are poorly reproducible; this aspect relates to such items as pore size of the adsorbent, solvent gradient along the plate, vapour phase composition, distribution of vapour phase composition in the chromatographic chamber, etc. CASPARIC (244) in 1968 still suggested the use of an internal standard for relative determination of R_F-values, even though it is known from experience that in comparing R_F-values from different sources, the method of relative R_F-values is invalid. Even a method using R_F-values in relation to a homologous series of compounds, in analogy to the successful use of KOVATS indexes in gas-chromatography, supposes a logarithmic relationship between R_F and chain length, which is seldom the case in chromatography on silica gel. It is a general belief that R_F-values in TLC are not reproducible in interlaboratory experiments, and that R_F-values measured in one laboratory are useless to workers in another.

[1] Dr. R. V. GOLOVNYA, Institute of Organo-Element Compounds Academy of Sciences USSR, Moscow.

However, surprisingly good results have often been obtained in the author's experience when the quite unsophisticated method of GALANOS and KAPOULAS (240) was used for calculating "standard" or "corrected" R_F-values.
In 1964, GALANOS and KAPOULAS suggested the use of two reference compounds for R_F-correction in *paper* chromatography. The method was later extensively tested with a variety of chromatographic systems (245—247), and we shall now discuss this method in some detail.
GALANOS and KAPOULAS showed that the following linear equation

$$R_F^c = aR_F + b \qquad \text{eq. 3.26.}$$

can be applied successfully in many cases. In this equation, R_F stands for the R_F-value of any compond in a particular chromatogram; R_F^c is the corrected R_F-value for that compound, and a and b are constants for the given chromatogram. To find the values for a and b, the R_F-values of two reference compounds in this chromatogram are compared with those found for the same two compounds in some previous chromatogram, defined as the chromatogram with standard conditions. Constants a and b then follow from the two equations

$$a = \frac{R_{FA}^0 - R_{FB}^0}{R_{FA} - R_{FB}} \quad \text{and} \quad b = R_{FA}^0 - aR_{FA} \qquad \text{eq. 3.27.}$$

where R_{FA}^0, R_{FA}, R_{FB}^0 and R_{FB} denote the R_F-values for the two reference substances, A and B, obtained in the chromatogram with standard conditions and in the given chromatogram, respectively.
The method is not applicable to spots close to the front or the starting line, where deviation from linearity becomes too appreciable. However, compounds with very high and low R_F values are usually discarded from consideration. Within this limitation, the two reference compounds are preferably chosen to differ as much as possible in R_F-value, so that the numerator and denominator in the quotient for the constant a may be known with greatest relative precision. The "chromatogram with standard conditions" is chosen arbitrarily, or it may even not exist at all, the R_F^0-values of the two reference compounds being taken from tabulated data. The aim of the method described above is solely to make R_F-values from different chromatograms intercomparable, by converting them all into R_F^c-values. A concrete example of calculation is given below.
Table 3-8 contains R_F-values for a selected number of compounds found in two laboratories, under the following conditions:

Lab. I: layer, Silica gel G (Merck); chamber saturation;
Lab. II: silica gel "Fertigplatte" (Merck), unsaturated chamber;
solvent in both cases: benzene:dioxane:acetic acid (90 + 25 + 4).
Arbitrarily, the R_F-values found in Lab. 1 are taken as the "true" values. For calculation of constants a and b the compounds having the lowest and the highest R_F-values are taken as the reference compounds: $R_{FA}^0 = R_{F(\text{dimethyl yellow})}^0 = 0.79$, and $R_{FB}^0 = R_{F(\text{protocatechualdehyde})}^0 = 0.37$.
$a = (0.79 - 0.37)/(0.90 - 0.39) = 0.823$; $b = 0.79 - 0.823 \times 0.90 = 0.048$
The complete calculation is shown in Table 3-9.

Table 3—8
R_F-values of some compounds determined in two laboratories

compounds	R_F found in lab.		
	I	II	ΔR_F
protocatechualdehyde	0.37	0.39	0.02
benzaldehyde	0.50	0.58	0.08
heliotropine	0.70	0.85	0.15
vanilline	0.56	0.70	0.14
dimethyl yellow	0.79	0.90	0.11

Table 3—9
Corrected R_F-values (R_F^c-values) of lab. II

compounds	R_F Lab. I	R_F^c Lab. II	$\|\Delta\| R_F$
protocatechualdehyde	0.37	0.37	0 (per definition)
benzaldehyde	0.50	$0.823 \times 0.58 + 0.05 = 0.52$	0.02
heliotropine	0.70	$0.823 \times 0.85 + 0.05 = 0.70$	0
vanilline	0.56	$0.823 \times 0.70 + 0.05 = 0.57$	0.01
dimethyl yellow	0.79	0.79	0 (per definition)

The correction procedure of GALANOS and KAPOULOS is even applicable under widely different conditions of types of silica gel, chamber saturation, etc. The independence of R_F^c from the type of silica gel is clearly demonstrated by DHONT and MULDERS-DIJKMAN (248).

All R_F values mentioned in this chapter have been obtained in our institute over several years. They have been corrected by the above-mentioned procedure.

3.4.3. Recipes for the preparation of chromatoplates

The recipes in this section are based on practical experience in the author's laboratory.

For basic concepts, such as sizes of plates and methods of application, the reader is referred to existing handbooks (236—238).

3.4.3.1. Recipes for non-impregnated plates

Silica gel: 25 g of silica gel is suspended in a mixture consisting of 20 ml of water and 40 ml of methanol by shaking in an Erlenmeyer flask. From this suspension, plates are made using a so-called applicator with the split-width set at about 0.25 mm. Before being used, the plates are allowed to dry in contact with the atmosphere for at least 3 hours. The plates can be stored indefinitely when kept, protected from dust and vapours, in a suitable compartment or desiccator.

Silanized silica gel: 25 g of silanized silica gel (Merck H_{254} brand) are suspended in a mixture of 5 ml of chloroform, 15 ml of water and 40 ml of methanol, using an electric mixer. Plates are made in the usual way. The plates can be used 1 hour after preparation and can be stored indefinitely.

Acetylcellulose: 15 g of acetylcellulose (Macherey & Nagel 300 G/Ac brand) are suspended in a mixture of 35 ml of ethanol, 15 ml of chloroform and 15 ml of water, using an electric mixer. From this suspension, plates are made in the usual way. The plates can be used about 2 hours after preparation and can be stored for several weeks, if protected from atmospheric influences. For a different method of preparation see 3.4.3.3.

3.4.3.2. Recipes for impregnated plates; layers impregnated with silver cations

Plates impregnated with non-volatile liquids: The most straightforward way to separate series of homologues by TLC is partition chromatography. Usually, partition chromatography is carried out on thin layers containing a non-volatile liquid as the immobile phase, while a volatile liquid acts as the mobile phase. The most widely used liquid phases have been collected in Table 3—10.

Table 3—10
Non-volatile liquids used as immobile phases in partition TLC

liquid	references
Carbowax	249
2-phenoxyethanol	250
3,3-oxydipropionitrile	251
paraffin oil	252
polyvinylacetate	253

Carbowax, 2-phenoxyethanol, 3,3'-oxydipropionitrile and polyvinyl acetate are polar phases used in conjunction with non-polar mobile phases in "normal" partition chromatography; paraffin oil is a nonpolar phase, to be used with polar mobile phases (mixtures of water with acetone, dioxane, methanol) in "reversed-phase" partition chromatography.
Our recipes for the preparation of these systems are as follows.

Carbowax, 3,3'-oxydipropionitrile, or 2-phenoxyethanol: 25 g of Kieselguhr G are mixed with 3 g of the stationary phase, 20 ml of water, and 40 ml of methanol, by shaking in an Erlenmeyer flask. The plates are left to dry in contact with the laboratory atmosphere and can be used 3 hours after preparation. Plates covered with Carbowax as the stationary phase can be stored indefinitely, if protected from dust and atmospheric vapours; plates containing 3,3'-oxydipropionitrile and 2-phenoxyethanol as the stationary phase should be used within 24 hours.

Polyvinyl acetate: 3 g of polyvinyl acetate (Mowilith 50 brand, Farbwerke Hoechst, Frankfurt, German Federal Republic) are dissolved in 40 ml of methanol and 15 ml of ethyl acetate. After dissolving, the volume is adjusted to 60 ml with methanol. In this solution are suspended 25 g of Kieselguhr G, by shaking in an Erlenmeyer flask. From this suspension, plates are made in the usual way. The plates can be used 3 hours after preparation and can be stored indefinitely if protected from the atmosphere. The apparatus used for the preparation of the plates should be cleaned immediately after use by rinsing with alcohol.

Remark: polyvinyl acetate dissolves very slowly. If no stirring is applied, the solution can best be prepared by standing overnight. With stirring, the material dissolves completely within 2—3 hours.

Paraffin oil: As a first step, a solution is prepared by dissolving 1 g of polyvinyl chloride in 100 ml of methyl ethyl ketone. Heating on a steam bath facilitates dissolution. 25 g of Kieselguhr G are suspended in 80 ml of the cooled solution of polyvinyl chloride. To the suspension are added 3 g of a good quality paraffin oil. The suspension is shaken in an Erlenmeyer flask before pouring it into the applicator, which should be cleaned with alcohol immediately after use.

Plates impregnated with volatile liquids: The impregnation of silica gel thin layers by vapour absorption was first described by URBACH (254), who impregnated the layers with methanol vapour. This method has been studied by several authors (for a literature survey see 255). Later, DHONT and MULDERS-DIJKMAN (248) described an arrangement (see Fig. 3—23) for the impregnation of silica gel with vapour of nitromethane and acetonitrile. The procedure currently applied in the author's laboratory is as follows.

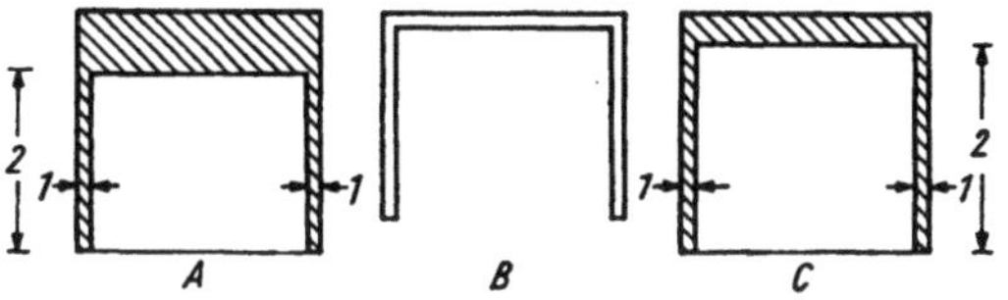

Fig. 3—23
Arrangement for vapour phase impregnation. *A* coverplate, *B* spacer, *C* normal thin layer plate. The shaded areas are scraped off.
1. 1.5—2.0 cm 2. ca. 15 cm.

A thick "cover plate", A, is impregnated with the liquid (nitromethane or acetonitrile), next the thin-layer plate C, on which the compounds to be separated have been spotted, is mounted in a sandwich arrangement with A, using strip B as the spacer between the plates. The plates are mounted in such a manner that the layers face each other. In the narrow gap that exists between the surfaces of the plates, the vapour given off by the liquid on the cover plate diffuses rapidly to the silica gel layer and is partially adsorbed by it. The sandwich is held together with clamps.
The layer of the cover plate is prepared as follows: 25 g of Kieselguhr G, 10 g of gypsum and 10 g of magnesium oxide are suspended in 120 ml of a 1:1 mixture of water and methanol by shaking in an Erlenmeyer flask. The suspension is spread out on glass plates of 20 × 20 cm, using standard equipment. If an adjustable applicator is used, a split width of approx. 1 mm is taken. After drying for about half an hour, part of the layer is scraped off as shown in Figure 3—23. The plates are left in contact with the atmosphere overnight; they can be used indefinitely.
For chromatography, a non-impregnated silica gel thin-layer plate is partially scraped to acquire the shape shown in Figure 3—23 and the spots applied at 1.5—2 cm from the bottom edge. Nitromethane or acetonitrile is poured on to the surface of the cover plate, the excess liquid is drained off, and the sandwich mounted as described. After

one minute, the sandwich is placed in the development tank containing 100 ml of mobile phase (hexane). A few words may here be added on the chromatographic separation itself. Because R_F-values are found to be quite low in the systems in question, multiple development is of advantage; after the first run, the sandwich is taken from the tank and dismounted, the cover plate is covered with a glass plate, and the hexane allowed to evaporate from the silica gel plate for a few minutes; the sandwich is mounted again and developed for the second time. Usually two or three developments will be sufficient for a good separation. The mechanism on which the separation is based seems to be pure liquid-liquid partition chromatography, as shown by the following: when ^1R_F-values are calculated from the ^3R_F-values using eq. 3.24., then these ^1R_F-values can be converted to R_M-values (eq. 3.25.) which, plotted against the number of carbon atoms, prove to yield a straight line (see Fig. 3—24), which according to theory is characteristic for liquid-liquid chromatography.

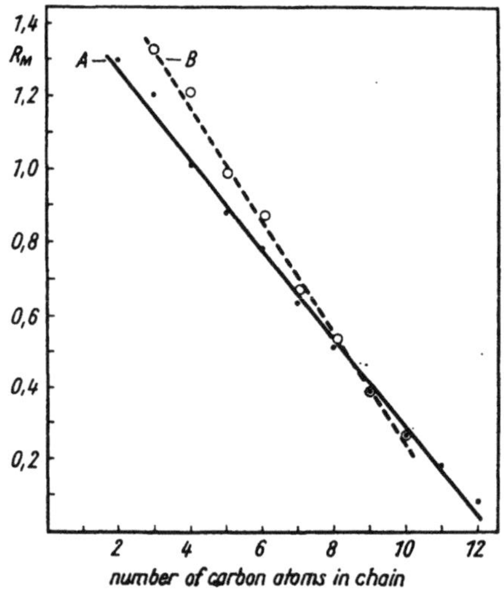

Fig. 3—24
R_M-values in relation to the number of carbon atoms in the aliphatic chain.
A. n-alkanal 2.4-dinitrophenyl hydrazones in the system acetonitrile-hexane.
B. n-alkanone 2.4-dinitrophenyl hydrazones in the system nitromethane-hexane.

Layers impregnated with silver cations: Layers impregnated with inorganic salts have been employed in TLC for a number of purposes. Well-known applications are the separation of sugars on layers impregnated with sodium borate and the separation of aromatic amines on layers containing cadmium or zinc ions.

The impregnation of layers with silver cations has gained considerable interest; it is used in the separation of aroma compounds according to their degree of unsaturation, in the separation of cis-trans-isomers and in the separation of positional isomers. In general, cis-olefinic structures complex more tightly with silver ions than do the trans-isomers; on a silver-impregnated layer, therefore, the cis-isomer is retained in respect to the trans-isomer. The R_F data of unsaturated compounds on these layers point to linear relationships with the amount of silver ions present in the layer. It is therefore advisable to estimate the optimum amount of silver ions for a given separation. There is a considerable amount of literature on this subject, see Table 3-11.

Table 3—11

Review of literature concerning the chromatography of unsaturated volatile compounds on silver impregnated layers

composition of thin layer	solvent system	compounds separated	reference
silica gel, 13% $AgNO_3$	benzene : acetone 95 + 5	sesquiterpene hydrocarbons	256
silica gel, 20% $AgNO_3$	benzene	monoterpene hydrocarbons	257
silica gel, 5—30% $AgNO_3$	toluene	isomeric octadecanoates	258
alumina, 28.5% $AgNO_3$	petroleum ether: ethylacetate 75 + 25	resin acid methylesters	259
silica gel, $AgNO_3$ content not specified	chloroform	tetracyclic triterpenes	260
silica gel, 26% $AgNO_3$	mixtures of benzene and methanol	allylic propenylic isomers	261
silica gel, 6—25% $AgNO_3$	benzene	monoterpene hydrocarbons	262
silica gel, 3% $AgNO_3$	dichlormethane : chloroform : ethylacetate : n-propanol 50 + 50 + 5 + 5	sesquiterpene alcohols	263
Kieselguhr, 30% $AgNO_3$	petroleum ether	2,4-DNPH's of unsaturated carbonyl compounds	249
aluminium oxide, 6.6% $AgNO_3$	mixtures of light petroleum-ether and chloroform	2,4-DNPH's of unsaturated carbonyl compounds	264
silica gel, 30% $AgNO_3$	benzene	2,4-DNPH's of cis-trans isomeric 2,6-nonadienals and 4-heptenals	265

3.4.3.3. Acetylated cellulose as stationary phase

General remarks. A systematic investigation of organic compounds on acetylated cellulose has not yet been made, although it is to be expected that partition chromatography on bound-ester phases such as cellulose acetate will be a useful addition to the chromatography on silica gel layers whether or not impregnated with liquid stationary phases. In this paragraph, results are discussed as obtained in our institute, when we used thin layers of acetylcellulose.

Dependence of R_F and R_F^c values upon acetyl content: Different batches of acetylcellulose of "MN 300G/Ac" brand, which contained different amounts of acetyl groups, were used in preparing TLC plates. The acetyl contents of the preparations used were determined as described by TANGHE et al. (266).
The test compounds were the 2,4-dinitrophenylhydrazones of methanal, ethanal, propanal, decanal, veratrumaldehyde and phenylpropanal. The solvent consisted of a mixture of ethyl acetate, methanol, di-n-butyl ether and hexane (85 + 5 + 5 + 5). The results are shown in Figure 3—25. The R_F^c-values of propanal and phenylpropanal

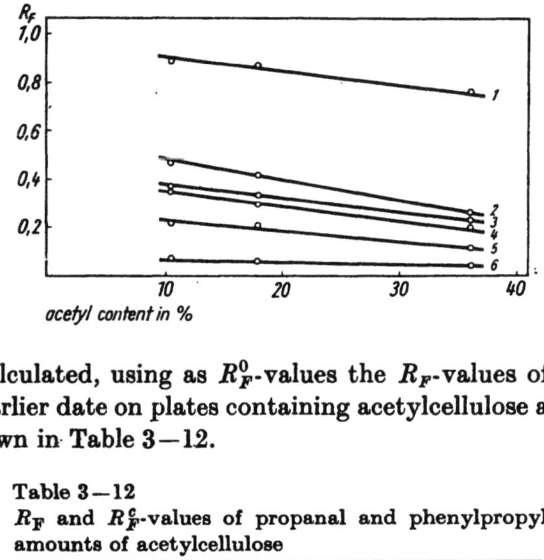

Fig. 3—25
R_F-values in relation to the acetyl content of the layer. 2.4-dinitrophenylhydrazones of
1. decanal, 2. propanal,
3. phenylpropylaldehyde,
4. ethanal, 5. methanal,
6. veratrumaldehyde.

were calculated, using as R_F^0-values the R_F-values of ethanal and decanal obtained at an earlier date on plates containing acetylcellulose acetylated to 10.0%. The results are shown in Table 3—12.

Table 3—12
R_F and R_F^c-values of propanal and phenylpropylaldehyde on plates with different amounts of acetylcellulose

compound	% acetylated					
	10.5		18.0		36.0	
	R_F	R_F^c	R_F	R_F^c	R_F	R_F^c
2,4-DNPH of propanal	0.47	0.68	0.42	0.68	0.25	0.65
phenylpropylaldehyde	0.37	0.62	0.33	0.63	0.22	0.63

SPOTSWOOD (267, 268) investigated the dependence of R_F-values of some polycyclic aromatic hydrocarbons on sheets of acetylated cellulose, on the degree of acetylation. From the R_F-values given by that author, the R_F^c-values can be calculated using 3,4-benzopyrene and phenanthrene as reference compounds. The results are shown in

Table 3—13
R_F- and R_F^c-values for some aromatic hydrocarbons on acetylated paper; solvent methanol:ether:water (4 + 4 + 1)

compound	% acetylated							
	21.2		23.7		25.1		28.0	
	R_F	R_F^c	R_F	R_F^c	R_F	R_F^c	R_F	R_F^c
chrysene	0.46	0.38	0.41	0.41	0.27	0.34	0.16	0.34
pyrylene	0.63	0.53	0.52	0.52	0.41	0.48	0.25	0.47
pyrene	0.77	0.65	0.63	0.63	0.55	0.63	0.37	0.63

Table 3—13. The conclusion is that the GALANOS-KAPOULAS correction procedure is a useful tool for obtaining better reproducible R_F-values on acetylated cellulose of varying acetyl content.

Improvements in spot shape: In our investigations, it was observed that spots sometimes became quite elongated. This phenomenon could be caused by the fibre structure of the acetylcellulose preparations used. Therefore, a method was worked out to prepare plates on which the thin-layer consists of Kieselguhr covered with acetylcellulose. The following method was found satisfactory.

Four grammes of fully acetylated cellulose (acetyl content 40—44.8%) are dissolved in 60—80 ml of a 1:1 mixture of methanol and dioxane. In this solution are suspended 25 g of Kieselguhr. The plates are prepared in the usual way. They can be stored for at least several weeks, and perhaps much longer. The layers are mechanically rugged, while the spots obtained during chromatography were always round and compact.

Separation of aliphatic homologous series on acetylcellulose: In the separation of the derivatives of the volatile members of the aliphatic series of carbonyl compounds, amines, etc., a mixture of hexane, methanol, ethyl acetate and di-n-butyl ether in the proportion 85 + 5 + 5 + 5 can be used as the solvent. This system has been the subject of systematic study in our laboratory. With the exception of the first two or three members of a series, all the series investigated show a linear relationship between R_M-value and the number of carbon atoms in the chain. In the calculation of R_M-values, R_F^c-values were used as obtained from the measured R_F-values, using the R_F-values of the 2,4-DNPH's of ethylvanillin ($R_F^0 = 0.064$) and decanal ($R_F^0 = 0.920$) as the reference values.

Linear equations established as the best fits for the data obtained, relating R_M-value and carbon number, are given in Table 3—14. Theoretically, the homologous increment

Table 3—14
Regression equations relating R_M-value and the carbon number for some homologous series on acetylcellulose

series	regression equation
alkylanilides	$R_M = -0.183\ C + 0.676$
7-alkyl-4-nitro-2-oxa-1,3-diazoles	$R_M = -0.179\ C + 1.61$
alkanal 2,4-dinitrophenyl hydrazones	$R_M = -0.166\ C + 0.367$
2,4-dinitrophenylmercaptides	$R_M = -0.170\ C + 0.372$

per$-CH_2-$ group should be constant irrespective of the nature of the functional group. Using the data from the table, the best estimate for this increment is the mean of the values found for the different series. Thus,

$$\Delta R_M(CH_2) = \frac{-0.183 + (-0.179) + (-0.166) + (-0.170)}{4} = -0.175\,.$$

A typical example showing the separation of homologues is given in Figure 3-26.

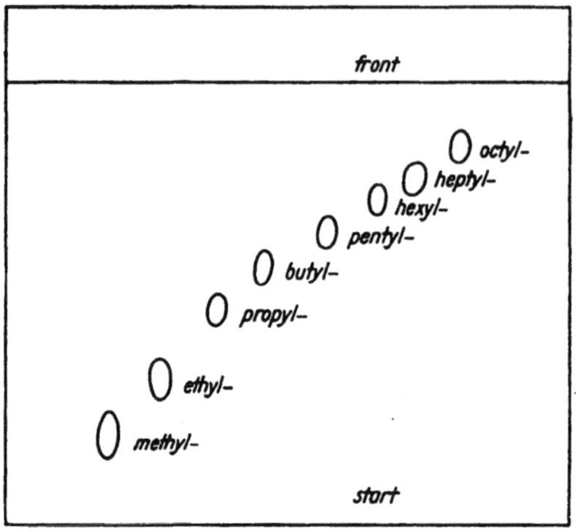

Fig. 3—26
Separation of dinitro-arylmercaptides on thin layers of acetylcellulose. Solvent: hexane, methanol, ethyl acetate, di-n-butyl ether (85 + 5 + 5 + 5).

Separation of 2,4-DNPH's of aromatic-type carbonyl compounds on acetylcellulose: Just as in the case of polyvinyl acetate, derivatives of compounds having an aromatic ring in the molecule require a more polar mobile phase than do the aliphatic-type compounds. For the separation of the 2,4-DNPH's of carbonyl compounds derived from benzaldehyde or benzophenone, the following system was found to give satisfactory results.

Stationary phase: acetylcellulose. Mobile phase: hexane:methanol:ethyl acetate:di-n-butyl ether (70 + 10 + 10 + 10).

Some results obtained with this system are recorded in Table 3-15.

Table 3—15
R_F-values for some 2,4-DNPH's on acetyl cellulose with hexane:methanol:ethylacetate:di-n-butylether (70 + 10 + 10 + 10) as the solvent

2,4-DNPH of:	R_F	2,4-DNPH of:	R_F
ethylvanilline	0.27	p-toluylaldehyde	0.69
decanal	0.96	hydratropaldehyde	0.79
syringa aldehyde	0.11	heliotropin	0.46
p-methoxyacetophenone	0.60	salicylaldehyde	0.49
cinnamaldehyde	0.55	benzylmethyl ketone	0.76
cuminaldehyde	0.76	benzophenone	0.80
a-pentylcinnamaldehyde	0.87	ethylstyryl ketone	0.76
iso-jasmone	0.93	m-methylacetophenone	0.70
anisaldehyde	0.43	methyl a-naphtylketone	0.69
benzaldehyde	0.54	methyl b-naphthyl ketone	0.56
vanilline	0.30	o-methylbenzaldehyde	0.52
p-hydroxybenzaldehyde	0.40		

3.4.3.4. Reversed-phase chromatography

In reversed-phase chromatography, the stationary phase is an apolar liquid such as paraffin oil, while the mobile phase is a mixture of polar liquids such as methanol, acetone and water.

There are several methods for applying the stationary phase to the plate. Besides the method described in 3.4.3.2., the paraffin oil film can be obtained by soaking a Kieselguhr or silica gel plate in a 3—6% solution of paraffin oil in petroleum ether or by allowing such a solution to rise into the thin layer by placing the plate in an ordinary developing chamber with a solution of paraffin in petroleum ether as the "solvent"; after evaporation of the petroleum ether, a thin film of paraffin oil remains on the sorbent layer.

In reversed-phase chromatography, the long-chain members of homologous series are more strongly retained in contrast to normal chromatography where the long-chain members always move faster than the members having a shorter carbon chain.

In aroma research, this kind of systems has not been used to any extent. McSweeney (269) described the separation of terpene alcohols on Kieselguhr G, covered with paraffin oil using a mixture of water and acetone (65 + 35) as the mobile phase.

In our laboratory, the separation of derivatives of volatile compounds has been studied. The plates used in this study were always made as described in 3.4.3.2., using 3 g of paraffin oil per 25 g of Kieselguhr. The mobile phase consisted of a mixture of acetone and water (60 + 40).

For the homologous series tested, linear relations could be set up between carbon number and R_M, in most cases with the exception of the first one or two members. The best-fitting linear equations (regression equations) are given in Table 3-16. The R_F-values were always standardized using the linear correction method of Galanos and Kapoulas (see 3.4.2.), two reference compounds being run on each plate to provide the reference R_F-values. From Table 3—16 it appears that the mean value

Table 3—16
Regression equations of some homologous series

series	equations
primary aliphatic amines (as alkyl anilides)	$R_M = 0.18\ C - 0.63$
primary aliphatic amines (as DANS-amides)	$R_M = 0.18\ C - 1.09$
n-fatty acids (as p-bromophenacyl esters)	$R_M = 0.20\ C - 1.88$
n-alkanals (as 2,4-DNPH's)	$RN = 0.18\ C - 0.51$

Stationary phase: paraffin oil
Mobile phase: acetone:water (60 + 40)
Number of developments: 1

for the homologous increment is 0.18. Figure 3—27 shows the separation of a number of "DANS" amides.

The silica gel surface of silica gel plates can be made hydrophobic by conventional methods of silylation. A silanized silica gel for reversed-phase chromatography can be obtained from the Merck company. On this material, the same kind of separations

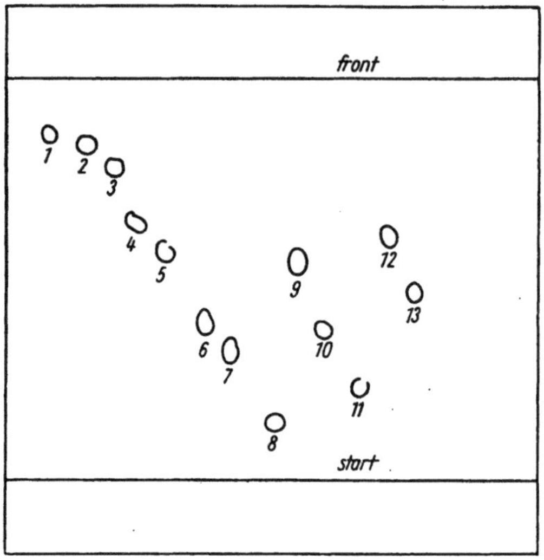

Fig. 3—27
Separation of some DANS-derivatives.
Derivative of 1. methyl-, 2. ethyl-, 3. propyl-, 4. pentyl, 5. hexyl-, 6. octyl-, 7. nonyl-, 8. dodecyl-, 9. diethyl-, 10. di-propyl-, 11. dibutyl, 12. cyclohexylamine, 13. piperidine.
Stationary phase: paraffin oil.
Mobile phase: acetone: water (6 + 4).

Table 3—17
Regression equations of some homologous series

series	equations
primary aliphatic amines (as alkyl anilides)	$R_M = 0.14\ C - 0.08$
n-fatty acids (as p-bromophenacyl esters)	$R_M = 0.14\ C - 0.20$
n-alkanals (as 2,4-DNPH's)	$R_M = 0.14\ C - 0.07$

Stationary phase: silanized silica gel (Merck)
Mobile phase: acetone:water (60 + 40)
Number of developments: 1

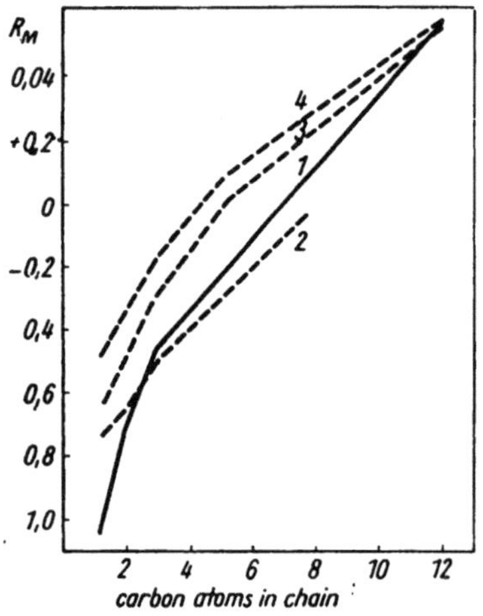

Fig. 3—28
Relation between R_M and number of C-atoms in chain for the homologous series of n-alkanal 2—4 DNPH's on different apolair phases. Stationary phase: 1. paraffin oil, 2. silicone rubber SE 52, 3. silicone oil MS 200, 4. silicone oil SF 96.
Mobile phase: acetone: water (75 + 25).

can be obtained as on paraffin oil. For good separation, however, multiple development is often necessary. The constants in the regression equations are definitely different from those obtained with paraffin oil, see Table 3—17.

Figure 3—28 shows the behaviour of the 2,4-DNPH's of a series of n-alkanals on different apolar phases (251). As can be seen, a marked deviation from linearity is shown by this series on the silicone oil phases.

3.4.4. Analysis of essential oils

One of the first applications of TLC occurred in the field of essential oil analysis. Essential oils may be regarded as aroma substances of sufficiently low volatility to be handled in micro amounts in the liquid (or solid) state. KIRCHNER et al. (270) used strips coated with silica gel for the separation of essential oil components, taking various organic liguids as the developing solvents. They applied different reagents for the visualization of the separated compounds, such as fluorescein, bromine, and concentrated sulphuric acid. This method was followed by MILLER and KIRCHNER (271) who developd identification methods for a number of compounds, using dehydratation, hydrolysis, and derivatization on the plate. REITSEMA (272) also used TLC for the characterization of essential oils.
One of the earliest solvent systems used in essential oil analysis which is still in use, is hexane admixed with 15—20% of ethyl acetate. It is worth noticing that the R_F-values in this system often show a remarkable constancy, irrespective of the silica gel preparation or the exact solvent composition; this is shown in Table 3—18.

Table 3—18
R_F-values of essential oil components obtained by different authors in the system; silica gel-hexane:ethylacetate (80 + 20 or 85 + 15)

compound	R_F as published by reference		
	251	270	271
geraniol	0.16	0.21	0.20
linalool	0.40	—	0.36
pulegone	0.52	0.49	0.58
carvone	0.48	0.45	0.45
cinnamaldehyde	0.31	0.31	0.31
1,8-cineole	—	0.48	0.48

Since the study by KIRCHNER and collaborators, a steady stream of publications on the analysis of aromatic oils by TLC began to appear. For good surveys of these studies the reader is referred to the literature (236—238). TLC has been found to be a rapid method for the separation of terpenoid isomers. The isomeric menthols were separated on silica gel by PETROWITZ (273); the solvent used was a mixture of hexane: ethanol (85 + 15). R_F-values have been determined and are to be found in Table 3—19. Isomeric thujyl alcohols were separated by NANO and MARTELLI (274) using the combination silica gel — benzene with multiple development. R_F-values found in our institute are given with their system in Table 3—20, while a chromatogram is shown in Figure 3—29.

TLC of the thujones and the thujylamines has been studied by BANTHORPE and TURNBULL (275). The influence of the double bond in some terpenoid compounds has been described by PETROWITZ (276). A typical result obtained by that author which has been confirmed in our institute, is shown in Table 3—21.

Table 3—19
R_F-values of isomeric menthols

compound	R_F
menthol	0.51
iso-menthol	0.36
neo-menthol	0.68
neo-iso-menthol	0.77

Stationary phase: silica gel
Mobile phase: hexane:ethanol (85 + 15)

Table 3—20
R_F^3-values of isomeric thujols and thujones

compound	R_F^3
thujol	0.40
iso-thujol	0.35
neo-thujol	0.50
neo-iso-thujol	0.62
a-thujone	0.80
b-thujone	0.83

Stationary phase: silica gel
Mobile phase: benzene
Number of developments: 3
Detection: phosphomolybdic acid

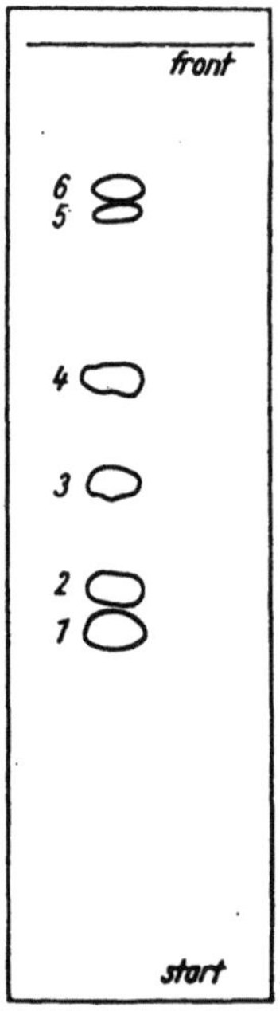

Fig. 3—29
Separation of thujone and thujyl alcohols by multiple development on silica gel.
Solvent: benzene, number of developments 3. Detection: phosphomolybdic acid.

Table 3—22
R_F-values of essential oil components

compound	R_F A	R^3_F B
α-ionone	0.54	0.29
β-ionone	0.54	0.25
germacrol	0.64	0.46
cedrol	0.36	0.18
a-cedrene	0,80	0,96
caryophyllene	0,87	0,91
methyl-α-ionone	0,66	0,44
methyl-β-ionone	0,65	0,40
carvacrol	0,47	0,47
phenol	0,26	0,48
geraniol	0.16	0.07
linaloöl	0.40	0.22
linalylacetate	0.68	0.46
a-terpineol	0.23	0.13
borneol	0.29	—
santonin	0.03	0.03
ascardiol	0.53	0.30
dihydrocarveol	0.29	0.41
benzylsalicylate	0.60	—
p-tert. butylphenol	0.33	—
b-methyl umbelliferone	0.05	0.41
methyl anthranilate	0.34	—
benzylmethyl ketone	0.33	—
benzophenone	0.50	0.58
thymol	0.43	0.46
eucalyptole	0.59	—
nerol	0.20	0.09
carvone	0.48	0.25
acetophenone	0.39	0.40
anisaldehyde	0.24	0.29
citral	0.44	0.24
citronellal	—	0.59
heliotropin	0.46	0.36
cuminaldehyde	0.60	0.55
methylheptenone	—	0.31
pulegone	0.52	0.25
p-toluylaldehyde	0.47	—
guajacol	0.39	0.40
orcin	0.08	0.01
a-naphthol	0.35	0.41
b-naphthol	0.29	0.28
farnesol	0.24	0.14

A. stationary phase: silica gel
mobile phase: hexane:ethylacetate (80 + 20)
number of developments: 1
B. stationary phase: silical gel
mobile phase: benzene
number of developments: 3

BAINES and JONES (277) compared the separation of terpene hydrocarbons on silica gel plates impregnated either with silver nitrate or with thallous nitrate. Benzene was the solvent for the silver nitrate plates while on the thallous nitrate plates the compounds were developed with hexane. From the R_F-values published by those authors it must be concluded that the silver nitrate impregnation gives a much better separation than does the impregnation with thallous nitrate.

Table 3—21
Influence of double bond on R_F-values for some terpenes

compound	R_F
piperitone	0.17
pulegone	0.22
isomenthone	0.28

Stationray phase: silica gel
Mobile phase: cyclohexane
Detection: 2,4 dinitrophenylhydrazine solution

Two-dimensional chromatography is often of advantage in the separation of complex mixtures. In our laboratory, silica gel layers are used which are first developed three times in benzene and, for the second dimension, once in hexane: ethyl acetate (80 + 20). Table 3—22 contains the R_F-values determined by the author.

3.4.5. Analysis of derivatives

In TLC, it is often desirable to treat the volatile concentrate with a reagent that reacts specifically with one chemical class of volatiles, to form non-volatile derivatives. Some of the most obvious advantages are the following:

- even highly volatile members of the class in question are amenable to TLC after derivatization; large volumes of solvents can be evaporated without loss of the compounds of the selected class;
- the non-derivatized volatiles can be discarded by evaporation, thus drastically simplifying the subsequent TLC separations;
- in the preceding isolation and concentration steps, the non-derivatizable volatiles can be disregarded so that these steps can be optimized specifically for the selected class;
- a reagent can be chosen that forms intensely coloured or fluorescing derivatives, so that the TLC spots are directly visible and can be estimated quantitatively.

An obvious disadvantage lies in the fact that only one class of volatiles can be handled at a time. In practice, however, the investigator often is interested in only one particular class of compounds.
The classes to be discussed in the following are 1) carboxylic acids, 2) primary and secondary amines, 3) aldehydes and ketones (referred to as "carbonyl compounds"), 4) alcohols. In the following the isolation, derivatization and the separation of these compound classes will be described.

3.4.5.1. Isolation by steam distillation

The isolation method is traditionally steam distillation at atmospheric pressure.

Carboxylic acids: The medium for steam distillation should be of low pH. In our laboratory, a number of experiments have been carried out on the isolation of volatile acids by steam distillation under standard conditions. In some experiments, the acids were distilled from a customary steam distillation set-up while in a second series, the distillation was carried out by the method and apparatus which ANTONACOPOULOS described (278). The results of our experiments are collected in Table 3—23.

Table 3—23
Comparison of "normal" steam distillation with the Antonacopoulos method (278)

compound	"normal" steam distillation		steam distillation according to ANTONACOPOULOS	
	ml of total distillate	acid distilled in %	ml of total distillate	acid distilled in %
acetic acid	500	96	150	100
butyric acid	300	100	—	—
stearic acid	300	16	200	12
benzoic acid	600	65	400	95
sorbic acid	850	95	600	100

From these experiments it is concluded that ANTONACOPOULOS' method is superior to "ordinary" distillation. Our procedure, using an ANTONACOPOULOS-type apparatus, is as follows: 10—20 g of the homogenized sample is placed in flask B (Fig. 3—30). To the sample in the flask are added 20 g $MgSO_4 \cdot 7 H_2O$ for a salting-out effect, approx.

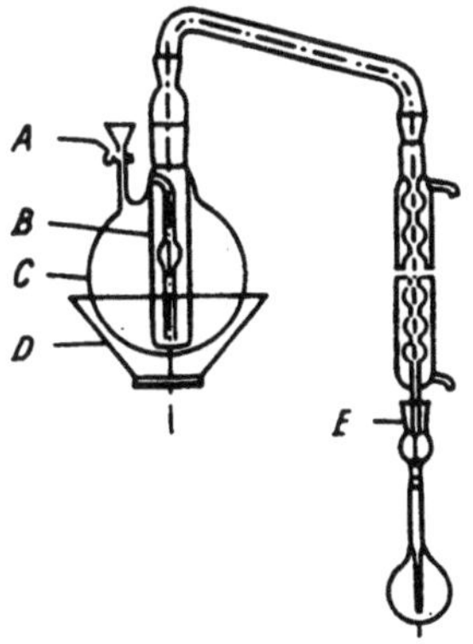

Fig. 3—30
ANTONACOPOULOS distillation apparatus.

2 g of tartaric acid (pH becomes approx. 1) and approx. 20 ml of water. Flask C is filled with about 1 litre of water and some boiling-chips are added. It is often of advantage to add 100—200 g of sodium chloride to flask *C* to raise the boiling point. The contents of this flask are heated, with stopcock *A* open, until the boiling point is reached; the stopcock is then closed and the distillate collected at *E*. At the end of the distillation it is necessary to open stopcock *A* before removing the heat source *D* Usually between 300 and 500 ml of distillate may be collected. To the distillate are added a few drops of 0.1% phenolphthalein solution in ethanol and the distillate is

neutralized by adding 0.1 N NaOH until a pink colour persists. It is then taken to dryness using a rotary-film evaporator. The required derivatives can be obtained in the residue.

Amines: Steam distillation from a slightly alkaline medium seems to be the most efficient method for the isolation of volatile amines from foods. According to NICHOLS and FOOTE (279), the pH should not be below 7.4 in order to recover quantitatively ammonia from solutions by steam distillation, nor much higher, in order to prevent decomposition of certain organic nitrogen compounds such as amino acids, urea, guanidine, acetamide and casein. WEURMAN and DE ROOY (280) investigated a number of compounds present in food samples for their behaviour during steam distillation at pH = 8.0.
The method followed in our institute is essentially the one described by STEIN VON KAMIENSKY (281) as modified by WEURMAN and DE ROOY (280). The details are as follows:
The sample, about 50 g, is mixed with 200—300 ml of 0.1 M K_2HPO_4 in an electric mixer, and the resulting suspension centrifuged. After decantation from the solid material, the remaining suspension is adjusted to pH approx. 8 with 0.1 N NaOH. The alkaline solution thus obtained is distilled with steam, a silicone anti-foam being used if required; the distillate is collected in 50 ml 0.1 N HCl. For qualitative purposes, distillation usually need not be continued beyond a volume of 100 ml. The acidified distillate is evaporated to dryness in a film evaporator and is then ready for derivatization.

Alcohols: Volatile alcohols can be isolated from bulk materials such as foods by steam distillation and subsequent extraction with ether; the ether extract is ready for derivatization. In some instances, ethanol and methanol are present in such large amounts that they would mask the presence of minor compounds in the mixture. It is then advisable to use an extractant that discriminates against the lower alcohols (see 2.2.6.), such as hexane. In this case it is of advantage to use the apparatus and procedure as described for carbonyl compounds in the next paragraph.

Carbonyl compounds: The isolation of volatile carbonyl compounds from foods and from plant material in general may be carried out by a combination of steam distillation and extraction. For routine analysis, the apparatus detailed in Figure 3—31 is used in our institute.
The procedure is as follows: in a flask connected to A the material is boiled with a sufficient amount of water and, if necessary, a small amount of anti-foam. The steam carrying the volatiles is condensed in D, while part B has been filled with 50 ml of hexane. As the condensate drops through the hexane layer, the volatiles are extracted from it, after which it returns to the distillation flask through C. After a sufficient distillation time, usually 1—3 hours, the hexane layer is collected through the stopcock and dried on sodium sulphate, after which it is ready for derivatization.

3.4.5.2. Derivatization procedures

Carboxylic acids: For the separation of mixtures of acids into their components, the following derivatives will be discussed in this chapter: a) p-nitrobenzyl esters; b) p-

bromophenacyl esters; c) p-phenylazophenacyl esters. The method of preparation, as practised in our institute, is as follows:
The dry residue obtained as described in 3.4.5.1. is dissolved in 0.5 ml of water and 10 ml of ethanol is added, followed by an excess of the reagent (p-nitrobenzyl bromide, p-bromophenacyl bromide or p-phenylazophenacyl bromide). The mixture is refluxed for one hour; if the contents of the flask do not dissolve completely, more alcohol is added through the condenser. The reaction mixture is then cooled and washed into

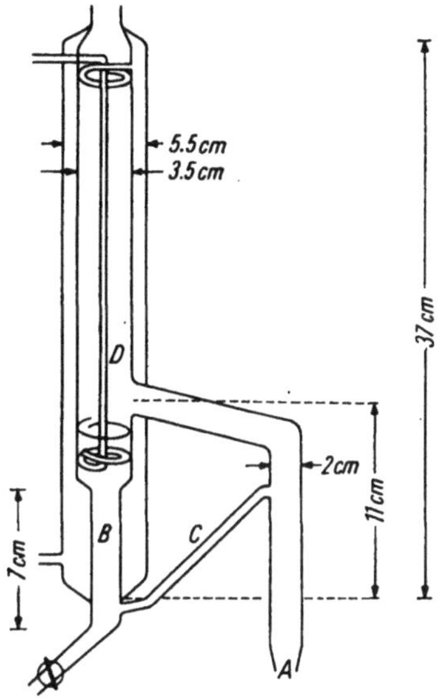

Fig 3—31·
Apparatus for continuous steam distillation and extraction.

a separating funnel with alcohol; 25—50 ml water is added, and the mixture extracted with ethyl acetate. The extract is dried on sodium sulphate and evaporated to a small volume. This solution is ready for TLC analysis.
Note: as p-phenylazophenacyl bromide is rather poorly soluble in ethanol, the excess of reagent should in this case be small.

Primary and secondary amines: The following derivatives will be discussed:

a. 2,4-dinitroanilides, formed by reaction of the amines with 2,4-dinitrochlorobenzene or 2,4-dinitrofluorobenzene (Fig. 3—32a).
b. the reaction product of amines with 7-chloro-4-nitrobenzo-2-oxa-1,3-diazole (NBD amines; the general structure is represented in Fig. 3—32b).
c. dimethylaminonaphthalenesulphonylamides, formed by reaction of the amines with 1-dimethylaminonaphthalene-5-sulphonyl chloride (DANS-amides; the reaction scheme is as represented in Fig. 3—32c).
d. products formed by raection of primary amines with Fluram. The reaction scheme is shown in Figure 3—32d.

We shall here present the methods of preparation of these derivatives as practised in our institute, without discussing the considerations behind each of them; for those the reader will be referred to the original literature. In each of the four derivatization procedures, the dry residue of amine salts obtained as described in 3.4.5.1. is first dissolved in one millilitre of water. The subsequent steps include adding the reagent, adding carbonate to liberate the amines for reaction with the reagent, and extraction of the reaction product. The details for derivatives a, b, c and d are as follows:

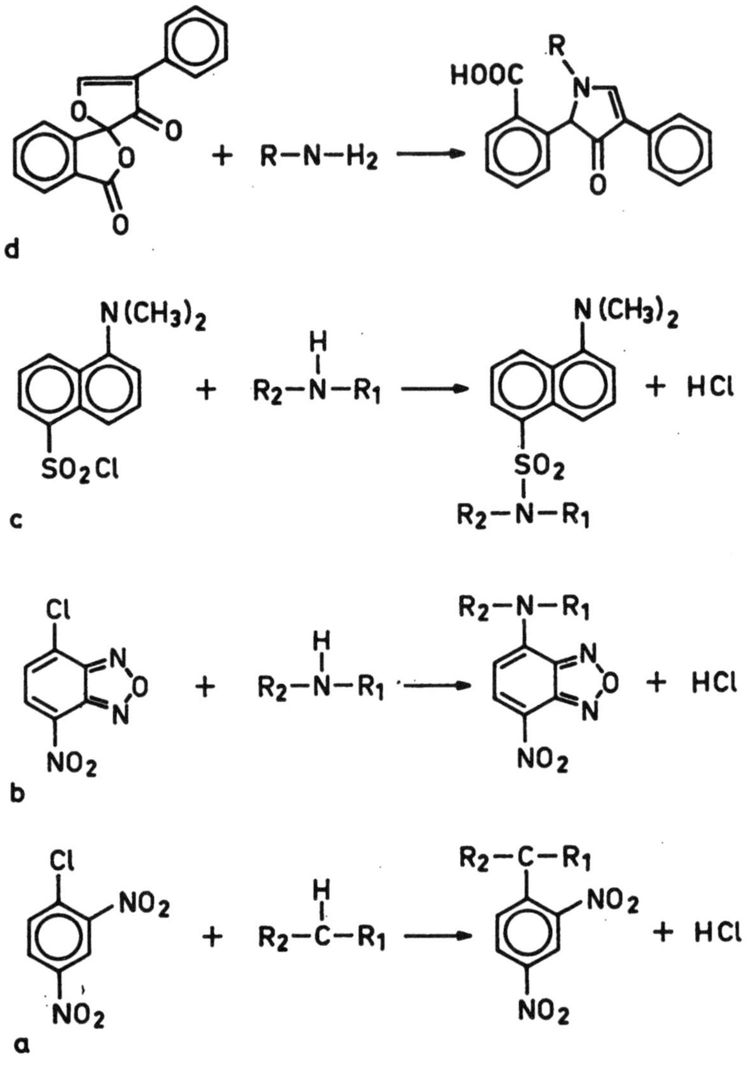

Fig. 3—32
Reaction schemes for the derivatization of volatile amines.

a. 2,4-dinitroanilides: 5 ml of a 0.6% ethanolic solution of the reagent is added in a small round-bottom flask to the aqueous solution of amine salts, followed by 2 ml of 0.1 M sodium bicarbonate. The flask is set aside at room temperature for 30 minutes and then heated during 10 minutes at 40—50 °C using a reflux condenser. Two millilitres of a 0.2 N solution of NaOH in 60% dioxane are added and heating continued for another 30 minutes. After cooling, the content of the flask is poured into 100 ml of water and the flask is washed several times with small amounts of alcohol which are added to the aqueous solution. This solution, which may be turbid, is extracted twice with 10 ml portions of cyclohexane. The cyclohexane extract is dried on sodium sulphate, filtered, evaporated to dryness under vacuum, and the residue taken up in 0.5—2 ml of ethyl acetate. See literature references 282—285.

b. NBD-amines: to the aqueous solution of the amine salts in a small round-bottom flask is added 1 ml of 0.1 M sodium bicarbonate and 5 ml of an 0.05% solution of the reagent in methanol. The mixture is heated to 50 °C for one hour using a reflux condenser. After cooling to room temperature, 1 ml of ethyl acetate is added to the mixture, followed by water until the ethyl acetate layer separates. See literature reference 286.

c. DANS-amides: to the aqueous solution of amine salts is added 5 ml of a 0.05% solution of the reagent in acetone. After standing for 2 hours at room temperature, 1 ml of a saturated sodium carbonate solution is added. After standing overnight, 2—4 ml of ethyl acetate is added and the derivatives are extracted by shaking. See literature references 287—289.

d. reaction products of Fluram: the aqueous solution of amine salts is transferred to a test tube and 2 ml of a saturated solution of sodium bicarbonate added, followed by 3 ml of a 0.01% solution of the reagent in acetone. After 5 minutes, the solution is acidified with 6 N HCl and 1 ml of ethyl acetate is added, followed by dropwise addition of water until the ester layer separates. Only primary amines are derivatized. See literature references 290, 291.

Of the ethyl acetate solutions obtained by each of the above procedures, only a few microlitres need be spotted if determination of R_F-values is the only purpose. If necessary, however, e.g. if the TLC separation is to be followed by spectroscopic analysis, the entire extract may be applied to the TLC plate as a streak.
The anilides can be detected by their strong absorption in UV; the other derivatives strongly fluoresce in filtered UV light.

Alcohols: 3,5-dinitrobenzoates (3,5-DNB's) are the only derivatives in general use. They can be prepared as follows:
The ether or hexane extract obtained from the steam distillate is dried on sodium sulphate. After decantation from the solid salt, approx. 0.5 g of 3,5-dinitrobenzoyl chloride is added, followed by 2 ml of dry pyridine. After standing overnight, the reaction mixture is washed into a separating funnel with 50 ml of water and the mixture set aside for one hour so that the excess of the chloride may be converted into the corresponding acid. After removal of the aqueous layer the organic layer is extracted with 5% sodium carbonate solution to remove the 3,5-dinitrobenzoic acid. The pyridine is removed by shaking the organic layer several times with 2 N HCl. After washing twice with water, the solution is taken to dryness using a rotary-film evaporator. After dissolving the residue in acetone, the solution is ready for application to the thin-layer plate.

Carbonyl compounds: The derivatization of carbonyl compounds usually consists of conversion into 2,4-dinitrophenylhydrazones (2,4-DNPH's). In our institute, this is carried out by the reaction column technique of SCHWARTZ et al. (292) with some minor modifications. In the following the procedure is described in detail.

Preparation of reaction column: 300 mg of 2,4-dinitrophenylhydrazine is mixed thoroughly with 2.5 ml of phosphoric acid and 0.25 ml of water, and the resulting suspension is filtered through a sintered glass filter. The filtrate is mixed with enough Celite 545 until a yellow damp powder is obtained. A chromatography tube of about 8 mm inner diameter is filled with the powder until a column is obtained of 4—8 cm height, the powder being poured into the tube in small amounts and tamped with a glass rod. Carbonyl-free hexane is now drained through the column at a rate of a few drops per second until the effluent becomes colourless. The column is now ready for use.

Conversion of carbonyls into 2,4-DNPH's: The dried hexane extract, obtained from the distillation apparatus as described in 3.4.5.1., is drained through the reaction column at a rate of a few drops per second. The column is washed with hexane and the washings added to the eluate. If carbonyl compounds were present in the hexane extract, the eluate appears yellow. This fraction will be called fraction *A*; it contains all 2,4-DNPH's that are soluble in hexane such as the derivatives of aliphatic type carbonyl compounds and most of the terpenoids. A number of 2,4 DNPH's, especially those derived from benzaldehyde or benzophenone derivatives, are very poorly soluble in hexane. During the conversion they remain on the reaction column in an undissolved state. These derivatives can be recovered, after the column has been washed with hexane, in the following way:
The outlet of the column is connected to a water suction pump and the excess hexane removed by suction of air through the column. After disconnection of the reaction column from the vacuum line, the excess reagent is removed by draining 2—4 N HCl through the column until the effluent becomes colourless. The residual HCl is then removed by washing with water. The outlet of the column is again connected to the vacuum line and the excess water removed as far as possible by suction. If the column appears white, no hexane insoluble 2,4 DNPH's are present. If such derivatives have been formed on the column, they show up as a band on the top of the reaction column. They are recovered as follows: A small amount of dioxane is drained through the column to dissolve and elute the compounds. The eluate is mixed with twice its volume of ethyl acetate, and dried on sodium sulphate. This fraction will be indicated as fraction *B*. The solvent is now removed from fractions *A* and *B* under vacuum and the dry residues are taken up in the minimum amount of ethyl acetate. To dissolve the residue from fraction *B* it is often of advantage to use a mixture of ethyl acetate and dioxane as the solvent.

Purification after conversion: If the fractions are obtained from aromatic plants or seeds, then fraction *A* often contains a considerable amount of non carbonyl compounds from the essential oil. Before applying the sample to a thin layer plate it is necessary to remove these compounds from the sample. Such removal of neutral compounds is also necessary from 2,4 DNPH's after separation on plates containing non volatile liquids as the stationary phase, because in such cases a considerable amount of stationary phase (paraffin oil, Carbowax, etc.) can be present besides the 2,4 DNPH's after elution from the plate. In our institute, the current method to purify DNPH samples consists in the following:
Two grammes of Dowex 1 × 8 (50—100 mesh) are transferred to a chromatographic tube (approx. 8 mm inner diameter). The column is washed with 1 N NaOH until the

effluent is strongly alkaline, followed by repeated washings with water until the effluent is neutral. The water is displaced by methanol, and the methanol is then displaced by an 8:2 mixture of benzene and methanol. The dry residue obtained from fraction *A* is dissolved in the minimum amount of this same mixture and allowed to suck into the column. The column is then washed with 4—6 times its own volume to remove the neutral compounds; the eluate and the washings are discarded. The 2,4 DNPH's remain adsorbed on the column and are now eluted with the same mixture but with 10% acetic acid added. The eluate is taken to dryness under vacuum. The residue is now ready for further analysis.

3.4.5.3. Separations of derivatives of acids, amines, and alcohols Acids

Since the majority of volatile acids belong to the homologous series of fatty acids, the method of choice to separate their derivatives is by partition chromatography, which will separate homologues according to their chain length. The following systems have been tested in our laboratory:

- Silica gel impregnated with acetonitrile or nitromethane by vapour phase adsorption; solvent, hexane;
- Kieselguhr covered with Carbowax 400 or 3,3'-oxydipropionitrile; solvent, hexane.

R_F values are collected in Table 3—24.

Table 3—24
R_F-values of p-bromophenacyl esters of n-fatty acids solvent: hexane

carbon number of acid	R_F-value on	
	Carbowax 400	3,3 oxydipropionitrile
1	0.11	0.11
2	0.24	0.22
3	0.45	0.39
4	0.59	0.53
5	0.71	0.65
6	0.81	0.76
7	0.87	0.86
8	0.92	0.89
9	0.95	0.93
10	0.96	0.96

Figure 3—33 shows the relation between the carbon number of the acid and the R_M-value in the system acetonitrile-hexane. The linear relationships in this figure are theoretical proof that in this particular system, the mechanism is essentially that of partition chromatography and that phenomena of adsorption on the silica gel layer are much less important; the silica gel seems to be fully deactivated by adsorption of the acetonitrile. An analogous picture is obtained when using nitromethane instead of acetonitrile as the impregnating vapour. For satisfactory resolution of the lower members of the homologous series, it is of advantage to use multiple development. Usually three developments will separate the homologous fatty acid esters $C_1—C_{10}$.

A good separation of the first eight members of the fatty acid series is obtained by using Carbowax 400 or 3,3'-oxydipropionitrile as the stationary phase (Table 3—24). In the literature, no communications on TLC of the derivatives described in this chapter have come to our attention. One paper deals with TLC of a related class of deriv-

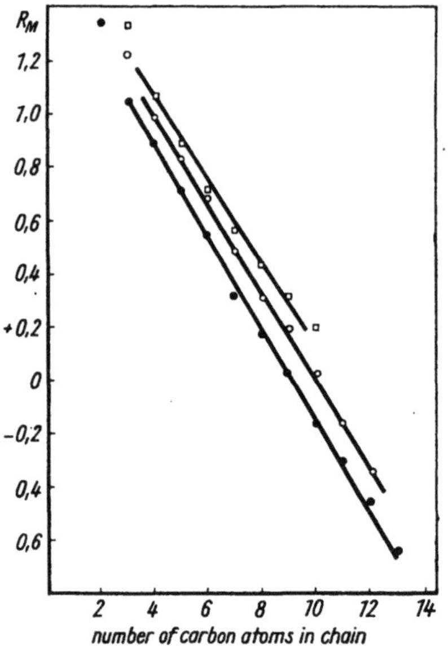

Fig. 3—33
Relation between R_M-values and carbon number for some derivatives of volatile acids. □ p-phenylazophenacyl esters, ○ p-nitrobenzyl esters, ● p-bromophenacyl esters. Stationary phase: silica gel-acetonitrile. Mobile phase: hexane.

atives, 2,4-dinitrobenzyl esters (293) and reports very nice separations on layers impregnated with paraffin oil, with a mixture of dimethylformamide and water (2 + 1) as the solvent.

Primary and secondary amines

2,4-Dinitroanilides: 2,4-dinitroalkylanilides, obtained by reacting primary or secondary alkyl amines with 2,4-dinitrofluorobenzene or 2,4-dinitrochlorobenzene, have been separated by ILERT and HARTMANN (282) using silica gel layers and pentane:iso-

Table 3—25
R_F^c-values calculated from different sources

2,4 dinitro anilide	R_F-value according to reference			mean R_F^c	R_M
	282	284	251		
n-propylamine	0.14	0.14	0.21	0.16	0.720
n-butylamine	0.29	0.33	0.30	0.31	0.347
n-amylamine	0.39	0.43	0.45	0.42	0.140
n-hexylamine	0.51	0.55	0.56	0.54	−0.070
n-heptylamine	0.64	0.68	0.70	0.67	−0.308
n-octylamine	0.76	0.79	0.81	0.79	−0.575
n-nonylamine	0.92	0.92	0.90	0.92	−1.061

amyl acetate:ammonia 70 + 29 + 1 or pentane:benzene:triethylamine 45 + 45 + 10 as the mobile phase.

Partition chromatography on Carbowax 400 with hexane or heptane as the mobile phase has been described by the same authors. SCHWARTZ et al. (284) used the same system, but with a mixture of Seasorb 43 and Kieselguhr as the carrier for the Carbowax layer. In our laboratory, Carbowax on Kieselguhr was used as the immobile phase and hexane as the mobile phase (251).

In these three cases, very different R_F-values were found by the respective authors. In Table 3—25, the R_F^c-values, calculated from the published results, are summarized. It is seen how well the R_F^c-values coincide. The R_M-values calculated from them fit a straight line, except for the first three members; their R_M-values cannot be calculated accurately because of the low R_F-values, and are expected to show deviations from linearity anyway.

Good separation, especially of the lower members of the series, is obtained on layers containing 3,3′-oxydipropionitrile as the stationary phase instead of Carbowax, and hexane as the mobile phase.

A separation of the derivatives of the primary amines from the symmetrical secondary amines having the same chain length, is difficult to attain. Table 3—26 shows the differences, in absolute values, between the R_F-values in a Carbowax system (251).

Table 3—26
R_F in the system Carbowax — hexane for amines having the same carbon number

2,4-dinitro anilide of secondary amine	R_F	2,4-dinitro anilide of primary amine	R_F	$\|\Delta R_F\|$
dimethylamine	0.07	ethylamine	0.10	0.03
diethylamine	0.21	butylamine	0.34	0.13
dipropylamine	0.60	hexylamine	0.56	0.04
dibutylamine	0.86	octylamine	0.78	0.08

The behaviour of the alkyl anilides on acetylated cellulose has been discussed in Section 3.4.3.3.

NBD-amines: These derivatives show intense fluorescence in UV-light. TLC separation of the aliphatic-type amines has been described by KLIMISCH and STADLER (286) using silica gel with two-dimensional development. For the first dimension, a mixture of cyclohexane and ethyl acetate (1 + 1) was used, and for the second dimension a mixture of petroleum ether, chloroform, ether and acetic acid (33 + 33 + 33 + 1). On polyamide layers, some separation was also obtained by these authors. Separation of series of homologues was, however, hardly possible. In our laboratory we obtained a very good separation in the system Carbowax/hexane:ethyl acetate (80 + 20) with two developments. The ^2R_F-values are recorded in Table 3—27.

Table 3—27
^2R_F-values of some NBD amines in the system Carbowax — hexane:ethyl acetate (80 + 20)

NBD of	^2R_F	NBD of	^2R_F
methylamine	0.07	amylamine	0.41
ethylamine	0.14	hexylamine	0.53
propylamine	0.22	heptylamine	0.62
butylamine	0.29	octylamine	0.72
		nonylamine	0.82

DANS-amides: These derivatives show a strong fluorescence in filtered UV-light, by which they can be detected. TLC of these derivatives has been described by SEILER and WIECHMANN (289, 294, 295). Those authors used silica gel as the stationary phase and various mixtures of organic liquids as the mobile phase. Since, however, the volatile amines occurring in foods mainly belong to the homologous series of aliphatic primary and secondary amines, and these are best separated by partition chromatography, it seems to be better to use partition chromatography in food research. Table 3-28 contains the R_F and R_M-values of the DANS derivatives of the primary aliphatic amines in the systems Carbowax/hexane and Carbowax/hexane: ethyl acetate (80 + 20) (251).

Table 3—28
R_F- and R_M-values of DANS amides from primary aliphatic amines on Carbowax layers

DANS amide derived from	Carbowax-hexane		Carbowax-hexane: ethyl acetate (80 + 20)	
	R_F	R_M	R_F	R_M
methylamine	0		0.14	0.79
ethylamine	0.01	2.00	0.26	0.45
propylamine	0.03	1.57	0.34	0.29
butylamine	0.04	1.40	0.42	0.14
amylamine	0.05	1.26	0.53	−0.05
hexylamine	0.08	1.05	0.61	−0.20
heptylamine	0.11	0.90	0.72	−0.41
octylamine	0.16	0.71	0.77	−0.52
nonylamine	0.24	0.82	0.82	−0.65
decylamine	0.29	0.39	0.84	−0.75

In contrast to the behaviour of the dinitroalkylanilides as described above under the heading "2,4-dinitroanilides", the DANS-amides of the symmetrical secondary amines are easily separated from the primary amines having the same chain. Table 3—29 summarizes the results in the Carbowax/hexane system.

Table 3—29
$|\Delta R_F|$ of primary and secondary DANS amides having the same carbon number

secondary amine	R_F	primary amine	R_F	$\|\Delta R_F\|$
diethylamine	0.50	butylamine	0.04	0.46
dipropylamine	0.73	hexylamine	0.08	0.65
dibutylamine	0.87	octylamine	0.16	0.71

Fluram derivatives: Since the reagent reacts only with primary amines, no interference will arise from secondary amines. Good separation of the homologous series of primary aliphatic amines is obtained using acetylcellulose as the stationary phase, see 3.4.3.3.

Alcohols:

The 3,5-dinitrobenzoic esters (3,5-DNB's) are practically the only derivatives of aliphatic alcohols used in TLC separations. We shall now first survey the literature on the chromatographic behaviour of these derivatives, which is very limited, and then present some of our unpublished work.
The chromatographic behaviour of 3,5-DNB's on silica gel plates was studied by DHONT and DE ROOY (296); the solvent mixtures used were the following: benzene: pe-

troleum ether (1 + 1), hexane:ethyl acetate (85 + 15) and toluene:ethyl acetate (90 + 10).

MEHLITZ et al. (297) separated a number of 3,5-DNB's on silica gel with the following solvent systems: 1) diisopropyl ether:petroleum ether (12 + 88), 2) diisopropyl ether:petroleum ether (6 + 94). The derivatives of the lower homologues were best separated using solvent system 1, while the higher members were best separated using system 2.

Two-dimensional chromatography on silica gel was described by SEVERIN (298). For the first dimension, the mixture was developed with a mixture of cyclohexane:ethyl acetate (90 + 10); after development and evaporation of the solvent, the plate was soaked in a 10% solution of paraffin oil in petroleum ether, and developed in the second dimension with acetic acid:water (80 + 20), i.e. in reversed-phase partition chromatography. Using this technique they were able to separate the derivatives of the complete homologous series of aliphatic alcohols from C_1 to C_{14}.

In our laboratory, some systematic work has been done on separations of 3,5-DNB's. Separations of the homologous series of the primary straight-chain aliphatic alcohols can be read from Table 3—30.

Table 3—30
R_F-values for the n-aliphatic alcohols (as 3,5-DNB's) in some partition systems

3,5 DNB of:	system*)			
	A R_F	B R_F	C R_F	D ^3R_F
methanol	0.34	0.28	0.11	0.05
ethanol	0.48	0.54	0.25	0.14
n-propanol	0.58	0.71	0.38	0.20
n-butanol	0.68	0.84	0.54	0.26
n-pentanol	0.74	0.91	0.64	0.37
n-hexanol	0.80	0.96	0.77	0.45
n-heptanol	0.85	0.99	0.86	0.53
n-octanol	0.89	—	0.90	0.61
n-nonanol	0.92	—	0.94	0.77
n-decanol	0.94	—	0.98	0.87

*) A. stationary phase: polyvinylacetate on Kieselguhr
solvent: hexane:methanol:ethylacetate:di-n-butylether (85:5:5:5)
number of runs: 1
B. stationary phase: carbowax on Kieselguhr
solvent: hexane
number of runs: 1
C. stationary phase: 3,3 dioxydipropionitrile on Kieselguhr
solvent: hexane
number of runs: 1
D. stationary phase: siliga gel impregnated with acetonitrile by vapour phase adsorption, see 3.4.3.2
solvent: hexane
number of runs: 3

The table refers to a number of partition sytems because separation on silica gel layers is usually insufficient.

The separation of the 3,5-DNB's of isomeric alkanols is a problem which, in the author's opinion, is still to be solved. Derivatives of branched-chain alkanols have practically the same R_F-values as have the derivatives of the corresponding n-alkanols, while those of the 2-alkanols have slightly higher R_F-values than the corresponding primary alcohols (see Table 3—31).
Terpene alcohols, at least the primary and secondary ones, also form 3,5-DNB's; these will not be discussed here.

Table 3—31
Differences in R_F-values of some 3,5 DNB's of primary and secondary alkanols

3,5 DNB of:	system*)	
	A	B
	$\Delta^3 R_F$	$\Delta^5 R_F$
butanol-1-butanol-2	—0.04	—0.07
pentanol-1-pentanol-2	—0.06	—0.04
hexanol-1-hexanol-2	—0.04	—0.04
heptanol-1-heptanol-2	—0.04	—0.08

*) A. stationary phase: silica gel impregnated with acetonitrile by vapour phase ad sorption
solvent: hexane
number of runs: 3
B. stationary phase: silica gel
solvent: benzene:hexane (1 + 1)
number of runs: 5

3.4.5.4. Separations of derivatives of carbonyl compounds

Review of literature on TLC of 2,4-DNPH's: Onoe was no doubt the first to describe the use of thin layers of silica gel for the separation of 2,4-DNPH's. The information about his work given here (299, 300) was mainly obtained from an abstract by Simamura.
Onoe prepared thin layers by slurrying 23 g of silica gel in 30 ml of a 2.5% solution of polyvinyl alcohol. Water-saturated benzene was used as the mobile phase. Onoe studied the separation of aliphatic saturated C_1 to C_{10} aldehydes and of acrolein, crotonaldehyde and citral.
After the TLC technique was streamlined by Stahl in 1956 (301), the number of papers describing the separation of 2,4-DNPH's rapidly increased. Dhont and de Rooy (302) published a paper concerning the separation of some 2,4-DNPH's derived from aromatic carbonyl compounds by using silica gel as the adsorbent and a mixture of light petroleum with benzene and of benzene and ethylacetate, as the mobile phases. Most papers that appeared since deal with the separation of the derivatives of aliphatic carbonyl compounds. Urbach (250) made a thorough investigation on the systematic separation of aliphatic-type 2,4-DNPH's. The individual members of homologous series were separated by partition chromatography using 2-phenoxyethanol on Kieselguhr and light petroleum as the immobile and the mobile phases, respectively. For the separation of the homologous series on 2-phenoxyethanol,

multiple development was found necessary. n-Alkanals, n-alkan-2-ones and n-alk-en-2-ones were separated from each other on aluminium oxide "G" with 4% ethyl ether in light petroleum as the solvent.
EDWARDS (303) observed the separation of syn and anti isomers of some aliphatic carbonyl compounds on plates of 2-phenoxyethanol. This phenomenon is not recorded in URBACH's paper. In our laboratory we did not observe the occurrence of double spots for the aliphatic carbonyls when using URBACH's system. The cause of these differences might be the method of preparation of the derivatives.
Separations similar to those of URBACH were obtained by BADINGS and WASSINK (249). For the separation into classes, they used activated zinc carbonate with a mixture of light petroleum, benzene and pyridine (7 + 1 + 2) as the solvent. Separation within the homologous series was obtained on Kieselguhr covered with Carbowax 400 as the immobile phase and light petroleum as the developing solvent. The advantage of using Carbowax instead of 2-phenoxyethanol lies in the fact that for the separation of a homologous series on Carbowax a single development suffices while the same separation on 2-phenoxyethanol can be obtained after 5 to 6 multiple runs. Unsaturated aldehydes were separated on thin layers of Carbowax by MEYBOOM (304). He observed that the separating efficiency rose by using 33.3% Carbowax in the suspension instead of the 25% used by BADINGS and WASSINK.
BADINGS and WASSINK could classify the 2,4-DNPH's according to the degree of unsaturation on Kieselguhr plates impregnated with 30% silver nitrate, which were developed with petroleum ether. SCHWARTZ and PARKS (305) described the separation into classes on plates of activated magnesium oxide with mixtures of hexane and chloroform as the solvent. In our experience magnesium oxide plates are much less sensitive to atmospheric moisture than are BADINGS and WASSINK's zinc carbonate plates.
Unsaturated 2,4-DNPH's with one or more isolated double bonds may as a class be separated from saturated 2,4-DNPH's and from 2,4-DNPH's containing conjugated double bonds, on alumina layers impregnated with silver nitrate, as described by DE JONG and MOSTERD (264). They found alumina layers impregnated with silver nitrate to be superior to silver nitrate-impregnated silica gel layers. Mixtures of light petroleum with ether or chloroform were used as the eluents. The R_F-value was found not only to depend upon the number of double bonds in the molecule but also upon the molecular geometry since cis-isomers moved slower than the trans-isomers. This effect was also demonstrated by MEYBOOM and JURRIENS (306), who succeeded in separating mixtures of isomeric hexenals and of isomeric heptenals. The separation of the radiolytic products of cyclohexanone was studied by DENTI and LUBOZ (307).
A number of different aliphatic aldehydic and ketonic 2,4-DNPH's were chromatographed on layers of silica gel and aluminium oxide with and without impregnation with silver nitrate.
The chromatographic behaviour of the derivavites of aliphatic aldehydes and ketones was also studied by ZAMOJSKI and ZAMOJSKA (308). These authors used silica gel layers in combination with a number of solvent systems. Among the most useful were: petroleum ether with ethyl acetate (7:3) and carbon tetrachloride with acetone (9:1). A study of the behaviour of aliphatic dicarbonyl compounds in the form of their osazones has been published by COBB (309). The osazones of these compounds can be separated into classes such as α-keto-alkanals and 2,3-diketones by TLC on

activated Seasorb 43 — silica gel and on Seasorb 43 — silica gel-calcium sulphate layers. For these separations, the following solvent systems were used: benzene saturated with ethanolamine + 8% methanol, and a mixture of chloroform, tetrahydrofuran and methanol (15:4:1). Later COBB et al. (310) published a method for the separation of 2,4-DNPH osazone homologues of vicinal dicarbonyl compounds. Separation was carried out, on silica gel plates impregnated with a mineral oil as the stationary phase and dioxane:water (6:4) as the mobile phase.

Low molecular weight 2,4-DNPH's were tried on several adsorbents by NANO (311) who found neutral alumina to be the best adsorbent for the separation of these compounds. The author describes a number of solvent systems most of which are mixtures of nitrobenzene with hydrocarbons or with carbon tetrachloride. Starch impregnated with dimethylformamide has been described by DAVIDEK (312) for the separation of some 2,4-DNPH's. The paper does not contain detailed information.

The separation of polar carbonyl compounds, as obtained from cellulose pyrolysis, has been described by BYRNE (313). These carbonyls have a —COOH or —OH group besides the carbonyl group. BYRNE used silica gel and alumina with solvent systems consisting of mixtures of light petroleum and ether, and of benzene and tetrahydrofuran with and without added glacial acetic acid.

Only a few communications deal with the TLC of aromatic-type 2,4-DNPH's. Substituted benzaldehydes were studied by FINLEY and GILMAN (314) with silica gel as the adsorbent and benzene as the developing solvent.

Carbonyl compounds found in wood degradation products can be chromatographed on silica gel, as is described by RUFFINI (315). A mixture of ethyl acetate and ligroin (1:2) was used as the solvent.

The separation of the 2,4-DNPH's of ketosteroids has been described by KENT and RAWITCH (316). These authors used silica gel with ethyl acetate in benzene (15 + 85). Steroid 2,4-DNPH's were also studied by WATSON and ROMANOFF (317) on silica gel and alumina, using a large number of solvents. In most of their systems, one spot occurred for each 2,4-DNPH. However, the geometrical isomers did separate in solvents consisting of mixtures of benzene and chloroform and of benzene and methylene chloride.

Some aromatic-type carbonyl 2,4-DNPH's besides aliphatic types were studied by JART and BIGLER (318). They tried a number of adsorbents such as silica gel, alumina and microcrystalline cellulose, and plates impregnated with dimethylformamide.

Terpene-type 2,4-DNPH's were studied to some extent by VASHIST and HANDA (319). Silica gel was used in combination with a number of solvents, most of which were mixtures of chloroform with carbon tetrachloride. The separation of a number of monoterpene carbonyls has been reported by NANO and SANCIN (320).

Reversed-phase chromatography of 2,4-DNPH's of higher fatty aldehydes was described by MAHADEVAN et al. (321). Silicone oil was used as the stationary phase with 85% aqueous acetone as the eluent. 2,4-DNPH's of alkyl substituted carbonyl compounds can be separated on alumina layers impregnated with silver nitrate as reported by SLOOT (322). The alkylsubstituted derivatives showed higher R_F-values than did the normal DNPH's having the same chain length.

A number of different types of 2,4-DNPH's were investigated by BRUMMER and MÜLLER-PENNING (323). They investigated aliphatic saturated, unsaturated and hydroxycarbonyl 2,4-DNPH's and a number of aromatic-type derivatives on silica gel with the mixture petroleum ether: chloroform: ethyl acetate (30 + 3 + 1) as the solvent. Circular TLC of 2,4-DNPH's obtained from carbonyls from butterfat was used by ABOUSTEIT (324).

Volatile carbonyl compounds from vegetables were separated as 2,4-DNPH's on thin layers of silica gel, using light petroleum and ethyl acetate as the solvent, by BALDRATI and GIANNONE (325).

Carbonyl compounds in leather were separated by BLOEM (326) using silica gel-Carbowax layers.

DHONT and DIJKMAN (327) described the separation of complex mixtures of 2,4-DNPH's into two groups; one containing the 2,4-DNPH's derived of aromatic-type carbonyl compounds and one containing the aliphatic and terpene-type carbonyl derivatives.

Two isomeric 2,6-nonadienals and two isomeric 4-heptenals were isolated by HOFFMAN and MEYBOOM (265) using silver nitrate-impregnated silica gel layers, and benzene as the solvent.

SCHWARTZ et al. (328) separated 2,4-DNPH's of monocarbonyl compounds on a mixture of magnesium oxide and Celite with hexane: chloroform (19 + 5) as the solvent, and the derivatives of dicarbonyls on a layer of the same composition but with chloroform: methanol (19 + 1) as the solvent.

The TLC behaviour of a number of 2,4-DNPH's derived from different classes of carbonyl compounds was studied by TUMLINSON et al. (329) on layers of silica gel, polyamide, and silica gel covered with Carbowax as the stationary phase.

Bis-2,4-DNPH's of dicarbonyl compounds were separated on silica gel with benzene: petroleumether: ethylacetate (34 + 5 + 1) as the solvent, by RONKAINEN (330).

A two-dimensional method combining adsorption chromatography on magnesium oxide with partition chromatography on Carbowax was described by CRASKE and EDWARDS (331).

Analysis of mixtures of 2,4-DNPH's: In practice, the mixture of 2,4-DNPH's obtained from foods is a complicated mixture consisting of the derivatives of saturated and unsaturated aliphatic aldehydes and ketones, terpenoids and aromatic-type carbonyl compounds. Before actual analysis by TLC it is, therefore, advisable to separate the mixture of 2,4-DNPH's into classes containing one type of carbonyl compounds. This can be done by adsorption chromatography on layers of Kieselguhr/magnesium oxide, which are prepared as follows: 15 g of magnesium oxide (Merck p.a., sieved to < 50 mesh) and 15 g of Kieselguhr G (Merck) are mixed thoroughly and suspended in an Erlenmeyer flask containing 80 ml of a mixture of equal parts of methanol and water; the suspension is spread onto glass plates using standard equipment, the thickness of the layer being taken at about 0.55 mm because the fractions to be obtained should be of sufficient quantity for the further analysis. The plates are left in contact to the atmosphere overnight and need not be 'activated'. To the bottom edge of such a plate, 100—200 µl of a concentrated solution of the mixture of 2,4-DNPH's in ethyl acetate is now applied as a streak. The plate is then developed in a solvent mixture consisting of hexane and chloroform (70 + 30). The relative positions attained by the different classes are shown in Figure 3—34. After development, the bands are scraped from the plate and the adsorbed 2,4-DNPH's are removed from the plate material by suspending the adsorbent in a small amount of ethyl acetate and filtering the solution from the solid particles. The ethyl acetate is then removed by distillation in vacuum. Since the different classes of 2,4-DNPH's have characteristic absorption maxima, the identity of each class separated out of the original

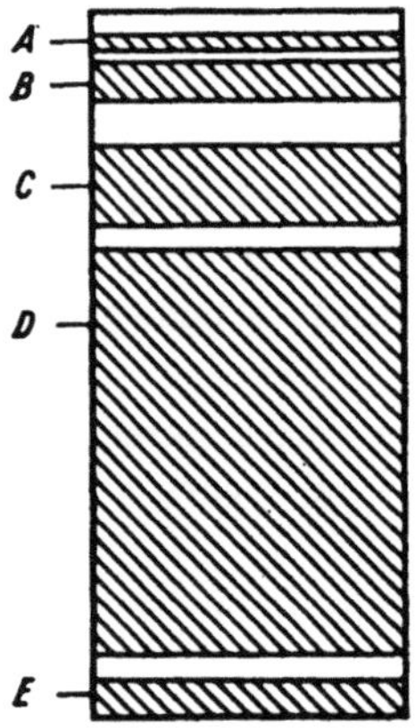

Fig. 3—34
Relative positions of the 2.4 DNPH's of some classes of carbonyl compounds on thin layers of magnesium oxide. Solvent: chloroform: hexane (7 + 3).
2.4 DNPH's of: *A* methyl ketones, *B* saturated aliphatic aldehydes, *C* unsaturated aliphatic aldehydes and methyl ketones, *D* monoderivatives of dicarbonyls and some aromatic ketones and aromatic aldehydes, *E* keto-acids, osazones, phenolic aldehydes.

Table 3—32
Absorption maxima of some 2,4-DNPH's in ethanol and chloroform

2,4-DNPH of:	λmax	
	in ethanol	in chloroform
2,4-dinitrophenylhydrazine	350	343
formaldehyde	348	348
saturated aliphatic aldehydes	356—360	358—361
methyl ketones	362—365	365—369
acrolein	366	367
2-enals	373	373
2-enones	376	380
2,4-dienals 2,4-dienones	379—395	388—407
2,4,6-trienals 2,4,6-trienones	395—410	400—415

Table 3—33
Maxima of some non-aliphatic-type carbonyl 2,4-DNPH's measured in chloroform

2,4-DNPH of:	λmax	2,4-DNPH of:	λmax
cinnamaldehyde	390	benzaldehyde	378
benzophenone	386	α-ionone	380
acetophenone	376	β-ionone	388
2-furaldehyde (yellow form)	379	5-methylfuraldehyde	389
2-furaldehyde (red form)	385	phenylacetaldehyde	355
phenylpropylaldehyde	355	p-toluylaldehyde	380
salicylaldehyde	377	vanillin	390

Table 3—34
R_F-values of some homologous series of 2,4-DNPH's in several partition chromatographic systems

2,4-DNPH of:	system[1]			
	A	B	C	D
	$^{2}R_F$	R_F	$^{3}R_F$	$^{3}R_F$
methanal	< 0.10	0.20	< 0.1	< 0.1
ethanal	0.16	0.27	< 0.1	0.13
n-propanal	0.23	0.37	< 0.1	0.17
n-butanal	0.35	0.45	0.13	0.24
n-pentanal	0.43	0.54	0.15	0.31
n-hexanal	0.51	0.61	0.21	0.37
n-heptanal	0.61	0.67	0.27	0.47
n-octanal	0.67	0.76	0.37	0.56
n-nonanal	0.73	0.78	0.46	0.65
n-decanal	0.80	0.84	0.57	0.73
n-undecanal	0.85	0.87	0.66	0.78
n-dodecanal	0.89	0.90	0.76	0.84
propanone-2	0.38	0.43	0.10	0.18
butanone-2	0.51	0.55	0.13	0.28
pentanone-2	0.61	0.61	0.19	0.35
hexanone-2	0.71	0.70	0.26	0.45
heptanone-2	0.79	0.78	0.34	0.53
octanone-2	0.83	0.81	0.43	0.63
nonanone-2	0.89	0.87	0.55	0.70
decanone-2	0.91	0.90	0.65	0.76
undecanone-2	0.95	0.94	0.73	0.81
propene-2-al-1	—	0.28	0.10	0.14
butene-2-al-1	0.20	0.32	0.10	0.16
pentene-2-al -1	0.27	0.41	0.14	0.25
hexene-2-al-1	0.38	0.52	0.15	0.30
heptene-2-al-1	0.45	0.59	0.18	0.42
octene-2-al-1	0.54	0.66	0.29	0.47
nonene-2-al-1	0.59	0.73	0.33	0.62
decene-2-al-1	0.67	0.75	0.46	0.63
undecene-2-al-1	0.83	—	0.52	0.81

[1] A. stationary phase: mixed layer of Carbowax 400 + 2-phenoxyethanol (1 + 1) see text
solvent: hexane
number of developments: 2

B. stationary phase: polyvinylacetate
solvent: hexane:methanol:ethylacetate:di-n-butylether (85 + 5 + 5 + 5)
number of developments: 1

C. stationary phase: silica gel impregnated with nitromethane by vapour phase adsorption see 3.4.3.2
solvent: hexane
number of developments: 3

D. stationary phase: silica gel impregnated with acetonitrile by vapour phase adsorption
solvent: hexane
number of developments: 3

Table 3—35
Values of some 2,4-DNPH's derived from terpene-type and aromatic carbonyl compounds

2,4-DNPH of:	system[1]			
	A	B	C	D
	[2]R_F	R_F	[3]R_F	[3]R_F
piperitone	0.81	0.81	0.43	0.64
α-phellandral	0.80	0.81	0.50	0.68
perillaaldehyde	0.46	0.57	0.20	0.39
menthone	0.87	0.86	0.55	0.73
iso-menthone	0.89	0.86	0.54	0.73
pulegone	0.81	0.83	0.57	0.64
iso-pulegone	0.84	0.84	0.41	0.56
carvotane acetone	0.80	0.79	0.49	0.66
a-thujone	0.87	0.84	0.46	0.66
b-thujone	0.87	0.84	0.46	0.66
carvone	0.71	0.73	0.31	0.54
citronellal	0.70	0.75	0.31	0.51
pseudoionone	0.81	0.80	0.40	0.60
dihydrocarvone	0.79	0.77	0.34	0.57
a-ionone	0.87	0.86	0.54	0.73
b-ionone	0.87	0.87	0.60	0.76
methyl-a-ionone	0.91	0.91	0.68	0.79
methyl-b-ionone	0.91	0.93	0.73	0.79
citral	0.51	0.64	0.23	0.44
methylheptenone	0.74	0.74	0.28	0.48

[1] for description of systems see Table 3—34

mixture can be confirmed by measuring the absorption maximum in the violet or near-ultraviolet region. Published data on the maxima of some typical classes are recorded in Table 3—32, while the absorption maxima of some other types of carbonyl compounds were determined in our institute (see Table 3—33).

Separations within the classes: The fractions from the magnesium oxide plate can now be separated into their components by other thin-layer systems. The bands containing the homologous aliphatic carbonyls and α-mono-unsaturated aldehydes are best separated by partition chromatography. Perfect separation is achieved by using a mixture of Carbowax and 2-phenoxyethanol as the stationary phase. The plates are prepared as follows: 1.5 g of Carbowax 400 and 1.5 g of 2-phenoxyethanol are dissolved in 60 ml of a mixture of equal parts of methanol and water, and the plates then prepared as described in 3.4.3.2. Other systems which gave good separations contain acetylcellulose or polyvinyl acetate as the stationary phase. Systems on the basis of silica gel impregnated with acetonitrile or nitromethane vapours can also be used to separate homologues; these latter systems have the advantage that the compounds can be scraped from the plates in a pure condition after separation, not accompanied by any stationary phase, because acetonitrile and nitromethane are volatile. R_F-values on acetylated cellulose are not tabulated but can be calculated by using the relation $R_M = -0.18\ C + 0.37$ (cf. 3.4.3.3.). R_F-values for a number of other 2,4-DNPH's are recorded in Tables 3—34, 3—35 and 3—36.

Table 3—36
$^{5}R_F$-values of some 2,4-DNPH's

2,4-DNPH	$^{5}R_F$	2,4-DNPH	$^{5}R_F$
methanal	0.23	propanone-2	0.21
ethanal	0.21	butanone-2	0.37
propanal	0.33	pentanone-2	0.41
butanal	0.39	hexanone-2	0.45
pentanal	0.44	heptanone-2	0.51
hexanal	0.48	octanone-2	0.53
heptanal	0.51	nonanone-2	0.55
octanal	0.53	decanone-2	0.58
nonanal	0.55	propene-2-al-1	0.35
decanal	0.57	butene-2-al-1	0.33
phenylacetealdehyde	0.30	pentene-2-al-1	0.37
heliotropin	0.21	hexene-2-al-1	0.44
cinnamaldehyde	0.31	heptene-2-al-1	0.46
cuminaldehyde	0.46	octene-2-al-1	0.49
furfural	0.16	nonene-2-al-1	0.51
	0.29	decene-2-al-1	0.53
m-methylacetophenone	0.48	anisaldehyde	0.19
p-toluylaldehyde	0.45	vanillin	0
benzaldehyde	0.40	ethylvanillin	0
piperitone	0.39	a-phellandral	0.65
perillaaldehyde	0.43	menthone	0.51
citronellal	0.49	iso-menthone	0.66
pulegone	0.46	a-thujone	0.53
iso-pulegone	0.51	b-thujone	0.63
carvone	0.61	carvotane acetone	0.67
pseudoionone	0.51	dihydrocarvone	0.57
α-ionone	0.60	citral	0.38
β-ionone	0.63	methylheptenone	0.44
methyl α-ionone	0.67		
methyl β-ionone	0.66		

layer: silica gel G
solvent: benzene:hexane (1 + 1)
number of developments: 5

Class separation after reaction with mercuric acetate: Further separations, especially in mixtures of aliphatic with terpenic carbonyls, may be achieved by reaction with mercuric acetate, based on the fact that 2,4-DNPH's containing non-conjugated double bonds readily react with mercuric acetate dissolved in methanol. The reaction is carried out as follows. At the bottom edge of the thin-layer plate (silica gel), two microlitres of a saturated solution of mercuric acetate in methanol are applied. After evaporation of the methanol, the 2,4-DNPH's, dissolved in a mixture of equal parts of methanol and ethyl acetate, are spotted onto the mercuric acetate. When the plate is developed in benzene:hexane (1 + 1) the 2,4-DNPH's which did not react with mercuric acetate move away from the start while 2,4-DNPH's derived from carbonyl compounds containing an isolated double bond remain at the start (Table 3—37).
This technique can be used in two-dimensional chromatography as is shown in Figure 3—35. The plate is first developed five times in benzene:hexane (1 + 1). After the fifth development, the plate is dried and covered with a glass plate leaving only

Table 3—37
Reaction of 2,4-DNPH's with mercuric acetate on thin layer plates

2,4-DNPH of:	reaction[1]	2,4-DNPH of:	reaction[1]
acrolein	—	carvotane acetone	—
carvone	+	citral	+
citronellal	+	methanal	—
benzaldehyde	—	methylheptenone	+
iso-pulegone	+	'phellandral	—
pulegone	—	a-ionone	—
b-ionone	—		

[1] — no reaction with mercuric acetate. Compound moves away from the start
+ positive reaction with mercuric acetate. Compound remains on the start.

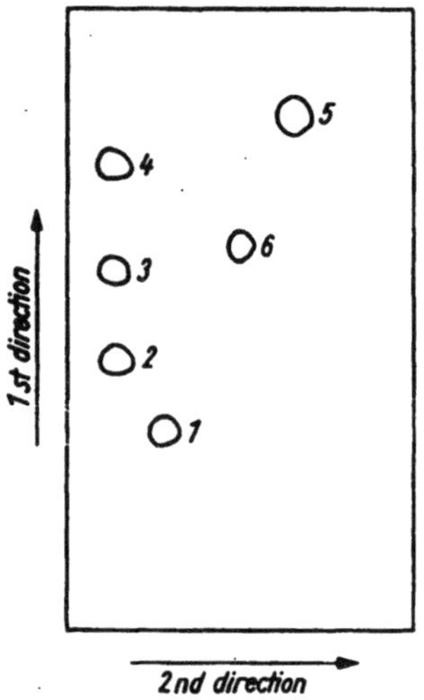

Fig. 3—35
Two-dimensional separation on silica gel of 2,4-DNPHs using reactions with mercuric acetate, 1st direction developed 5 times in hexane: benzene (1+1), 2nd direction developed in the same solvent mixture. 2,4-DNPH of 1 methanol, 2 citral,3 is opulegone, 4 carvone, 5 carvotane acetone, 6 pulegone.

the track of the separated compounds uncovered. The uncovered part is next sprayed with a saturated solution of mercuric acetate in methanol. After evaporation of the methanol, the plate is developed in the second dimension using one development in the same solvent mixture as before.

3.4.6. Identification and quantitative determination

3.4.6.1. Identification

Although a TLC separation procedure may in itself be called an identification method if R_F-values are found to coincide with reference values, yet unambiguous identification calls for further analysis of the separated compound. To this end, the eluate in the centrifuge tube (see 3.4.6.2.) can be used. Köhler (332) evaporates the eluate to dryness directly in the sample holder of a mass spectrometer. Stahl and Schild (333) evaporate the eluate in the presence of KBr for infra-red examination. A similar method has been described by Clemett (334).

In the author's laboratory, MS identification is usually carried out after evaporating the eluate to a small volume; the concentrated eluate is then injected into a GC-MS combination. Derivatized compounds, such as the 2,4-DNPH's of carbonyl compounds, and high-boiling compounds in general, are handled in a different way. The concentrated eluate is evaporated into the tip of a small piece of glass fibre paper, as shown in Figure 3-36.

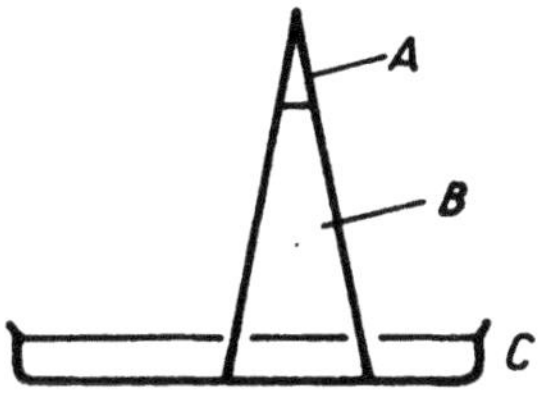

Fig. 3—36
Concentrating a component on the tip of a piece of glass fibre paper for identification by mass spectrometry.

This is effected by placing the glass fibre paper (*B*) in a small dish (*C*) containing the concentrated eluate. When the original solvent has completely evaporated, fresh solvent is added to the dish until the compound has been concentrated on the top(*A*) of the glass fibre paper. This top is cut off and placed into the sample holder of the mass spectrometer. It should be emphasized that for this operation all solvents must be freshly distilled. The plate should be washed before use by placing it into a chamber filled with distilled acetone. Glasswork and the glass fibre paper should also be washed beforehand.

3.4.6.2. Quantitative determination

There are essentially two different methods for the quantitative determination of compounds separated by thin-layer chromatography:

a) measurement of the spot on the plate;
b) measurement after elution from the plate.

All the derivatives discussed in the preceding sections are either intensely coloured or show intense fluorescence in UV light, which facilitates determination by optical methods.

Measurement of the spot on the plate: The amount of material in the spot is found either by estimating the spot size visually or by measuring the concentration in the spot by a scanning device, measuring the transmittance, the reflectance, the fluorescence or the absorption of the spot.
Measurement of the spot area should be considered a semi-quantitative method. Both the spot area (335) and its square root (336) have been linearly related with the log of the amount of substance. For direct optical scanning methods, a number of mathematical theories and many sophisticated contrivances have been worked out (337—342). Some investigations have been concerned with a scanning method using pyrolysis with detection of the pyrolytic products by flame ionization detectors (343).

Measurement after elution of the compounds from the plate: Determination of a compound after elution from the plate material is a much simpler method than are direct scanning methods. In many cases, spectrophotometric, colorimetric, polarographic or other methods can be applied to the solution obtained from thin-layer plates. Such methods are at least as accurate as the direct scanning methods. Usually the spot or band is scraped from the plate by some type of pipe-line filter device (344). In the writer's laboratory, the device shown in Figure 3-37a is used. *A* is a glass tube with a constriction in the middle. A cotton plug, B, is placed against the constriction. Glass tube *D* is connected to *A* by a piece of silicone tubing, *C*; *E* is connected to a vacuum line. The spot or band is scraped from the plate with the help of mouthpiece F while the liberated plate material is sucked against cotton plug *B*. After all the adsorbent has been collected, tube A is disconnected from *D* and placed into a centrifuge tube as shown in Figure 3—37b. To elute the compound from the adsorbent, 100—200 μl of a suitable solvent is placed on the top of the adsorbent column (*B*) in the centrifuge tube (*D*) and the solvent is forced through the adsorbent by centrifug-

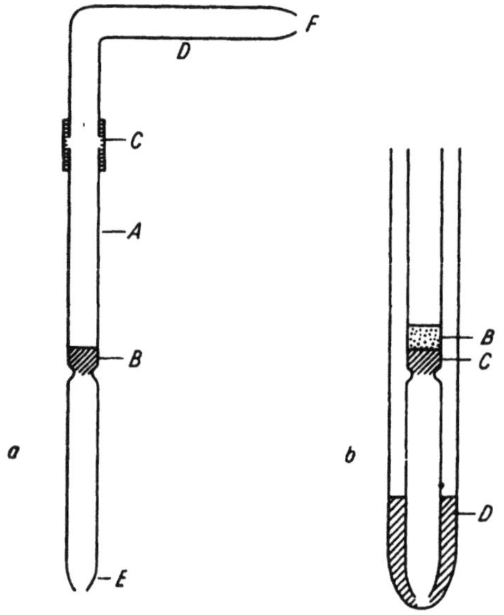

Fig. 3—37
a. Small 'vacuum cleaner' for isolating separated compounds from the plate material. *A*. glass tube, *B*. cotton wool, *C*. silicone rubber sleeve, *D*. glass tube bent to 90°, *E*. connection to pump, *F*. opening for aspiration of plate material
b. Part A from figure 3.—37a placed into a centrifuge tube for elution. *C* cotton wool, *B* aspirated adsorbent material, *D* centrifuge tube with eluate at the bottom of the tube.

ing. This elution is repeated once. After this operation, the eluate is found at the bottom of the centrifuge tube. The eluted compound can now be determined by any suitable analytical method.

References

1. Keulemans, A. I. M., Gas Chromatography, 2nd ed., Reinhold Publishing Corporation, New York 1959, pp. 234
2. Dean, J. A., Chemical separation methods, Van Nostrand Reinhold Company, New York 1969, pp. 398
3. Ettre, L. S., Open tubular columns in gas chromatography, Plenum Press, New York 1965, pp. 164

4. KAISER, R. E., Chromatographie in der Gasphase, I Gas-Chromatographie 1960, pp. 205; II Kapillar-Chromatographie 1961, pp. 156; III Tabellen 1962, pp. 216; IV Quantitative Auswertung 1965, pp. 278, Bibliographisches Institut, Mannheim 1960—1965
5. KAISER, R. E., Neuere Ergebnisse zur Anwendung der Gas-Chromatographie, Z. Anal. Chem. **189**, 1—14 (1962)
6. PALFRAMAN, J. F. and E. A. WALKER, Techniques in gas chromatography. I. Choice of solid supports. A review, Analyst (London) **92**, no. 1091, 71—82 (1967)
7. OTTENSTEIN, D. M., Column support materials for use in gas chromatography, J. Gas Chromatogr. **1**, 11—23 (1963)
8. OTTENSTEIN, D. M., The chromatographic support, Adv. Chromatogr. **3**, 137—196 (1966)
9. HISHTA, C., J. BOMSTEIN and W. D. COOKE, Advances in the technology of lightly loaded glass bead columns, Adv. Chromatogr. **9**, 215—242 (1970)
10. ROHRSCHNEIDER, L., Eine Methode zur Charakterisierung von gaschromatographischen Trennflüssigkeiten, J. Chromatogr. **22**, 6—22 (1966)
11. MCREYNOLDS, W. O., Characterization of some liquid phases, J. Chromatogr. Sci. **8**, 685—691 (1970)
12. ŘEHÁK, V. and E. SMOLKOVÁ, Chemically bonded stationary phases for gas and high performance liquid chromatography, Chromatographia **9**, 219—229 (1976)
13. LITTLE, J. N., W. A. DARK, P. W. FARLINGER and K. J. BOMBAUGH, GC packings with chemically-bonded stationary phases, J. Chromatogr. Sci. **8**, 647—652 (1970)
14. PESEK, J. J. and J. E. DANIELS, Investigation of the retention mechanism of chemically bonded stationary phases in gas chromatography, J. Chromatogr. Sci. **14**, 288—291 (1976)
15. CRAMERS, C. A., J. A. RIJKS and P. BOCEK, Potentialities of micropacked columns: some applications in petroleum chemistry, J. Chromatogr. **65**, 29—37 (1972)
16. RIJKS, J. A., C. A. CRAMERS and P. BOCEK, Permeability and preparation of micropacked columns, Chromatographia **8**, 482—485 (1975)
17. BRUNER, F., P. CICCIOLI, G. BERTONI and A. LIBERTI, Gas chromatographic characteristics and analytical applications of micropacked columns, J. Chromatogr. Sci. **12**, 758—762 (1974)
18. BRUNER, F., P. CICCIOLI, E. BRANCALEONI and A. LONGO, The application of glass micropacked columns in gas liquid solid chromatography to the GC-MS analysis of polar compounds, Chromatographia **8**, 503—506 (1975)
19. GOLAY, M. J. E., Theory and practice of gas-liquid partition chromatography with coated capillaries, in "Gas Chromatography 1957", eds. COATES, V. J., H. J. NOEBELS and I. S. FAGERSON, Academic Press, New York 1957, p. 1—13
20. GOLAY, M. J. E., Theory of chromatography in open and coated tubular columns with round and rectangular cross-sections, in "Gas Chromatography 1958", ed. DESTY, D. H., Butterworths Scientific Publications, London 1958, p. 36—55
21. BERTSCH, W., F. SHUNBO, R. C. CHANG and A. ZLATKIS, Preparation of high resolution nickel open tubular columns, Chromatographia **7**, 128—134 (1974)
22. GOSTECNIK, G. F. and A. ZLATKIS, Computer evaluation of gas chromatographic profiles for the correlation of quality differences in cold pressed orange oils, J. Chromatogr. **106**, 73—81 (1975)
23. ETTRE, L. S., E. W. CIEPLINSKI and W. AVERILL, Application of open tubular (Golay) columns with larger diameter, J. Gas Chromatogr. **1**, 7—16 (1963)
24. TERANISHI, R. and T. R. MON, Large-bore capillary and low-pressure-drop packed columns, Anal. Chem. **36**, 1490—1492 (1964)
25. TERANISHI, R., I. HORNSTEIN, P. ISSENBERG and E. L. WICK, "Gas chromatographic separations" in "Flavor Research, Principles and Techniques", Marcel Dekker, Inc., New York 1971, p. 77—106
26. MON, T. R., Preparation of large bore open tubular columns for GC, Res. Devel. **22**, 14—17 (Dec. 1971)
27. LOWRY, R. R. and I. J. TINSLEY, Open tubular columns for gas liquid chromatography, cleaning and recoating procedures, J. Am. Oil Chem. Soc. **52**, 298—299 (1975)
28. RYBA, M., Adsorption properties of stainless-steel capillaries used in the preparation of open tubular gas chromatographic columns, J. Chromatogr. **123**, 317—325 (1976)

29. METCALFE, L. D. and R. J. MARTIN, Technique for coating gas-liquid chromatography capillary columns using long chain quaternary ammonium compounds, Anal. Chem. **39**, 1204—1205 (1967)
30. VERBEEK, A. M. and H. MAARSE, The honeycomb spool: a modified method for winding capillary columns, J. Chromatogr. **84**, 389—391 (1973)
31. ETTRE, L. S. and J. E. PURCELL, Porous-layer open tubular columns — Theory, practice and applications, Adv. Chromatogr. **10**, 1—97 (1974)
32. NOVOTNY, M. and A. ZLATKIS, Glass capillary columns and their significance in biochemical research, Chromatogr. Rev. **14**, 1—44 (1971)
33. ALEXANDER, G., G. GARZÓ and G. PÁLYI, Method for preparing glass capillary columns for gas chromatography, J. Chromatogr. **91**, 25—37 (1974)
34. JENNINGS, W. G., K. YABUMOTO and R. H. WOHLEB, Manufacture and use of the glass open tubular column, J. Chromatogr. Sci. **12**, 344—348 (1974)
35. NOVOTNY, M. und K. TESAŘIK, Behandlung der Innenoberfläche von Glaskapillaren für die Gas-Chromatographie, Chromatographia **1**, 332—333 (1968)
36. BADINGS, H. T., J. J. G. VAN DER POL and D. G. SCHMIDT, Some observations on the hydrochloric acid etching of soda-lime glass in the preparation of capillary columns for gas chromatography, Chromatographia **10**, 404—411 (1977)
37. FRANKEN, J. J., G. A. F. M. RUTTEN and J. A. RIJKS, Preparation of glass capillary columns coated with polar phase for high-temperature gas chromatography. I. Study of roughening of the glass surface, J. Chromatogr. **126**, 117—132 (1976)
38. ALEXANDER, G. and G. A. F. M. RUTTEN, Preparation of polar phase coated open tubular columns for steroid analysis, Chromatographia **6**, 231—233 (1973)
39. ALEXANDER, G. and G. A. F. M. RUTTEN, Surface characteristics of treated glasses for the preparation of glass capillary columns in gas-liquid chromatography, J. Chromatogr. **99**, 81—101 (1974)
40. BADINGS, H. T., J. J. G. VAN DER POL and J. G. WASSINK, Wide-bore glass capillary columns for gas chromatography. Their preparation and application, Chromatographia **8**, 440—448 (1975)
41. KRUPČÍK, J., M. KRISTÍN, M. VALACHOVIČOVÁ and Š. JANIGA, Effect of etching with gaseous hydrogen chloride on the quality of glass capillary columns, J. Chromatogr. **126**, 147—160 (1976)
42. TESAŘIK, K. und M. NOVOTNY, Behandlung der Innenoberfläche von Glaskapillaren für die Gas-Chromatographie, VI Symposium über Gas Chromatographie, ed. Struppe, H. G. 1968, p. 575—584
43. SCHIEKE, J. D., N. R. COMINS and V. PRETORIUS, Whiskers: a new support for glass open tubular columns in gas chromatography, Chromatographia **8**, 354 (1975)
44. SCHIEKE, J. D., N. R. COMINS and V. PRETORIUS, Whisker-walled open tubular glass columns for gas chromatography. Techniques for inducing whisker growth, J. Chromatogr. **112**, 97—107 (1975)
45. SCHIEKE, J. D., N. R. COMINS and V. PRETORIUS, An improved technique for observing the inner surface of open-tubular columns for gas chromatography using scanning electron microscopy, J. Chromatogr. **114**, 190—192 (1975)
46. SCHIEKE, J. D., N. R. COMINS and V. PRETORIUS, Surface-modified open tubular columns for gas chromatography. Observation and analysis of the surface using scanning electron microscopy (SEM) and energy-dispersive X-ray analysis, J. Chromatogr. **115**, 373—381 (1975)
47. SCHIEKE, J. D. and V. PRETORIUS, Whisker-walled open tubular glass columns in gas chromatography, I. Deactivation, J. Chromatogr. **132**, 217—222 (1977)
48. SCHIEKE, J. D. and V. PRETORIUS, Whisker-walled open tubular columns in gas chromatography, II. Chromatographic performance, III. Applications, J. Chromatogr. **132**, 223—236 (1977)
49. SANDRA, P. and M. VERZELE, Surface treatment, deactivation and coating in GC[2] (Glass Capillary Gas Chromatography), Chromatographia **10**, 419—425 (1977)
50. ONUSKA, F. I. and M. E. COMBIA, Preparation and application of surface-modified high-resolution wall-coated open tabular columns, J. Chromatogr. **126**, 133—145 (1976)
51. GROB, K., Polare Imprägnierung von Glaskapillaren für die Gas-Chromatographie, Helv. Chim. Acta **48**, 1362—1370 (1965)

52. GORETTI, G. C., A. LIBERTI and G. NOTA, A general procedure for preparing high resol utio glass capillary columns, Chromatographia **8**, 486—490 (1975)

53. VIDAL-MADJAR, C., S. BEKASSY, M. F. GONNORD, P. ARPINO and G. GUIOCHON, Glass-solid and gas-liquid chromatography using porous layer open tube columns made with graphitized thermal carbon black, Anal. Chem. **49**, 768—772 (1977)

54. DEELDER, R. S., J. J. M. RAMAEKERS, J. H. M. VAN DEN BERG and M. L. WETZELS, Study on the efficiency of support-coated open-tubular columns for steroid analysis, J. Chromatogr. **119**, 99—107 (1976)

55. GERMAN, A. L., C. D. PFAFFENBERGER, J.-P. THENOT, M. G. HORNING and E. C. HORNING, High resolution gas chromatography with thermostable glass open tubular capillary columns, Anal. Chem. **45**, 930—935 (1973)

56. GERMAN, A. L. and E. C. HORNING, Thermostable open tube capillary columns for the high resolution gas chromatography of human urinary steroids, J. Chromatogr. Sci. **11**, 76—82 (1973)

57. CRONIN, D. A., Preparation and properties of porous layer open tubular borosilicate and soda-glass columns possessing a permanently fixed support layer, J. Chromatogr. **101**, 271—279 (1974)

58. MCKEAG, R. G. and F. W. HOUGEN, Dynamic coating of glass capillaries with polar phases and Silanox, J. Chromatogr. **136**, 308—310 (1977)

59. GROB, K. and G. GROB, A new, generally applicable procedure for the preparation of glass capillary columns, J. Chromatogr. **125**, 471—485 (1976)

60. GROB, K., G. GROB and K. GROB JR., The barium carbonate procedure for the preparation of glass capillary columns; further informations and developments, Chromatographia **10**, 181—187 (1977)

61. CRAMERS, C. A., E. A. VERMEER, L. G. VAN KUIK, J. A. HULSMAN and C. A. MEIJERS, Quantitative determination of underivatized anti-convulsant drugs by high resolution gas chromatography with support-coated open tubular columns, Clin. Chim. Acta **73**, 97—107 (1976)

62. CRAMERS, C. A., E. A. VERMEER and J. J. FRANKEN, Preparation and evaluation of polar stabilized phase open tubular (SPOT) columns, Chromatographia **10**, 412—418 (1977)

63. SCHULTE, E., Belegung von Kapillarsäulen aus Glas nach Aufbringen von kolloidaler Kieselsäure, Chromatographia **9**, 315—320 (1976)

64. MALEC, E. J., Improved open tubular gas chromatographic columns for use at 250 °C and above, J. Chromatogr. Sci. **9**, 318—320 (1971)

65. RUTTEN, G. A. F. M. and J. A. LUYKEN, Analysis of steroids by high resolution gas-liquid chromatography, I. Preparation of apolar columns, J. Chromatogr. **74**, 177—193 (1972)

66. BLOMBERG, L., Deactivation of glass capillary columns for gas chromatography, J. Chromatogr. **115**, 365—372 (1975)

67. AUE, W. A., C. R. HASTINGS and S. KAPILA, On the unexpected behaviour of a common gas chromatographic phase, J. Chromatogr. **77**, 299—307 (1973)

68. CRONIN, D. A., The preparation of stable glass capillary columns coated with Carbowax 20 M, J. Chromatogr. **97**, 263—266 (1974)

69. GROB, K., Are noble metals the ideal material for interfaces in capillary GC?, Chromatographia **9**, 509—512 (1976)

70. DIJKSTRA, G. and J. DE GOEY, The use of coated capillaries as columns for gas chromatography, in "Gas Chromatography 1958", ed. DESTY, D. H. Butterworths Scientific Publications, London 1958, p. 56—68

71. BOUCHE, J. and M. VERZELE, A static coating procedure for glass capillary columns, J. Gas Chromatogr. **6**, 501—505 (1968)

72. NOVOTNY, M., K. D. BARTLE and L. BLOMBERG, Dependence of film thickness on column radius and coating rate in preparation of capillary columns for gas chromatography, J. Chromatogr. **45**, 469—471 (1969)

73. NOVOTNY, M., L. BLOMBERG and K. D. BARTLE, Some factors affecting the coating of open tubular columns for gas chromatography, J. Chromatogr. Sci. **8**, 390—393 (1970)

74. SCHOMBURG, G. and H. HUSMANN, Methods and techniques of gas chromatography with glass capillary columns, Chromatographia **8**, 517—530 (1975)

75. Venema, A. and L. G. J. van der Ven, Factors affecting film thickness in glass capillary chromatography columns coated by the mercury drop method, Proc. 2nd Int. Symp. on Glass Cap. Chrom., ed. Kaiser, R. E. Institute of Chromatography, Bad Dürkheim 1977, p. 385—395
76. Ilkova, E. L. and E. A. Mistryukov, A simple versatile method for coating of glass capillary columns, J. Chromatogr. Sci. **9**, 569—570 (1971)
77. Jennings, W. G., personal communication, 1977
78. Kaiser, R. E., Trenncassetten zur Gas-Chromatographie. Ein neues technisches System zur Erleichterung normaler und extrem schwieriger Analysen, besonders in der Umwelt-Analytik, Chromatographia **7**, 688—695 (1974)
79. Berg, P. M. J. van den und Th. P. H. Cox, Ein Ganzglas-System zur Dosierung von Festproben auf Trennkapillaren in der Gas-Chromatographie, Chromatographia **5**, 301—305 (1972)
80. Schulte, E. und L. Acker, Gas-Chromatographie mit Glaskapillaren bei Temperaturen bis zu 320 °C und ihre Anwendung zur Trennung von Polychlorbiphenylen, Z. Anal. Chem. **268**, 260—267 (1974)
81. Hartigan, M. J. and L. S. Ettre, Questions related to gas chromatographic systems with glass open-tubular columns, J. Chromatogr. **119**, 187—206 (1976)
82. Grob, K. and H. J. Jaeggi, Techniques of capillary gas chromatography. Possibilities for full utilization of high-performance columns. II. Manipulation and operation, Chromatographia **5**, 382—391 (1972)
83. Grob, K. Glaskapillaren für die Gas-Chromatographie. Verbesserte Erzeugung und Prüfung stabiler Trennflüssigkeitsfilme, Helv. Chim. Acta **51**, 718—737 (1968)
84. Grob, K. and G. Grob, An acidity test as part of the quality tests of a capillary column, Chromatographia **4**, 422—424 (1971)
85. Jentzsch, D. und E. Otte, Detektoren in der Gas-Chromatographie 1st ed., Akademische Verlagsgesellschaft, Frankfurt am Main 1970, pp. 454
86. David, D. J., Gas chromatographic detectors, 1st ed., Wiley-Interscience, New York 1974, pp. 295
87. Sevcik, J., Detectors in gas chromatography, 1st ed., Elsevier Scientific Publishing Company, Amsterdam 1975, pp. 192
88. Lawson, A. E. and J. M. Miller, Thermal conductivity detectors in gas chromatography, J. Gas Chromatogr. **4**, 273—284 (1966)
89. Driscoll, J. N., Evaluation of a new photoionization detector for organic compounds, J. Chromatogr. **134**, 49—55 (1977)
90. Driscoll, J. N. und J. B. Clarici, Ein neuer Photoionisationsdetektor für die Gas-Chromatographie, Chromatographia **9**, 567—570 (1976)
91. Bruhn, W., Natural substances or artifacts, Dragoco Rep. **6**, 115—121 (1974)
92. Weurman, C., Developments in head space food odour analysis, Proc. 2nd Intern. Congress on Food Science and Technol., Warsaw 1966, p. 289—301
93. Drozd, J. and J. Novak, Quantitative head-space gas analysis by the standard additions method, J. Chromatogr. **136**, 37—44 (1977)
94. Kepner, R. E., J. Strating and C. Weurman, Quantitative determination of esters in beer by gas chromatographic analysis of head space vapours, J. Inst. Brew. (London), **69**, 399—405 (1963)
95. Kunitake, N., Gas chromatographic determination of low-boiling volatiles in beer without any concentration steps, Bull. Brew. Sci. **11**, 1—9 (1965)
96. Weurman, C., Gas-liquid chromatographic studies on the enzymatic formation of volatile compounds in raspberries, Food Technol. (Chicago) **15**, 531—536 (1961)
97. —, unpublished results, Central Institute for Nutrition and Food Research TNO, Zeist
98. Davis, P. L., A simple method to prevent loss of volatiles during headspace analysis, J. Chromatogr. Sci. **8**, 423—425 (1970)
99. Gilliver, P. J. and H. E. Nursten, Loss of volatiles during headspace analysis through absorption by rubber septa and its avoidance, Chem. Ind. (London), 541—542 (1972)
100. Kepner, R. E., H. Maarse and J. Strating, Gas chromatographic head space techniques for the quantitative determination of volatile components in multicomponent aqueous solutions, Anal. Chem. **36**, 77—82 (1964)

101. HARRISON, G. A. F., Headspace analysis of beer, 4th Wilkens Gas Chrom. Symposium 1966, p. 31—38
102. JANSEN, H. E., J. STRATING and W. M. WESTRA, The gas chromatographic determination of traces of hydrogen sulphide and alkanethiols in beer, J. Inst. Brew. **77**, 154—159 (1971)
103. HARRISON, G. A. F., W. J. BYRNE and E. COLLINS, Applications of electron capture chromatography to brewery problems, Eur. Brew. Conv., Proc. Congr. Stockholm, 1965, p. 352—359
104. BUTTERY, R. G. and R. TERANISHI, Gas-liquid chromatography of aroma of vegetables and fruit. Direct injection of aqueous vapours, Anal. Chem. **33**, 1439—1441 (1961)
105. GASCÓ, L., R. BARRERA and F. DE LA CRUZ, Gas chromatographic investigations of the volatile constituents of fruit aromas, J. Chromatogr. Sci. **7**, 228—238 (1969)
106. BANDYOPADHYAY, C., A. N. SRIRANGARAJAN and A. SREENIVASAN, Studies on flavour components of onion. I. Thin-layer chromatographic investigation of onion oil, J. Chromatogr. **47**, 400—407 (1970)
107. HOUGEN, F. W., M. A. QUILLIAM and W. A. CURRAN, Headspace vapors from cereal grains, J. Agric. Food Chem. **19**, 182—183 (1971)
108. PALO, V., Contribution to the identification of easily volatile compounds in milk products analysed by using GLC (headspace), Chromatographia **4**, 55—58 (1971)
109. BASSETTE, R., S. OZERIS and C. H. WHITNAH, Direct chromatographic analysis of milk, J. Food Sci. **28**, 84—90 (1963)
110. GILBERT, J. and H. E. NURSTEN, Volatile constituents of horseradish roots, J. Sci. Food Agric. **23**, 527—539 (1972)
111. PAILLARD, N. M. M., Nature et origine de l'émission organique volatile des pommes (*Pirus Malus* L.), Thesis, l'Université Pierre et Marie Curie, Paris 1975, pp. 88 and XXX
112. HURST, R. E., Device for sampling headspace from canned food, Anal. Chem. **47**, 1221—1223 (1975)
113. EVANS, C. D. and E. SELKE, Headspace gas analysis of salad oils by mass spectrometry, J. Am. Oil Chem. Soc. **49**, 106—110 (1972).
114. JENTZSCH, D., H. KRÜGER, G. LEBRECHT, G. DENCKS und J. GUT, Arbeitsweise zur automatischen gas-chromatographischen Dampfraum-Analyse, Z. Anal. Chem. **236**, 96—118 (1968)
115. MIETHKE, H., Anwendung der automatischen Dampfraum-Gaschromatographie in der Lebensmittelüberwachung, Dtsch. Lebensm. Rundsch. **65**, 379—382 (1969)
116. NAWAR, W. W., Some considerations in interpretation of direct headspace gas chromatographic analysis of food volatiles, Food Technol. (Chicago) **20**, 115—117 (1966)
117. LUNTEREN, G. VAN, S. VAN STRATEN and C. WEURMAN, Irregularities in gas chromatographic direct vapour analysis, Troisième Colloque International sur la chimie des Cafés, Trieste 1967, p. 191—196
118. BUTTERY, R. G., L. C. LING and D. G. GUADAGNI, Volatilities of aldehydes, esters and ketones in dilute water solution, J. Agric. Food Chem. **17**, 385—389 (1969)
119. DREWS, B., G. BÄRWALD und H. J. NIEFIND, Gaschromatographische Bestimmung flüchtiger Schwefelverbindungen, Monatsschr. Brau. **22**, 140—144 (1969)
120. BÄRWALD, G., Head space-Analyse in der Brauereibetriebskontrolle, Brauwissenschaft **24**, 343—347 (1971)
121. KREJČÍ, M., K. HÁNA and M. ROUDNÁ, Anomalous sorption in the injection port of the gas chromatograph, J. Chromatogr. **41**, 145—155 (1969)
122. FRANKEN, J. J. and H. L. VADER, Open hole tubular columns in pesticide analysis, Chromatographia **6**, 22—27 (1973)
123. KOLLOFF, R. H., Septum bleeding in flame ionization programmed temperature, gas liquid chromatography, Anal. Chem. **34**, 1840—1841 (1962)
124. TAMSMA, A., F. E. KURTZ, N. RAINEY and M. J. PALLANSCH, Preconditioning of septums to reduce eluate contamination and anomalous peaks during temperature programming, J. Gas Chromatogr. **5**, 271—272 (1967)
125. CALLERY, I. M., Septum performance in gas chromatography, J. Chromatogr. Sci. **8**, 408—411 (1970)
126. SMITH, E. D. and K. E. SORRELLS, A quantitative comparison of gas chromatograph septum bleed, J. Chromatogr. Sci. **9**, 15—17 (1971)

127. PURCELL, J. E., H. D. DOWNS and L. S. ETTRE, Extraneous peaks in gas chromatography; their source and elimination, Chromatographia **8**, 605—616 (1975)
128. SMITH, E. D., K. E. SORRELLS and R. G. SWINEA, Gas chromatographic septum bleed studies, J. Chromatogr. Sci. **12**, 101—104 (1974)
129. DRESSMAN, R. C., Elimination of "memory" peaks encountered in aqueous-injection gas chromatography, J. Chromatogr. Sci. **8**, 265—271 (1970)
130. MIZUNO, G., D. JARVIS and J. R. CHIPAULT, Spurious peaks in temperature programmed gas chromatography, J. Gas Chromatogr. **6**, 179—180 (1968)
131. TUCKNOTT, O. G. and A. A. WILLIAMS, Injection port modification to prevent septa bleeding in programmed temperature gas-liquid chromatography, Anal. Chem. **41**, 2086—2088 (1969)
132. RENSBURG, J. F. J. VAN, P. L. MOUTON and V. PRETORIUS, A water-cooled inlet system for gas chromatographs, J. Chromatogr. Sci. **10**, 580—581 (1972)
133. GROB, K. and G. GROB, Techniques of capillary gas chromatography. Possibilities of the full utilization of high-performance columns. I. Direct sample injection, Chromatographia **5**, 3—12 (1972)
134. BARTEL, E. E. and S. J. VAN DER WALT, Reduction of plug length in capillary gas chromatography using an on-column micro-trap, J. Gas Chromatogr. **6**, 396—398 (1968)
135. EVRARD, E. and G. GUIOCHON, A splitless sampling system without a septum for high temperature gas chromatography on packed capillary columns, Chromatographia **5**, 587—593 (1972)
136. JENNINGS, W. G., Glass inlet splitter for gas chromatography, J. Chromatogr. Sci. **13**, 185—187 (1975)
137. JENNINGS, W. G. and S. ADAM, Gas chromatography: elution temperature, speed of analysis, and separation efficiency as influenced by rate of temperature programming and carrier gas velocity in open tubular glass capillary columns, Anal. Biochem. **69**, 61—69 (1975)
138. CRAMERS, C. A. and M. M. VAN KESSEL, Direct sample introduction system for capillary columns, J. Gas Chromatogr. **6**, 577—581 (1968)
139. GROB, K. and G. GROB, Splitless injection on capillary columns I. The basic technique; steroid analysis as an example. II. Conditions and limits, practical realization, J. Chromatogr. Sci. **7**, 584—591 (1969)
140. KARLSEN, J. and A. BAERHEIM SVENDSEN, Gas liquid chromatography of the monoterpene hydrocarbons of the volatile oil of separate needles and pieces of branches from *Juniperus communis* L., Med. Nor. Farm Selsk. **30**, 1—5 (1968)
141. MAARSE, H., Composition of the volatile oil of *Origanum vulgare* L. ssp. vulgare during the growth of the plant (in Dutch), Thesis, University of Groningen 1971
142. SCHOMBURG, G., R. DIELMANN, H. HUSMANN and F. WEEKE, New developments and experience on glass capillary column production and sampling techniques, in "Proc. 2nd Int. Symp. on Glass Cap. Chromatogr.", ed., KAISER, R. E. Institute of Chromatography, Bad Dürkheim, 359—383 (1977)
143. MEYER, A. S., A solid sample injector and capillary effluent splitter providing a completely metal-free gas chromatographic system, Anal. Biochem. **33**, 447—453 (1970)
144. WONG, N. P. and D. P. SCHWARTZ, A versatile injector for the gas chromatograph, J. Chromatogr. Sci. **7**, 569—571 (1969)
145. HARDING, R. J., H. E. NURSTEN and J. J. WREN, Basic compounds contributing to beer flavour, J. Sci. Food Agric. **28**, 225—232 (1977)
146. BADINGS, H. T., J. J. G. VAN DER POL and J. G. WASSINK, Wide-bore capillary columns for gas chromatography. Their preparation and applications. Proc. 1st Int. Symp. on Glass Cap. Chrom., ed. R. E. KAISER, Institut für Chromatographie, Bad Dürkheim, Germany 1975, p. 193—217
147. KLIMES, I., W. STÜNZI and D. LAMPARSKY, Nanogramm-Verfahren. I. Das verlustlose Einspritzen auf eine Kapillarkolonne. II. Präparative Gaschromatographie im Mikromaßstab, J. Chromatogr. **136**, 13—36 (1977)
148. DROZD, J., Chemical derivatization in gas chromatography, J. Chromatogr. **113**, 303—356 (1975)
149. HASHIMOTO, N., Studies on volatile carbonyl compounds in beer, Rept. Res. Lab. Kirin Brew. Co. Ltd, no. 9, 1—10 (1966)

150. FEDELI, E. and M. CIRIMELE, The separation of phenylhydrazones of volatile carbonyl compounds by vapour phase chromatography, J. Chromatogr. **15**, 435 (1964)
151. SOUKUP, R. J., R. J. SCARPELLINO and E. DANIELCZIK, Gas chromatographic separation of 2,4-dinitrophenylhydrazone derivatives of carbonyl compounds, Anal. Chem. **36**, 2255—2256 (1964)
152. KALLIO, H., Gas-liquid chromatographic analysis of 2,4-dinitrophenylhydrazones of carbonyl compounds, J. Chromatogr. **65**, 355—360 (1972)
153. HOSHIKA, Y. and Y. TAKATA, Gas chromatographic separation of carbonyl compounds as their 2,4-dinitrophenylhydrazones using glass capillary columns, J. Chromatogr. **120**, 379 to 389 (1976)
154. PIAS, J. B. and L. GASCÓ, Gas-liquid chromatography of 2,4-dinitrophenylhydrazones of carbonyl compounds; retention indices and molecular structure, Chromatographia **8**, 270 to 273 (1975)
155. KALLIO, H., Volatile monocarbonyl compounds of arctic bramble at various stages of ripeness, Z. Lebensm. Unters. Forsch. **153**, 23—30 (1973)
156. KOROLCZUK, J., M. DANIEWSKI and Z. MIELNICZUK, Gas chromatographic determination of carbonyl compounds as their phenylhydrazones, J. Chromatogr. **88**, 177—182 (1974)
157. TEITELBAUM, C. L., Isolation of carbonyl compounds under neutral conditions using the Girard reagent, J. Org. Chem. **23**, 646—647 (1958)
158. MAARSE, H. and J. SCHAEFER, Quantitative headspace analysis, total and specific group analysis, in "Analysis of Foods and Beverages, Head space techniques", ed. CHARALAMBOUS, G. Academic Press, New York, San Francisco, London 1978, p. 17—35
159. GALETTO, W. G., R. E. KEPNER and A. D. WEBB, Gas chromatographic analysis of some alcohol 3,5-dinitrobenzoates with application to analysis of acetals, Anal. Chem. **38**, 34—36 (1966)
160. WINTER, M., B. WILLHALM, M. HINDER, E. PALLUY and E. SUNDT, Chromatography on columns and paper in the study of the volatile fruit flavour of strawberries, Perfum. Essent. Oil Rec. **49**, 250—255 (1958)
161. SCHORMÜLLER, J. und W. GROSCH, Untersuchungen über Aromastoffe von Lebensmitteln III. Ein Beitrag zur Analytik in Tomaten vorkommender flüchtiger Alkohole, Z. Lebensm. Unters. Forsch. **126**, 188—193 (1964)
162. SEIDENSTÜCKER, W., Gas-chromatographische Bestimmung höherer Terpenalkohole nach Silylierung, Z. Anal. Chem. **237**, 280—289 (1968)
163. TULLBERG, L., I. B. PEETRE and B. E. F. SMITH, Structural investigation of phenols and alcohols using silylation and gas chromatography, J. Chromatogr. **120**, 103—113 (1976)
164. CRATHORNE, B. and M. J. SAXBY, The separation and identification of some isomeric alkylamines as their trifluoroacetyl derivatives, J. Chromatogr. **82**, 373—376 (1973)
165. WESTON, R. E. and B. B. WHEALS, The determination of cyclohexylamine by electron-capture gas chromatography, Analyst (London) **95**, 680—682 (1970)
166. GASCÓ, L. and R. BARRERA, The use of derivatives for the gas-chromatographic identification of alcohols, primary and secondary amines, and thiols in food aromas, Anal. Chim. Acta **61**, 253—264 (1972)
167. HOSHIKA, Y., Gas chromatographic determination of the lower aliphatic primary amines as their Schiff bases, J. Chromatogr. **115**, 596—601 (1975)
168. HOSHIKA, Y., Gas chromatographic separation of lower aliphatic amines, Anal. Chem. **48**, 1716—1717 (1976)
169. HOSHIKA, Y., Gas chromatographic separation of lower aliphatic primary amines as their sulphur-containing schiff bases using a glass capillary column, J. Chromatogr. **136**, 253—258 (1977)
170. —, Gas chromatographic determination of fatty acid composition of vegetable and animal fats and oils and of technical fatty acids (in Dutch), Netherlands Standard no 3428, Febr. 1972
171. ALLEN, G. R. and M. J. SAXBY, Gas chromatography of isomeric fatty acid methyl esters, J. Chromatogr. **37**, 312—314 (1968)
172. MLEJNEK, O., Methylation of dibasic acids with diazomethane in gas chromatography, J. Chromatogr. **70**, 59—65 (1972)
173. KRUPČÍK, J., J. HRIVŇÁK and J. JANÁK, Analytical aspects of capillary gas chromatography of lower fatty acids (up to C_{18}), J. Chromatogr. Sci. **14**, 4—17 (1976)

174. KAISER, R. E., Retention data and retention indices in gas chromatography, Chromatographia 3, 127—133 (1970)
175. KOVÁTS, E., Gas-chromatographische Charakterisierung organischer Verbindungen, Teil 1: Retentionsindices aliphatischer Halogenide, Alkohole, Aldehyde und Ketone, Helv. Chim. Acta 41, 1915—1932 (1958)
176. ETTRE, L. S., The retention index system; its utilization for substance identification and liquid phase characterization, Chromatographia 6, 489—495 (1973)
177. ETTRE, L. S., The interpretation of analytical results; qualitative and quantitative analysis, in "The practice of gas chromatography", eds. ETTRE, L. S. and A. ZLATKIS, Interscience Publishers, New York 1967, p. 373—406
178. ANDERSEN, N. H. and M. S. FALCONE, The identification of sesquiterpene hydrocarbons from gas-liquid chromatography retention data, J. Chromatogr. 44, 52—59 (1969)
179. MAARSE, H. and F. H. L. VAN OS, Volatile oil of *Origanum vulgare* L. ssp. vulgare. Qualitative composition of the oil, Flavour Ind., 477—484 (1973)
180. DOOL, H. VAN DEN and P. Dec. KRATZ, A generalization of the retention index system including linear temperature programmed gas-liquid partition chromatography, J. Chromatogr. 11, 463—471 (1963)
181. MCREYNOLDS, W. O., Gas chromatographic retention data, Preston Technical Abstracts Co. Evanston 1966
182. SCHUPP, O. E. and J. S. LEWIS, Compilation of gas chromatographic data, ASTM data series series publication no. DS 25 A, American Society for Testing and Materials, Philadelphia, 1967
183. GUIOCHON, G., Retention indices in programmed temperature gas chromatography, Anal. Chem. 36, 661—663 (1964)
184. HABGOOD, H. W. and W. E. HARRIS, Retention indices in programmed temperature gas chromatography, Anal. Chem. 36, 663—665 (1964)
185. HARRIS, W. E. and H. W. HABGOOD, Programmed temperature gas chromatography, John Wiley & Sons, Inc., New York 1966, pp. 305
186. SCHOMBURG, G. and G. DIELMANN, Identification by means of retention parameters, J. Chromatogr. Sci. 11, 151—159 (1973)
187. SCHOMBURG, G., Identification by retention and response values, Adv. Chromatogr. 6, 211—245 (1968)
188. ETTRE, L. S., The retention index system; its utilization for substance identification and liquid phase characterization, Part II: Correlation between retention index, structure and analytical characteristics, Chromatographia 7, 39—46 (1974)
189. HAKEN, J. K., Retention indices in gas chromatography, Adv. Chromatogr. 14, 367—407 (1976)
190. GOLOVNYA, R. V., Zur gaschromatographischen Analyse von Aminen als Aromakomponenten von Lebensmitteln, Ernährungsforschung 15, 333—347 (1970)
191. GOLOVNYA, R. V. and YU. N. ARSENYEV, A standardless gas-chromatographical method for analysis of homologous series of mercaptans, sulphides and disulphides (in Russian), Izv. Akad. Nauk. SSSR, Ser. Khim. no. 6, 1402—1404 (1972)
192. GOLOVNYA, R. V., YU. N. ARSENYEV and T. V. KURGANOVA, Accuracy of a determination of the thermodynamic functions of dissolution from retention indexes (in Russian), Izv. Akad. Nauk SSSR, Ser. Khim. no 3, 632—635 (1972)
193. GOLOVNYA, R. V. and V. P. URALETS, Monocarbonyl compounds in the volatile components of Rossiiskii cheese, flesh salmon, sturgeon and salmon caviar, Nahrung 16, 497—506 (1972)
194. GOLOVNYA, R. V., M. B. TERENINA and V. P. URALETS, Separation and gas chromatographic analysis of free fatty acids in salmon fillet (in Russian), Izv. Akad. Nauk SSSR, Ser. Khim. no. 10, 2351—2354 (1973)
195. GOLOVNYA, R. V. and I. L. ZHURAVLEVA, Gas chromatographic method of identification of n-aliphatic amines through the use of donor-acceptor interaction with phosphate, Chromatographia 6, 508—513 (1973)
196. GOLOVNYA, R. V., M. B. TERENINA and V. P. URALETS, Gas chromatographic identification of free fatty acids in trace amounts in natural materials, J. Chromatogr. 91, 127—131 (1974)

197. Golovnya, R. V. and V. G. Garbuzov, Analogy in the gas chromatographic behaviour of sulphur and oxygen containing compounds and possibilities to use it for identification, Chromatographia 8, 265–269 (1975)
198. Drawert, F. und A. Rapp, Gas-chromatographische Untersuchungen pflanzlicher Aromen I. Anreicherung, Trennung und Identifizierung von flüchtigen Aromastoffen in Traubenmosten und Weinen, Chromatographia 1, 446–457 (1968)
199. Drawert, F. und A. Rapp, Über Inhaltsstoffe von Mosten und Weinen VII. Gaschromatographische Untersuchungen der Aromastoffe des Weines und ihre Biogenese, Vitis 5, 351–355 (1966)
200. Badings, H. T. and J. G. Wassink, A multipurpose device for the collection of fractions separated by gas chromatography, J. Chromatogr. 18, 159–161 (1965)
201. Burson, K. R. and C. T. Kenner, A simple trap for collecting gas chromatographic fractions for mass spectrometric analysis, J. Chromatogr. Sci. 7, 63–64 (1969)
202. Soedarmo Moeljohardjo, D., On the volatile flavour compounds of cooked trassi, a cured shrimp paste condiment of the Far East, Thesis, Agricultural University of Wageningen 1972
203. Cronin, D. A. and J. Gilbert, Hydrogenation and ozonolysis of submicrogram quantities of unsaturated organic compounds eluted from gas chromatographic columns, J. Chromatogr. 87, 387–400 (1973)
204. Clark, R. G. and H. E. Nursten, The sensory analysis and identification of volatiles from walnut (*Juglans regia* L.) headspace, J. Sci. Food Agric. 28, 69–77 (1977)
205. Straten, S. van, Fl. de Vrijer and J. C. de Beauveser, Volatile compounds in food, 4th ed., Central Institute for Nutrition and Food Research TNO, Zeist, The Netherlands 1977, pp. 325
206. Groenen, P. J., R. J. G. Jonk, C. van Ingen and M. C. ten Noever de Brauw, Determination of eight volatile nitrosamines in thirty cured meat products with capillary gas chromatographic-high-resolution mass spectrometry: the presence of nitrosodiethylamine and the absence of nitrosopyrrolidine, I.A.R.C. Sci. Publ. 14, 321–331 (1976)
207. Larry, D., Gas chromatographic determination of methylsalicylate, safrole, and related compounds in non-alcoholic beverages, J. Assoc. Off. Anal. Chem. 52, 481–485 (1969)
208. Larry, D., Gas-liquid chromatographic determination of ß-asarone, a component of oil of Calamus, in flavors and beverages, J. Assoc. Off Anal. Chem. 56, 1281–1283 (1973)
209. Snyder, L. R., Principles of adsorption chromatography, Marcel Dekker, Inc., New York 1968, pp. 413
210. Kirkland, J. J., Modern practice of liquid chromatography, John Wiley, New York 1971
211. Baker, D. R., R. C. Williams and J. C. Steichen, A comparison of photometric detectors for high speed liquid chromatography, J. Chromatogr. Sci. 12, 499–505 (1974)
212. Berry, L. and B. L. Karger, Pumps and injectors for modern liquid chromatography, Anal. Chem. 45, 819A–828A (1973)
213. Chandler, C. D., High pressure liquid chromatography equipment, J. Chromatogr. Sci. 11, 468–475 (1973)
214. Kirkland, J. J., Preferred experimental conditions for trace analysis by modern liquid chromatography, Analyst (London) 99, 859 –885 (1974)
215. Lawrence, J. F. and R. W. Frei, Chemical derivatization in liquid chromatography, J. Chromatogr. Libr. 7, 1976, pp. 213
216. Leitch, R. E. and J. J. De Stefano, Column packings for modern liquid chromatography, J. Chromatogr. Sci. 11, 105–113 (1973)
217. Locke, D. C., Chemically bonded stationary phases for liquid chromatography, J. Chromatogr. Sci. 11, 120-128 (1973)
218. McNair, H. M. and C. D. Chandler, High pressure liquid chromatography equipment II, J. Chromatogr. Sci. 12, 425–432 (1974)
219. McNair, H. M., High performance liquid chromatography equipment III, J. Chromatogr. Sci. 14, 477–487 (1976)
220. Fleet, B. and C. J. Little, Design and evaluation of electrochemical detectors for HPLC, J. Chromatogr. Sci. 12, 747–752 (1974)
221. Lankelma, J. and H. Poppe, Design and characterization of a coulometric detector with a glassy carbon electrode for HPLC, J. Chromatogr. 125, 375–388 (1976)

222. Leman, M., P. Versaud et M. Porthault, Détection par spectrométrie infrarouge en chromatographie liquide à haute performance, J. Chromatogr. **132**, 295—301 (1977)
223. McLafferty, F. W., R. Knutti, R. Venkataraghavan, P. J. Arpino and B. G. Dawkins, Continuous mass spectrometric monitoring of a liquid chromatograph with subnanogram sensitivity using an on-line computer, Anal. Chem. **47**, 1503—1505 (1975)
224. O'Hara, J. R., M. S. Chin, B. Dainius and J. H. Kilbuck, Determination of benzo(a)pyrene in smoke condensates by high pressure rapid liquid-liquid chromatography, J. Food Sci. **39**, 38—41 (1974)
225. Lee, S. C. and S. S. Chang, Use of high-pressure liquid chromatography for the fractionation of less volatile flavor compounds. J. Agric. Food Chem. **23**, 337—339 (1975)
226. Komae, H. and N. Hayashi, Separation of essential oils by liquid chromatography, J. Chromatogr. **114**, 258—260 (1975)
227. Schmit, J. A., R. G. Williams and R. A. Henry, Applications of combined gel permeation chromatography and high speed liquid chromatography for the separation of complex flavor mixtures, J. Agric. Food Chem. **21**, 551—556 (1973)
228. Ross, M. S. F., Analysis of cinnamon oils by high-pressure liquid chromatography, J. Chromatogr. **118**, 273—275 (1976)
229. Charalambous, G., K. J. Bruckner, W. A. Hardwick and A. Linnebach, Determination of beer flavor compounds by high pressure liquid chromatography, Tech. Q. Master Brew. Assoc. Am. **10**, 74—78 (1973)
230. Charalambous, G., K. J. Bruckner, W. A. Hardwick and A. Linnebach, Separation, identification and determination of beer flavor compounds by high pressure liquid chromatography II, Tech. Q. Master Brew. Assoc. Am. **11**, 150—154 (1974)
231. Schaefer, J., unpublished results
232. Pelick, N., H. R. Bolliger and H. K. Mangold, The history of thin-layer chromatography, Adv. Chromatogr. **3**, 85—118 (1966)
233. Consden, R., A. H. Gordon and A. J. P. Martin, Qualitative analysis of proteins: partition chromatographic method using paper, Biochem. J. **38**, 224—232 (1944)
234. Izmailov, N. A. and M. S. Schraiber, Farmatsiya, **3**, 1 (1938) Translation of this paper is to be found in ref. 232
235. Stahl, E., Thin-layer chromatography, Pharmazie, **11**, 633—637 (1956)
236. —, Dünnschichtchromatographie ed., Stahl, E., Springer Verlag, Berlin 1967, pp. 534
237. Randerath, K., Thin-layer chromatography, Verlag Chemie, Weinheim, 1963, pp. 250
238. Kirchner,. J. G., Thin-layer chromatography, Interscience publishers, New York 1967, pp. 788
239. —, Spices and condiments — Thin layer chromatography and determination of stabilized R_F-values (in Dutch), NEN 5380 Nederlands Normalisatie-Instituut, Rijswijk 1976, pp. 5
240. Galanos, D. S. and V. M. Kapoulas, The paper-chromatographic identification of compounds using two reference compounds, J. Chromatogr. **13**, 128—138 (1964)
241. Thoma, J. A., A quantitative approach to unidimensional multiple chromatography, J. Chromatogr. **12**, 441—452 (1963)
242. Jeanes, A., C. S. Wise and R. Dimler, Improved techniques in paper chromatography of carbohydrates, Anal. Chem. **23**, 415—420 (1951)
243. Thoma, J. A., Application and theory of unidimensional multiple chromatography, Anal. Chem. **35**, 214—224 (1963)
244. Casparic, J., Determination and publication of R_F-values in the case of large groups of compounds, J. Chromatogr. **33**, 291—293 (1968)
245. Dhont, J. H., C. Vinkenborg, H. Compaan, F. J. Ritter, R. P. Labadie, A. Verweij and R. A. de Zeeuw, Application of R_F correction in thin-layer chromatography by means of two reference R_F-values, J. Chromatogr. **47**, 376—381 (1970)
246. Dhont, J. H., C. Vinkenborg, H. Compaan, F. J. Ritter, R. P. Labadie, A. Verweij and R. A. de Zeeuw, Application of R_F correction in thin-layer chromatography by means of two reference R_F-values II. Results obtained with a polar multicomponent solvent system, J. Chromatogr. **71**, 283—289 (1972)
247. Dhont, J. H., C. Vinkenborg, H. Compaan, F. J. Ritter, R. P. Labadie, A. Verweij and R. A. de Zeeuw, Application of R_F correction in thin-layer chromatography by means of two reference R_F-values III. Results obtained in reversed-phase thin-layer chromatography, J. Chromatogr. **130**, 205—208 (1977)

248. DHONT, J. H. and G. J. C. MULDERS-DIJKMAN, The resolution of mixtures of aliphatic and terpene-type 2,4-dinitrophenylhydrazones by thin-layer chromatography on silica gel. Analyst (London) **94**, 1090—1094 (1969)
249. BADINGS, H. T. and J. G. WASSINK, Separation and identification of aliphatic aldehydes and ketones by thin-layer chromatography, Neth. Milk Dairy J. **17**, 132—149 (1963)
250. URBACH, G., Thin-layer chromatography of aliphatic 2,4-dinitrophenylhydrazones, J. Chromatogr. **12**, 196—207 (1963)
251. DHONT, J. H., unpublished results
252. DHONT, J. H., Standardization of R_F-values in thin-layer partition chromatography, J. Chromatogr. **60**, 265—269 (1971)
253. DHONT, J. H., G. J. C. MULDERS-DIJKMAN, J. C. DE BEAUVESER and G. KUYPERS, Polyvinyl acetate as a stationary phase in thin-layer chromatography, J. Chromatogr. **52**, 429—435 (1970)
254. URBACH, G., Thin-layer chromatography of aliphatic γ- and δ-lactones, J. Am. Oil Chem. Soc. **42**, 927—930 (1965)
255. DHONT, J. H., Vapour pre-adsorption thin-layer chromatography, J. Chromatogr. **90**, 203—207 (1974)
256. GUPTA, A. S. and D. SUKH, Thin-layer chromatography of olefins, J. Chromatogr. **12**, 189—195 (1963)
257. SCHANTZ, M. V., S. JUVONEN und R. HEMMING, Trennung von Monoterpenen mittels Silbernitrat-Kieselgeldünnschichten, J. Chromatogr. **20**, 618—620 (1965)
258. MORRIS, L. J., D. M. WHARNY and E. M. HAMMOND, Chromatographic behaviour of isomeric long-chain aliphatic compounds, J. Chromatogr. **31**, 69—76 (1967)
259. ZINKEL, D. F. and J. W. ROWE, Thin-layer chromatography of resin acid methyl esters, J. Chromatogr. **13**, 74—77 (1964)
260. IKAN, P., Thin-layer chromatography of tetracyclic triterpenes, J. Chromatogr. **17**, 591—593 (1965)
261. NANO, G. M. and A. MARTELLI, The separation of some pairs of allylic-propenylic isomers by TLC on silica impregnated with silver nitrate, J. Chromatogr. **21**, 349 (1966)
262. LAWRENCE, B. M., The use of silver nitrate impregnated silica gel layers in the separation of monoterpene hydrocarbons, J. Chromatogr. **38**, 535—537 (1968)
263. STAHL, E. und H. VOLLMANN, Trennung von Terpen- und Sesquiterpenalkoholen auf Silbernitrat-Schichten, Talanta, **12**, 525—528 (1965)
264. JONG, K. DE, K. MOSTERD and D. SLOOT, Separation of unsaturated dinitrophenylhydrazones into classes, Rec. Trav. Chim. Pays Bas, **82**, 837—842 (1963)
265. HOFFMANN, G. and P. W. MEYBOOM, Isolation of two isomeric 2,6-nonadienals and two isomeric 4-heptenals from beef and mutton tallow, J. Am. Oil. Chem. Soc. **45**, 468—470 (1968)
266. TANGHE, L. F., L. B. GENUNG and J. W. MENCH, Determination of acetyl content and degree of substitution on cellulose acetate, in "Methods in carbohydrate chemistry", eds. WHISTLER, R. L., J. W. GREEN, J. N. BEMILLER and M. L. WOLFROM, Ac. Press, New York 1963, vol. III, p. 202—204
267. SPOTSWOOD, T. M., The chromatography of polycyclic aromatic hydrocarbons on acetylated paper, J. Chromatogr. **2**, 90—94 (1959)
268. SPOTSWOOD, T. M., The chromatography of polycyclic aromatic hydrocarbons on partially acetylated cellulose. Part II, J. Chromatogr. **3**, 101—110 (1960)
269. MCSWEENEY, G. P., Two methods for the thin-layer chromatographic separation of terpene alcohols and identification by a colour reaction, J. Chromatogr. **17**, 183—185 (1965)
270. KIRCHNER, J. G., J. M. MILLER and G. J. KELLER, Separation and identification of some terpenes by a new chromatographic technique, Anal. Chem. **23**, 420—425 (1951)
271. MILLER, J. M. and J. G. KIRCHNER, Chromatostrips for identifying constituents of essential oils, Anal. Chem. **25**, 1107—1109 (1953)
272. REITSEMA, R., Characterization of essential oils by chromatography, Anal. Chem. **26**, 960—963 (1954)
273. PETROWITZ, H. J., Silica gel thin-layer chromatography of stereo isomers of menthol, Angew. Chem. **72**, 921 (1960)

274. NANO, M. and A. MARTELLI, Thin-layer chromatography of thujones and thujyl alcohols, Gazz. Chim. Ital. **94**, 816 (1964)
275. BANTHORPE, D. V., K. W. TURNBULL, Thin-layer chromatography of thujyl compounds, J. Chromatogr. **37**, 366—368 (1968)
276. PETROWITZ, H. J., Zur Praxis der Dünnschicht-Chromatographie, Riechst. Aromen Koerperpflegem. **16**, 345—348 (1966)
277. BAINES, D. A. and R. A. JONES, An evaluation of the use of $TlNO_3$-SiO_2 thin-layer chromatographic plates for the analysis of monoterpene hydrocarbons, J. Chromatogr. **47**, 130—132 (1970)
278. ANTONACOPOULOS, N., Verbesserte Apparatur zur quantitativen Destillation wasserdampfflüchtiger Stoffe, Z. Lebensm. Unters. Forsch. **113**, 113—116 (1960)
279. NICHOLS, M. S. and M. E. FOOTE, Distillation of free ammonia nitrogen from buffered solutions, Ind. Eng. Chem: Anal. Ed. **3**, 311—313 (1931)
280. WEURMAN, C. and C. DE ROOY, Volatile amines in the odors of food, J. Food Sci. **26**, 239—243 (1961)
281. STEIN VON KAMIENSKY, E., Die flüchtigen Amine der Pflanzen, Planta **50**, 291—315 (1957)
282. ILBERT, H. I. und TH. HARTMANN, Dünnschichtchromatographie der 2,4-dinitrophenylderivate wasserdampfflüchtiger Amine, J. Chromatogr. **71**, 119—125 (1972)
283. ZEMAN, A. und I. P. G. WIROTAMA, Identifizierung von Aminen, Z. Anal. Chem. **247**, 155 — 158 (1969)
284. SCHWARTZ, D. P., R. BREWINGTON and O. W. PARKS, Separation of homologous series of dinitrophenyl amines by thin-layer chromatography, Microchem. J. **8**, 402—406 (1964)
285. MCINTIRE, F. C., L. M. CLEMENTS and M. SPROULL, Fluoro-2.4-dinitrobenzene as a quantitative reagent for primary and secondary amines, Anal. Chem. **25**, 1757—1758 (1953)
286. KLIMISCH, H. J. and L. STADLER, Mikroquantitative Bestimmung von aliphatischen Aminen mit 7-Chlor-4-nitro-benzo-2-oxa-1,3-diazol, J. Chromatogr. **90**, 141—148 (1974)
287. GRAY, W. R. and B. S. HARTLEY, A fluorescent end-group reagent for proteins and peptides, Biochem. J. **89**, 59P (1963)
288. HARTLEY, B. S., Amino-acid sequence of bovine chymotrypsinogen-A., Nature **201**, 1284 (1964)
289. SEILER, N. und M. WIECHMANN, Quantitative Bestimmung von Aminen und von Aminosäuren als 1-Dimethylaminonaphtalin-5-sulfonsäureamide auf Dünnschichtchromatogrammen, Z. Anal. Chem. **220**, 109—127 (1966)
290. UDENFRIEND, S., S. STEIN, P. BÖHLER and W. DAIRMAN, Fluorescamine: A reagent for assay of amino acids and primary amines in the picomole range, Science **178**, 871—872 (1972)
291. WEIGELE, M., S. DE BERNARDO, J. P. TENGI and W. LEIMGRUBER, A novel reagent for the fluorometric assay of primary amines, J. Am. Chem. Soc. **94**, 5927—5928 (1972)
292. SCHWARTZ, D. P., A. R. JOHNSON and O. W. PARKS, Thin-layer chromatography of carbonyl compounds, Microchem. J. **6**, 37—44 (1962)
293. CHURACEK, J. und H. PECHOVA, Dünnschichtchromatographie von Fettsäuren als 2,4-dinitrobenzylester und N,N-Dimethyl-p-aminobenzolazophenacylester, J. Chromatogr. **48**, 250—258 (1970)
294. SEILER, N. und M. WIECHMANN, Zur Chromatographie einiger 1-Dimethylaminonaphtalin-5-sulfonyl-derivate auf Kieselgel-G-Schichten, J. Chromatogr. **28**, 351—362 (1967)
295. SEILER, N. und M. WIECHMANN, Zum Nachweis von Aminen im 10^{-10} Mol-Maßstab, Experientia **21**, 203—204 (1965)
296. DHONT, J. H. and C. DE ROOY, Thin-layer chromatography of 3,5-dinitrobenzoates, Analyst (London) **86**, 527—530 (1961)
297. MEHLITZ, A., K. GIERSCHNER and TH. MIRAS, Thin-layer chromatographic separation of 3,5-dinitrobenzoates, Chem Ztg. **89**, 175—177 (1965)
298. SEVERIN, M., Chromatographie sur couche mince des 3,5-dinitrobenzoates, J. Chromatogr. **26**, 101—110 (1967)
299. ONOE, K., The separation and identification of aliphatic aldehyde 2,4-dinitrophenylhydrazones by the chromatostrip method, J. Chem. Soc. Japan, Pure Chem. Section **73**, 337—339 (1952)
300. SIMAMURA, The separation and identification of aliphatic aldehyde 2,4-dinitrophenylhydrazones by the chromatostrip method, Chem. Abstr. **47**, 3757^d (1953)

301. STAHL, E., Thin-layer chromatography, Chem. Ztg. **82**, 323—329 (1958)
302. DHONT, J. H. and C. DE ROOY, Chromatographic behaviour of 2,4-dinitrophenylhydrazones on chromatoplates, Analyst (London) **86**, 74—76 (1961)
303. EDWARDS, H. M., Thin-layer chromatography of stereo isomeric 2,4-dinitrophenylhydrazones of aliphatic aldehydes, J. Chromatogr. **22**, 29—35 (1966)
304. MEYBOOM, P. W., Identification of straight-chain unsaturated aldehydes by thin-layer partition chromatography, J. Chromatogr. **24**, 429—431 (1966)
305. SCHWARTZ, D. P. and O. W. PARKS, Thin-layer chromatography of carbonyl compounds, Microchem. J. **7**, 403—406 (1963)
306. MEYBOOM, P. W. and G. JURRIENS, Separation of positional and geometrical isomers of monoenoic aldehydes via the dinitrophenylhydrazones, J. Chromatogr. **18**, 424—426 (1965)
307. DENTI, E. and M. P. LUBOZ, Separation of 2,4-dinitrophenylhydrazones of carbonyl compounds by thin-layer chromatography, J. Chromatogr. **18**, 325—330 (1965)
308. ZAMOJSKI, A. and F. ZAMOJSKA, Identification of volatile aldehydes and ketones as 2,4-dinitrophenylhydrazones by thin-layer chromatography, Chem. Anal. (Warsaw) **9**, 589—596 (1964)
309. COBB, W. Y., Separation of 2,4-dinitrophenylosazones of vicinal dicarbonyls into classes by thin-layer chromatography, J. Chromatogr. **14**, 512—513 (1964)
310. COBB, W. Y., L. M. LIBBEY and E. A. DAY, Thin-layer chromatography of 2,4-dinitrophenylosazone homologs of vicinal dicarbonyls, J. Chromatogr. **17**, 606—607 (1965)
311. NANO, G. M., Thin-layer chromatography of 2,4-dinitrophenylhydrazones of aliphatic carbonyl compounds, Symp. on TLC, Rome, 2—3 May 1963, ed. MARINI-BETTOLO, G. B., Elsevier Publ. Comp. 1964, p. 138—143
312. DAVIDEK, J., Chromatography on thin-layers of starch with reversed phases, Symp. on TLC, Rome, 2—3 May 1963, ed. MARINI-BETTOLO, G. B., Elsevier Publ. Comp. pp. 117
313. BYRNE, G. A., The separation of 2,4-dinitrophenylhydrazones by thin-layer chromatography, J. Chromatogr. **20**, 528—540 (1965)
314. FINLEY, K. TH. and R. E. GILMAN, Thin-layer chromatography of substituted benzaldehyde 2,4-dinitrophenylhydrazones, J. Chromatogr. **22**, 36—40 (1966)
315. RUFFINI, G., Thin-layer chromatography of 2,4-dinitrophenylhydrazones of aromatic aldehydes and ketones, J. Chromatogr. **17**, 483—487 (1965)
316. KENT, R. and A. B. RAWITCH, Thin-layer chromatography of ketosteroid dinitrophenylhydrazones, J. Chromatogr. **20**, 614—615 (1965)
317. WATSON, D. J. and E. B. ROMANOFF, Resolution of steroid 2,4-dinitrophenylhydrazones isomers by thin-layer chromatography, J. Chromatogr. **22**, 192—194 (1966)
318. JART, A. and A. J. BIGLER, Improved methods for thin-layer chromatographic separation of 2,4-dinitrophenylhydrazones, J. Chromatogr. **23**, 261—266 (1966)
319. VASHIST, V. N. and K. L. HANDA, Separation of 2,4-dinitrophenylhydrazones of oxo-terpenes by thin-layer chromatography, J. Chromatogr. **18**, 412—413 (1965)
320. NANO, M. and P. SANCIN, The determination of α- and β-thujone, Ann. Chim. Roma **53**, 677—689 (1963)
321. MAHADEVAN, V., C. V. VISWANATHAN and W. V. LUNDBERG, Thin-layer chromatography of homologues of higher fatty aldehydes, dimethylacetals, and 2,4-dinitrophenylhydrazones, J. Chromatogr. **24**, 357 (1966)
322. SLOOT, D., Thin-layer chromatography of 2,4-dinitrophenylhydrazones of α-alkyl substituted carbonyl compounds, J. Chromatogr. **24**, 451—452 (1966)
323. BRUMMER, J. M. and F. J. MULLER-PENNING, Dünnschichtchromatographische Untersuchungen von Carbonylverbindungen als 2,4-Dinitrophenylhydrazone, J. Chromatogr. **27**, 290—295 (1967)
324. ABOUSTEIT, O., Ein Beitrag zur quantitativen Trennung der 2,4-Dinitrophenyl-Hydrazone mit Hilfe der Dünnschicht-Chromatographie, Fette Seifen Anstrichm. **69**, 1—5 (1967)
325. BALDRATI, G. and L. GIANNONE, Analysis of volatile carbonyl compounds by thin-layer chromatography, Industria Conserve **42**, 252—254 (1967)
326. BLOEM, E., Thin-layer chromatography of aldehydes, J. Chromatogr. **35**, 108—110 (1968)
327. DHONT, J. H. and G. J. C. DIJKMAN, The qualitative analysis of complex carbonyl mixtures by thin-layer chromatography, Analyst (London) **92**, 431—435 (1967)

328. SCHWARTZ, D. P., J. SHAMEY, C. R. BREWINGTON and O. W. PARKS, Methods for isolation and characterization of constituents of natural products, Microchem. J. 13, 407—417 (1968)
329. TUMLINSON, J. H., J. P. MINYARD, P. A. HEDIN and A. C. THOMPSON, Reaction chromatography, J. Chromatogr. 29, 80—87 (1967)
330. RONKAINEN, P., Chromatographic identification of carbonyl compounds, J. Chromatogr. 27, 380—383 (1967)
331. CRASKE, J. D. and R. A. EDWARDS, A two-dimensional thin-layer chromatographic technique for the resolution of monocarbonyl dinitrophenylhydrazones, J. Chromatogr. 51, 237—241 (1970)
332. KÖHLER, M., Massenspektrometrische Identifizierung dünnschichtchromatographisch getrenner Substanzen, Chromatographia 8, 685—689 (1975)
333. STAHL, E. und W. SCHILD, Schnell-Transfer Dünnschichtchromatographie-Infrarotspektroskopie, J. Chromatogr. 53, 387—389 (1970)
334. CLEMETT, C. J., Simple device for transferring thin-layer chromatographic fractions for spectroscopic examination, Anal. Chem. 43, 490 (1971)
335. SHER, A., Eine neue Methode zur qualitativen und quantitativen Analyse synthetischer Antioxydantien, Nahrung 4, 466—478 (1960)
336. PURDY, S. J. and E. V. TRUTER, The weight-area relationship in thin film chromatography, Chem. Ind. (London), 506 (1962)
337. MACNEIL, J. D. and R. W. FREI, Quantitative thin-layer chromatography of pesticides, J. Chrom. Sci. 13, 279—285 (1975)
338. EBEL, S. und H. KUSSMANE, Auswertung von Dünnschicht-Chromatogrammen über die Kubelka-Munk Funktion, Chromatographia 7, 197—199 (1974)
339. EBEL, S., G. HAROLD und J. HOCKE, Vollautomatische Auswertung von Dünnschicht-Chromatogrammen, Chromatographia 8, 573—576 (1975)
340. SCHMIDTMANN, W., L. RESCHKE, L. BAUMEISTER und M. KOCH, Erfahrungen mit dem Densitometer TLD 100, Chromatographia 3, 163—169 (1970)
341. KYNAST, G., Einsatz elektronischer Kleinrechner bei der direkten quantitativen Auswertung von Dünnschicht-Chromatogrammen, Chromatographia 3, 425—430 (1970)
342. SAMUELS, S., Direct quantification of thin-layer chromatograms by empirical methods, J. Chromatogr. 96, 1—13 (1974)
343. GOTGREAVE, T. and A. J. LYNES, Quantitative analysis by thin-layer chromatography using a flame ionization detector, J. Chromatogr. 30, 117—124 (1967)
344. HAY, A. W. M., A. J. SINCLAIR and W. G. RAY, Some useful techniques in thin-layer chromatography, Lab. Pract. 23, 483—484 (1974)

4. Identification

4.1. Introduction

R. BELZ

After an aroma has been isolated from a matrix and separated into its constituents, one will naturally be interested to find out the chemical formulas of these constituents, i.e. to identify them. It depends on the purpose of the investigation what effort the investigator is prepared to make in this respect. As has been pointed out in Chapter 2.1. one tends to be selective nowadays, and concentrate on those compounds that sensory apprasial has proved to be essential. We shall not discuss this question here, but merely indicate the investigator's various tools for identification.

From the preceding chapters it will have become clear that, after separation by chromatography, the aroma components are generally available in fairly pure condition, though in very small amounts; usually well below the milligramme range. In the early days of GLC, no systematic procedures existed for identification of such samples; the most one could do was guess the identity of a component by its smell or by some other piece of information, and inject the authentic compound to see whether the retention times agreed.

As techniques improved and new ones became available, two tendencies may be said to have developed: one consists in extending and perfecting the retention data approach, and the other in developing methods for spectroscopic characterization.
The "retention data approach" has the advantage that no extra equipment and specialism are required beyond those used in chromatography. Its successful application calls for strict standardization of chromatography conditions in order that a retention value may be sharply defined. Moreover, it is now general practice to add to the mixture a few selected known compounds whose retention values serve as points of reference between which to interpolate the unknowns. This makes chromatographic data even better inter-comparable. Thus, in GLC, the series of n-alkanes has found widespread application as reference compounds — see Chapter 3.2. —, while in TLC the employment is advocated of two known reference substances, selected according to the conditions of the experiment, whose R_F-values serve to adjust the R_F-values of all other spots on the chromatogram (see Section 3.4.2.). With these improvements, a retention value is a no less sharp criterion than, say, a melting or boiling point; it can be compared with tabulated reference data without the need for checking with the authentic sample, which may not be available to the investigator.

One retention value is not sufficient to identify a substance, as agreement with the reference may be sheer coincidence. This chance is much diminished, however, if two or more retention values agree. In identification work by retention data, it is therefore customary to use at least two judiciously chosen different systems. In GLC, this could be two columns of different polarity; in TLC, two-dimensional development. Reliability of identification is further enhanced if the number of possibilities can be narrowed down beforehand; this can be achieved by such expedients as group separation, selective detection and reaction chromatography. The first of these has been discussed in the preceding chapter and the other will be in the present chapter.

The "retention data approach" in GLC has been applied in aroma research by Golovnya and co-workers, who have been especially successful with the members of homologous series, whose retention values can be predicted (Refs. 190—197 of Chapter 3).

As regards the "spectrometric" approach, this is probably used by the majority of investigators. The argument is that a complete spectrogram of a substance carries far more data than can be provided by a few numerical values such as retention indices or R_F-values. Among the spectrometric methods used to characterize chromatographic fractions, mass and infra-red spectrometry have attained greatest popularity thanks to their ready applicability to micro samples, and to the following facts: every organic compound possesses a characteristic mass and an infra-red spectrum, and there exist extensive spectra collections for reference purposes. If no matching IR or mass spectrum (MS) is found in the reference files, some information still follows from the spectra. In particular, IR will characterize functional groups in the molecule, while MS will usually reveal the molecular weight and some structural features. If high-resolution MS is available, the molecular weight determination can be sufficiently precise to find the molecular formula. At the present stage, the tendency is to regard mass and infra-red spectra as the indispensable "fingerprints" in the identification of chromatographically separated components, retention data being treated as additional bits of information. This view-point may be further specified by Table 4—1.

Table 4—1
Rough classification of the reliability of identifications using some major identification methods.
(+) = not reliable; + = rather raliable; ++ = very reliable

Spectroscopic evidence	Confirmation by chromatography		
	Number of retention values by GC, TLC or LC with reference values		
	none	one	two*
IR without reference spectrum	(+)	(+)	+
MS without reference spectrum	(+)	+	+
IR with reference spectrum	+, ++**	++	++
MS with reference spectrum	+	++	++
IR and MS with both reference spectra	++	++	++

* GC, polar and apolar stationary phase; TLC and LC, two different systems.
** First members of a homologous series, ++; higher members of a homologous series, +.

Table 4—1 is for illustrative purposes only; it is of necessity crude. The reliability of an identification in practice depends on many factors, such as: the nature of the compound, the quality of a spectrogram, etc. However, it is generally true that the qualification "very reliable" implies the availability of at least one reference spectrum. Therefore, if a novel substance is encountered, organic synthesis of test substances is a must, and the investigator should be guided by intelligent guessing based on the spectral data obtained from the unknown, until he hits upon the substance giving the identical spectrum. An idealized example of such a procedure is presented pictorially in Figure 4—1; here, the PMR (Proton Magnetic Resonance) spectrum is presented besides IR and MS. More realistic examples of identification procedures can be found throughout the literature (see e.g. 1—7).

Of the two spectrometric methods, infra-red and mass, the latter is in several ways the better adapted to chromatographic fractions and especially to GLC. In fact, its speed and sensitivity permit us to record mass spectra while the GC peaks emerge from the column, even if only in sub-microgramme amounts. Moreover, a mass spectrum, by its nature of peaks occurring at discrete regular intervals, is very suited to computer processing; this fact has led to specialized applications such as trying a "first shot" at an aroma problem by running a chromatogram directly from the original sample and, aided by an on-line computer, making a search in this chromatogram for pre-selected MS features that would correspond to a component or class of components suspected to be present. These techniques have found wide application outside the aroma field, too, and will be discussed in Section 4.2.3 and in Chapter 6.

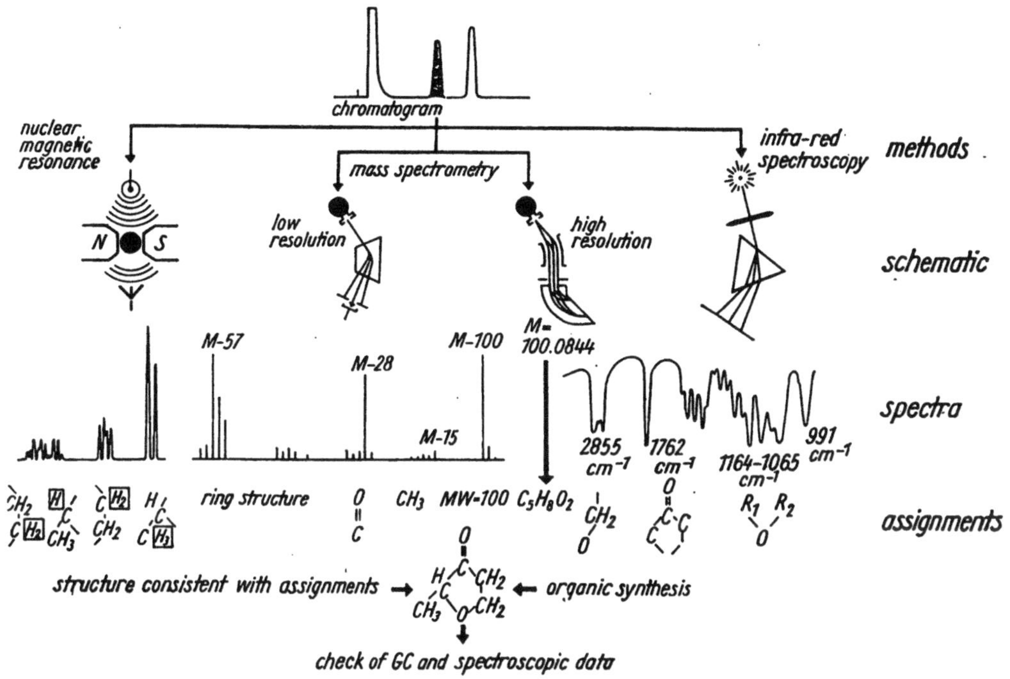

Fig. 4—1
Example of identification by combined spectroscopic methods.
Top to bottom: symbolic representations of techniques; spectrograms; evidence collected from spectrograms; chemical structure consistent with evidence.

In infra-red spectrometry, an "on-the-fly" spectrum of reasonable quality can only be obtained with GC peaks that are several orders of magnitude larger than with MS. It is still most common practice to trap a component issuing from the GC column and, next, transfer it to the IR cell, where its spectrum can at leisure be taken for maximum quality. The same applies *a fortiori* to PMR.

4.2. Mass spectrometry

M. C. TEN NOEVER DE BRAUW
C. VAN INGEN

4.2.1. Introduction

In the early days of aroma research, identification of volatile components occurring in food products was a very difficult and time-consuming task. Even identification of the two or three major components, using classical analytical methods, required considerable time. The amounts of substance (g-mg range) needed for these methods were, moreover, a severe restriction. From the early sixties, aroma investigators have begun to appreciate mass spectrometry; it had already been successfully applied as a powerful analytical tool for qualitative and quantitative analyses in the field of organic chemistry.

Mass spectrometry has its origin in the mass spectrometer first designed by ASTON and DEMPSTER (8, 9). It separates chemical elements into their isotopes and is based on the principle that ions accelerated to a certain kinetic energy by an electric field, describe, in a subsequent magnetic field, paths that differ according to their mass to charge ratios. Nowadays there exist mass spectrometers with mass analyzers other than the magnetic one (quadrupole, time of flight), but magnetic mass analyzers are still the most widely used. When it is attempted to ionize organic molecules for mass spectrometry, generally not only molecular ions are formed but also a large number of different fragment ions. The fragmentation pattern found is characteristic for the identity of the organic molecules ionized, and this has been the basis for the application of mass spectrometry in the identification of organic compounds.

To obtain useful information from a mass spectrum it is necessary that the substance introduced into the mass spectrometer be pure. In the past however, separation and purification procedures were very inefficient and required considerable time, apart from the risks of changes in structure of the individual compounds during the preparation of the sample. Furthermore, ionization techniques were practically limited to substances of reasonable volatility. Nowadays, chromatographic procedures and ionization techniques have become so diversified that practically every organic compound can be isolated and made to yield a mass spectrum. Introduction of so-called double-focusing techniques increased the resolving power of the mass spectrometer, so that not just the mass number but also the exact mass could be determined and the elemental composition of a molecular ion or fragment ion established unequivocally, giving another dimension to mass spectrometric identification.

With the advent of the combined approach using gas chromatography and mass spectrometry (GCMS), the number of applications of mass spectrometry in the field of aroma research went up considerably. To handle the amount of mass spectral data obtained with GCMS efficiently, the computerization of the system became indispensable. With such computerized GCMS systems it is possible to perform mass spectrometric sampling of a total GC effluent and store all the mass spectral information into the computer. With such a facility, analyses of complex mixtures are nowadays performed within a few hours' time.

In this chapter only those GCMS applications that are frequently applied in the field of aroma research will be discussed. For other techniques and details for each of the mass spectrometric disciplines, the reader is referred to the textbooks (10—16).

4.2.2. Combination of gas chromatography and mass spectrometry

4.2.2.1. General

Before the advent of GCMS combinations, mass spectrometric identification procedures of organic components were performed with the so-called "batch inlet system"; an example is shown in Figure 4—2. The compound trapped from the effluent of a GC column was injected into a gas reservoir of a volume of 1 to 3 litres, and let into the mass spectrometer through a molecular leak. However, trapping components from a complex mixture is not an easy task (Chapter 3.2.). In some cases it is still employed in aroma research because:

- other spectra (NMR, IR, UV) of the separated GC components may be needed;
- it enables selection of compounds based on their organoleptic properties.

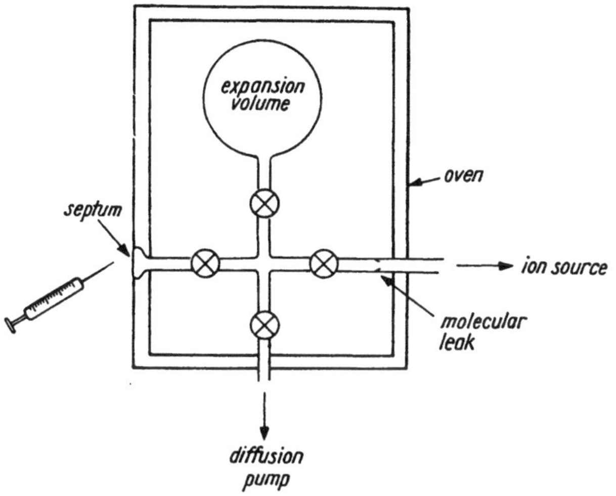

Fig. 4—2
Batch inlet system for mass spectrometry sample introduction (10).

However, fractionation of complex mixtures is often also needed, even if GCMS is used, since trace compounds have to be concentrated and separated from interfering main compounds. Furthermore, separation on two columns with different stationary phases may be needed because otherwise mixed spectra are obtained, and these cannot be interpreted satisfactorily.

Many articles (17—21) scrutinize the application of MS and GCMS in qualitative and quantitative analysis of odorous compounds. Therefore, we here restrict ourselves to mentioning the main requirements for an optimal GCMS combination:

a) **differential pumping system:** A mass spectrometer must be kept at high vacuum for proper functioning because otherwise ion-molecule reactions and peak broadening would occur. Therefore it has to be equipped with a differential pumping system; one diffusion pump for the mass analyzer and another for the ion source. The latter is the most critical one because it has to pump away the carrier gas, thereby maintaining an acceptable vacuum (10^{-5} Torr). The flow rate of carrier gas accepted by the mass spectrometer is 0.5—5 ml/min.

b) **pressure-linearity of the ion source:** The intensity of the ion current must vary linearly with the amount of sample introduced.

c) **dynamic range:** In order to cover the GC sample amount ranges, the dynamic range of the mass spectrometer should be in terms of six or seven orders of magnitude. Actually, the maximum ion current that can be recorded by the detector without saturation effects should be 10^6 or 10^7 times as large as the minimum ion current, discernible above noise.

d) **scanning cycle time:** In order to obtain sufficient information about the purity of a GC peak, and good representative spectra, the mass spectrometer should be able to record a mass spectrum within less than 1/3 of the half-width time of the GC peak.

e) **computer compatibility:** In order to handle the amount of mass spectral information obtained with GCMS analysis efficiently, connection with a computer system is indispensable.

f) **GCMS interface:** The GCMS interfaces have to provide the pressure drop between the GC exit and the mass spectrometer. Properties of some GCMS interfaces will be discussed in 4.2.2.2.

g) **stationary phases:** The stationary phases should have low volatility and good thermal stability properties in order to yield low bleed intensity, so that mass spectra are not mixed up with background masses.

h) **chromatographic resolution:** To prevent mixed spectra of non-separated compounds, maximum chromatographic resolution is necessary.

At present, commercial GCMS combinations exist which meet the mass spectrometric requirements. The GC requirements depend on the type of investigation; they have to be carefully considered by the investigator.

4.2.2.2. GCMS interfaces

When combining a gas chromatograph and a mass spectrometer, one is confronted with two major problems. Firstly, a GC column normally operates under atmospheric exit pressure (760 Torr) whereas the operating conditions for a mass spectrometer are high vacuum (10^{-6} Torr). Secondly, the GC effluent contains only small concentrations of organic compounds, the remainder being carrier gas. GCMS interfaces should overcome the pressure drop of seven or eight orders of magnitude and deliver the organic sample into the mass spectrometer, while removing as much of the carrier gas as possible so that an acceptable vacuum in the ion source is maintained. In this context, two properties are very important: the enrichment factor and the efficiency. The first is defined as the relative increase in the concentration of the compounds in the carrier gas after passing the separator. The second is the percentage of the quantity of the compound in the GC effluent that enters the mass spectrometer.
The amount of carrier gas (0.5—5 ml/min) accepted by the mass spectrometer, entirely depends on its vacuum equipment (pumping capacity and/or differential pumping system). The properties of an ideal GCMS interface system can be summarized as follows:

- as much sample and as little carrier gas as possible enter the mass spectrometer;
- no dependence of the enrichment factor and the efficiency on molecular weight, temperature and nature of carrier gas;
- no chemical changes of the sample;
- no discrimination against compounds with particular functional groups;
- no adsorption and memory effects.

Attempts to combine a GC and an MS were first reported in 1959 by GOHLKE (22). The GC was connected to the mass spectrometer through an open split system, so that only a few percent of the GC effluent was admitted to the mass spectrometer. The maximum admissible amount was dictated by the performance of the vacuum

system. It is obvious that the efficiency of this interface was low (1—10%), while no enrichment was achieved. With this coupling method, GC conditions are not affected; this technique is suitable for all types of GC columns. At present, with the improvement of the mass spectrometer pumping systems, there are more and more GCMS applications with a modified version of this type of interface (23). Other coupling methods were applied by selecting GC conditions for minimum gas flow (24, 25). A narrow bore capillary (i.d. 0.25 mm) can be coupled directly to the mass spectrometer through a capillary restriction. Total GC effluent then enters the mass spectrometer, so the efficiency is 100% but the enrichment factor is only one. The GC column exit operates under vacuum conditions, so the resolution properties of the column may be affected. The application of the direct coupling technique is restricted by limited capacity of the vacuum pumping equipment (0.5—5 ml carrier gas/min).
These simple interfaces were not sufficient for all GCMS applications. To achieve other coupling systems, enrichment devices or so-called molecular separators were constructed; the enrichment processes can be divided in three categories:

(1) fractionation of gases in an expanding jet stream (jet separator);
(2) selective effusion through fine pores (BIEMANN-WATSON separator) or through a narrow slit (slit separator);
(3) preferential diffusion of carrier gas or sample through a semipermeable membrane (teflon membrane separator, silicone membrane separator).

At the moment, a variety of different enrichment devices exist; some of them are appropriately and frequently applied. An excellent textbook (26) and review articles (27, 28) are available; they give technical details and applications of these interface systems. Table 4—2 briefly reviews some molecular separator systems and their main properties.

Table 4—2
Review of some molecular separator systems and their main properties

Type of interface	carriergas flowrange (ml/min)*	efficiency (%)**	enrichment factor**	maximal temperature (°C)	decomposition and adsorption effects***
Open split coupling	0—50	5— 80	negl.	400	—
Direct coupling	0— 5	100	negl.	400	—
Jet separator (1 or 2 stages)	10—80	20— 50	20— 80	400	+
Biemann Watson separator (1 or 2 stages)	10—80	20— 60	20—400	400	++
Membrane separator (1 or 2 st.)	1—50	60— 95	10^2—10^5	220	+
Variable slit separator	0—80	10—100	10— 60	400	+

* depends on type of column and on 1 or 2 stages type of separator used.
** depends on flow range and molecular weight.
*** — = no effects; + = small effects; ++ = pronounced effects.

From the properties mentioned in Table 4—2 it will be obvious that there does not exist any ideal GCMS interface system. Every investigator has to visualize his particular problem and then choose the appropriate coupling device.

Ionization techniques

Electron impact: The method of ionization most widely applied in organic mass spectrometry is electron impact ionization (E.I.). With this type of ionization, a heated tungsten or rhenium wire emits electrons, which are accelerated to an anode. The organic compound is introduced into the ion source at a pressure of 10^{-6} Torr and is bombarded with electrons having an energy of 20—100 eV. Molecular ions and fragment ions are then formed; they generally carry one positive charge, resulting in a mass spectrum. From the fragmentation pattern, structural information can be obtained because the pattern is characteristic for the molecule ionized. A very important piece of information which can be obtained from a mass spectrum is the molecular weight of the compound. Mass spectra of organic compounds obtained with electron impact ionization, however, do not always contain abundant molecular ions; sometimes the molecular ion is of such low "abundance" as to go unnoticed. This is clearly illustrated in Figure 4—3, showing a spectrum of caryophyllene oxide (mol. weight 220).

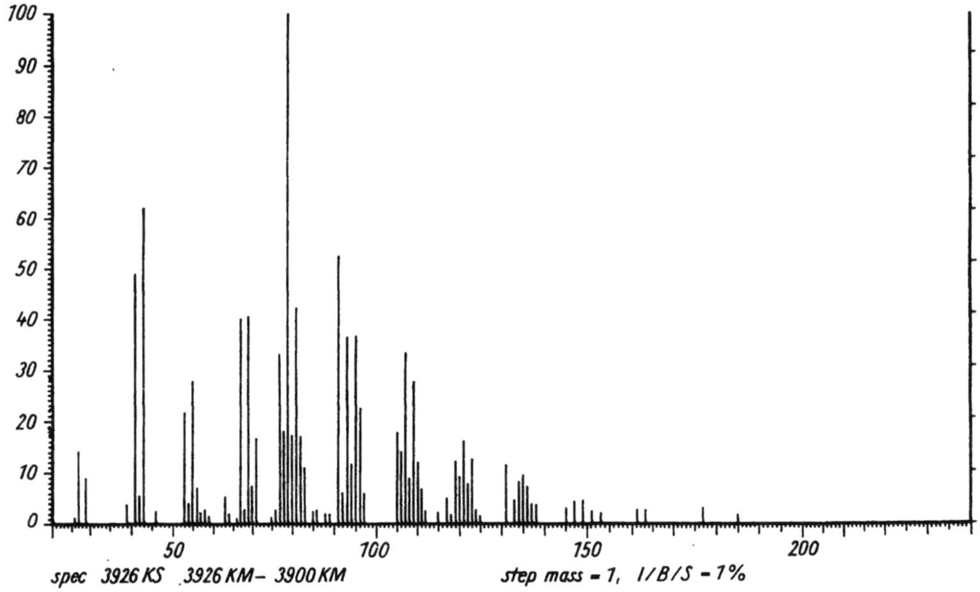

Fig. 4—3
Electron impact spectrum of caryophyllene oxide.

Consequently, new techniques for lower energy ionization and thus reduced fragmentation have been developed, such as chemical ionization, atmospheric pressure ionization, field ionization and field desorption.

Chemical ionization: The chemical ionization (C.I.) technique was introduced in 1966 by Munson and Field (28). In the C.I. mode, the ionization chamber is filled with a reagent gas (methane, propane, iso-butane, water, ammonia, etc.) at a pressure of about 0.5—1 Torr. The reagent gas is ionized by electron impact at 70 to 400 eV, whereby primary and secondary reactant ions are formed. The most stable ions can react with the sample introduced, transferring their charge to a sample molecule.

During these ion-molecule reactions, little energy is transferred; this results in a limited fragmentation. On the other hand, chemical ionization spectra often contain pseudo-molecular ions (particles made up of a molecule + reactant ion). In Figure 4—4, the C.I. spectrum of caryophyllene oxide is presented; it clearly shows the $(M + H)^+$ ion (mass 221).

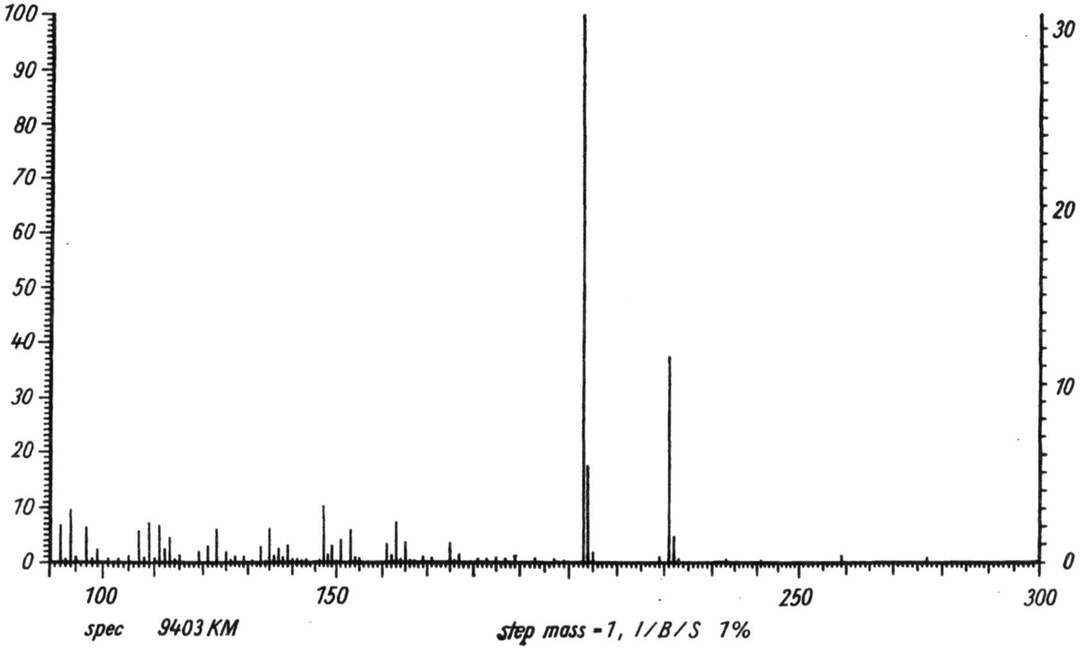

Fig. 4—4
Chemical ionization spectrum of caryophyllene oxide.

Atmospheric pressure ionization: In 1973, HORNING et al. (33) described the atmospheric pressure ionization (A.P.I.) technique. Electrons are generated from radioactive ^{63}Ni in a reaction chamber at atmospheric pressure. The organic compound, together with the carrier gas, is introduced into this reaction chamber where the carrier gas is ionized, followed by ion-molecule reactions with the organic sample. This technique, though still in development, is promising for trace analysis in the sub-femtogram range.

Field ionization: Field ionization (F.I.) can occur when an atom or molecule is acted upon by an electric field of the order of 10^7—10^8 V/cm, so an electron is being removed from the molecule based on the quantum-mechanical tunnel effect (31). In this ionization method, only a small amount of energy is transferred to the molecule. Due to molecule-ion reactions in the ion source, field ionization spectra mostly contain $(M + H)^+$ ions, so that the molecular weight can be easily established.

Field desorption: Field desorption (F.D.), introduced in 1969 by BECKEY et al. (32) is comparable to field ionization, but especially adopted to the analysis of thermolabile and non-volatile compounds.

Thanks to the availability of the various ionization techniques discussed, qualitative and even quantitative analyses of mixtures of organic compounds are now possible. Electron impact ionization and chemical ionization are the techniques mostly applied in aroma research; some applications will be discussed in 4.2.4.

4.2.2.3. Detection and recording of mass spectral signals

In mass spectrometry, ion-current values in the range of 10^{-10}—10^{-18} Amp. must be detected. With certain mass spectrometric geometries (e.g. HERZOG-MATTAUCH, 37) it is possible to apply *photoplate detection.* The accumulating feature of this detection technique, and the fact that it records all masses simultaneously, make it especially valuable in those cases that a sample is only available for a very short time (e.g. flash evaporation, laser induced mass spectrometry or GCMS-high resolution mass spectrometry). This technique, however, has many disadvantages (39); the principal one is the complicated and expensive processing of the photoplate itself. Consequently it is avoided whenever possible. At present the *electron multiplier* is universally used in organic mass spectrometry. This detection device has a gain-factor of about 10^6—10^7. To record mass spectral signals in the fast sequences (1—10 ms) in which they occur in GCMS work, fast response times are indispensable. The main recording systems which meet this requirement are the oscillographic recorder and the computer. With the *oscillographic recorder,* a complete mass spectrum recorded on U.V. photographic paper can be obtained within 1 to 2 s. In the case of a simple gas chromatogram, this recording system will be satisfactory, it is still being applied nowadays. However, with complex gas chromatograms a few hundred or more mass spectra have to be recorded, marked and calibrated. The investigator will be drowned in the amount of information and the difficulties to handle it. Figure 4—5 shows a capillary chromatogram and mass spectra of a mixture of sec-butanol, isoamyl alcohol, ethylbenzene and n-amyl alcohol. Spectra 1, 2 and 3 recorded of GC peak 2 are identical; this means that GC peak 2 contains only one compound. However, spectra 4, 5 and 6 recorded of GC peak 3 display a different pattern. In spectrum 4, masses 91 and 106, characteristic for ethylbenzene, are abundant. In spectrum 5 and 6 these masses decrease, while masses 42, 55 and 70, characteristic for amyl alcohol increase; this means that GC peak 3 contains 2 compounds (ethylbenzene and n-amyl alcohol). This example illustrates the necessity of taking more than one spectrum per GC peak, in order to obtain sufficient information about the composition of the GC peak. In practice the investigator will often be confronted with complex chromatograms. It is almost impossible to obtain the required information, if the spectra are recorded with an oscillographic recorder. A solution to this problem is the introduction of a *computer system* which can perform the marking and calibration routines accurately and automatically. Most automatic GCMS data processing is performed by dedicated minicomputers with limited ability for performance outside the field of dedication (34, 35). With such a computerized GCMS system, a mass spectrum of the GC effluent can be taken every few seconds and the analog mass spectral signals can be digitized, calibrated and stored on disk or magnetic tape. Upon completion of the analyses, all mass spectrometric information of the GC effluent is available.

A further description and applications of such a computerized GCMS system are given in the following section and in Chapter 6.

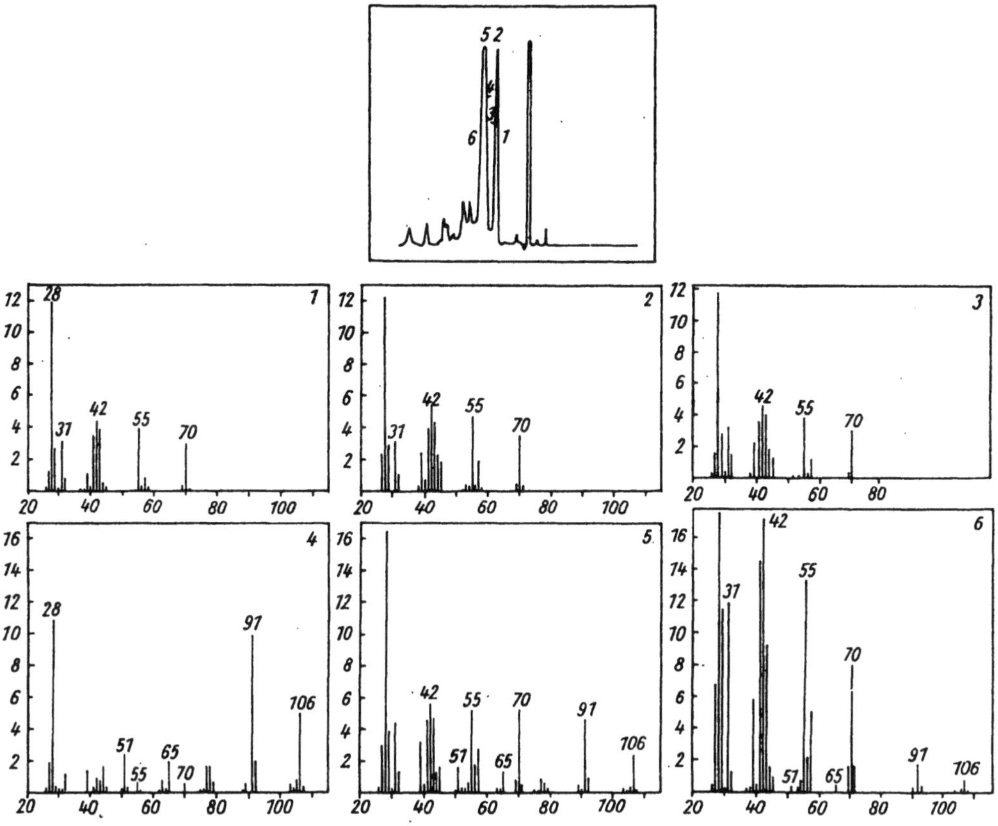

Fig. 4—5
Capillary gas chromatogram and mass spectra of a mixture of sec.-butanol, isoamyl alcohol, ethyl benzene and n-amylalcohol.

4.2.3. Methods and applications

As regards the application of GCMS in the field of aroma research, two different aims must be considered. First, the investigator may be generally interested in the identities of those compounds in a product which contribute to its aroma and, secondly, he may have to determine and prove the identity of a specified, well-defined compound or group of compounds. Figure 4—6 schematically shows the different procedures which can be followed by the investigator.

In the first case (Fig. 4—6 — route 1), spectra must be recorded of every GC peak of interest. Figure 4—7 schematically shows a GCMS combination used for such an analysis.

The GC is connected to the mass spectrometer through an interface system. The ion source in this mass spectrometer is equipped with two electron-impact ionization chambers: one chamber, working at 70 eV, yields the ions for the mass spectra, and

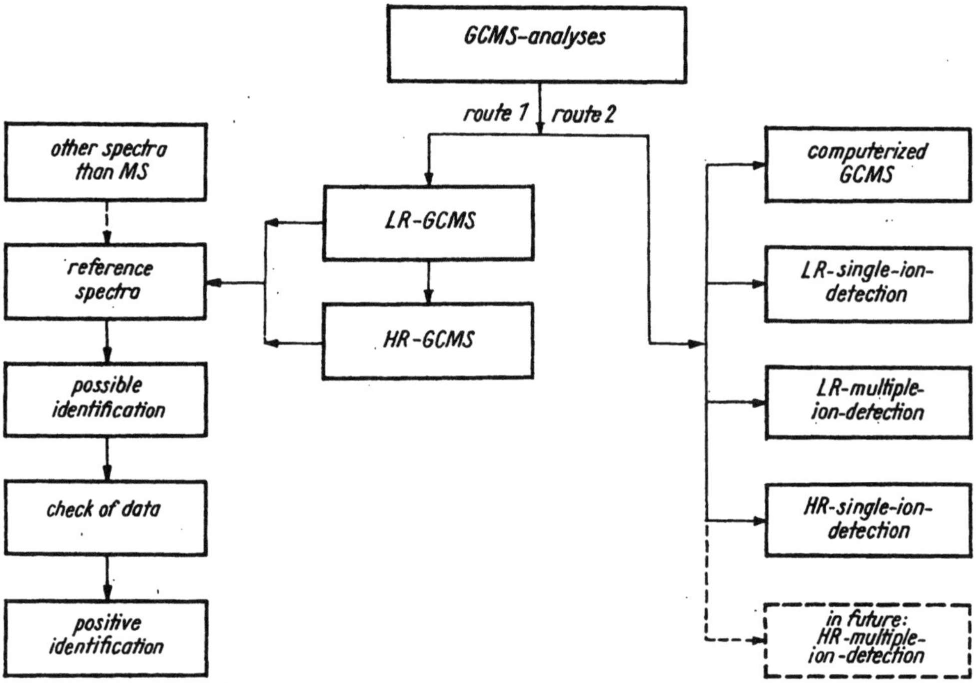

Fig. 4—6
Scheme of the different procedures in GCMS analysis.

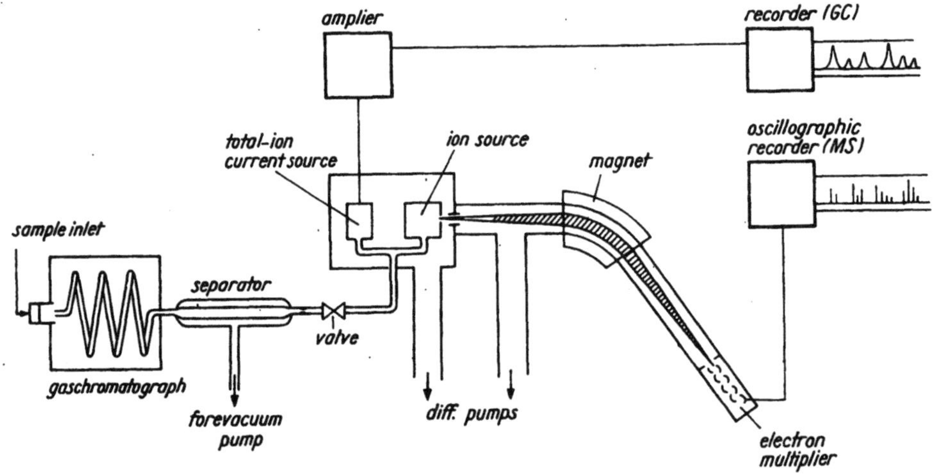

Fig. 4—7
Schematic presentation of a GCMS combination.

the other at 20 eV, provides total-ion-current signals. The electron energy of 20 eV is chosen to avoid ionization of the carrier gas, helium, which has an ionization potential of 24.4 eV. The ions from the 70 eV source enter the magnetic field, ion beams of different mass-to-charge ratio sweep across the exit slit of the mass analyzer as the field is varied, and are detected by an electron multiplier. This electrical signal produces the mass spectra, which can be recorded by the various types of recording systems. With computerized GCMS systems, these analyses can today be performed in such a way that a maximum number of spectra are recorded automatically. Figure 4—8 schematically shows such a computerized GCMS system. The mass spec-

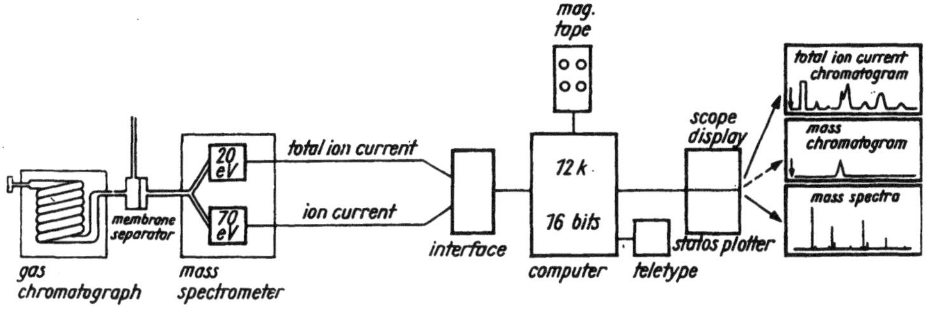

Fig. 4—8
Schematic view of a computerized GCMS system.

trometer can operate in the cyclic scanning mode, so that every few seconds a mass spectrum and a total-ion-current signal are recorded from the GC effluent. These spectra and total-ion-current signals are stored by the spectrosystem, on disk or magnetic tape. After completion of the analyses, the investigator can ask the computer to produce the required spectra, mass-fragmentograms and total ion-current chromatograms. Individual mass spectra, available at any retention time of interest, can be corrected by a subtraction technique for interfering GC peaks and, of course, for column bleeding (40—47). The mass spectra have to be interpreted by comparing the spectra manually or automatically with spectra in a reference library (34, 48).
The following example is given to illustrate the consecutive steps shown in Figure 4-6, route 1.

A pentane extract of vermouth was analyzed on a stainless steel capillary column coupled with a single focusing mass spectrometer (Varian MAT CH_4) through a single

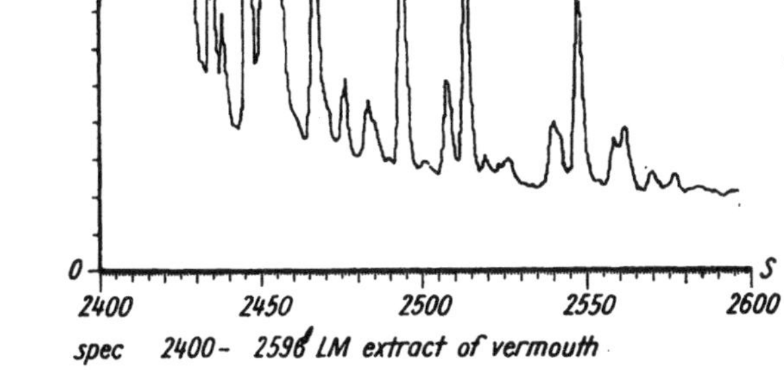

Fig. 4—9
Total-ion-current chromatogram of an extract of vermouth. GC conditions: 150 m × 0.75 mm (i. d.) stainless steel column, stationary phase Carbowax 20 M. Temperature of column and injection port were 100 °C and 120 °C, respectively. Helium flow 7.5 ml/min.

stage silicone membrane separator. The application of this separator has an essential advantage in the field of aroma research, since one is able to smell at the separator bypass capillary, whereby spectroscopic and organoleptic data can be obtained simultaneously. The efficiency of the membrane separator depends on flow rate, solubility of the compound in the membrane and temperature (49). The mass spectrometer was controlled by a Varian Spectrosystem SS100, such that every 4 seconds the mass spectrometric information was recorded from the GC effluent and stored in the computer. This resulted in about 200 spectra. The total ion-current chromatogram is shown in Figure 4—9. Next, one is able to retrieve mass spectra of every GC peak of interest. Figure 4—10 presents the spectrum of the GC peak at location 2548 (see

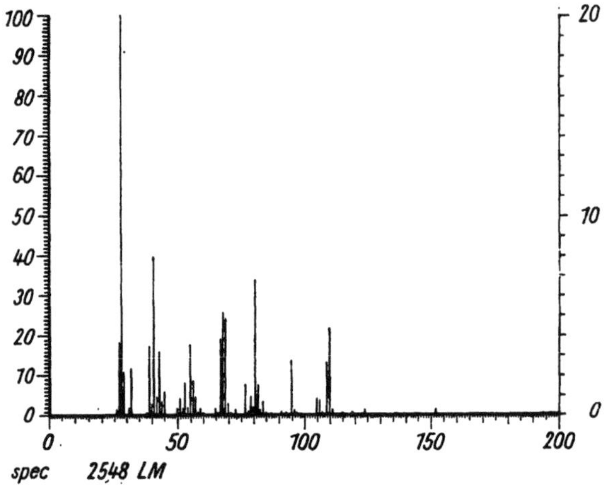

Fig. 4—10
Mass spectrum at location 2548 in the chromatogram of an extract of vermouth (see Fig. 4—9).

Figure 4—9). To determine the maximum concentration of the compound with molecular weight 152, and possible fragmentations 81 and 110, a mass fragmentogram was plotted. Figure 4—11 shows the fragmentograms of these masses. From this plot one

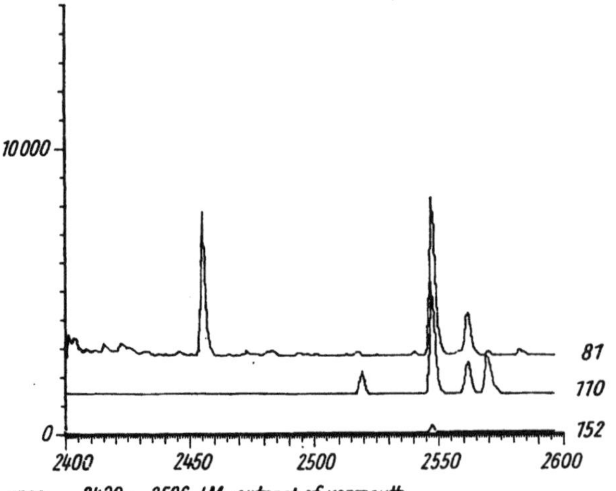

Fig. 4—11
Mass fragmentograms of m/e values 152, 110 and 81 (see also Figs. 4—9 and 4—10).

can see that the maximum concentration of the compound is at location 2548, and that the selected masses belong to one compound. Now one can correct the spectrum for neighbouring GC peaks and column bleeding. Figure 4—12 shows the corrected spectrum. This spectrum was identified as α-thujone by manually comparing with spectra in the authors' reference library, and by library file searching with the Varian Spectrosystem SS100 (Table 4—3).

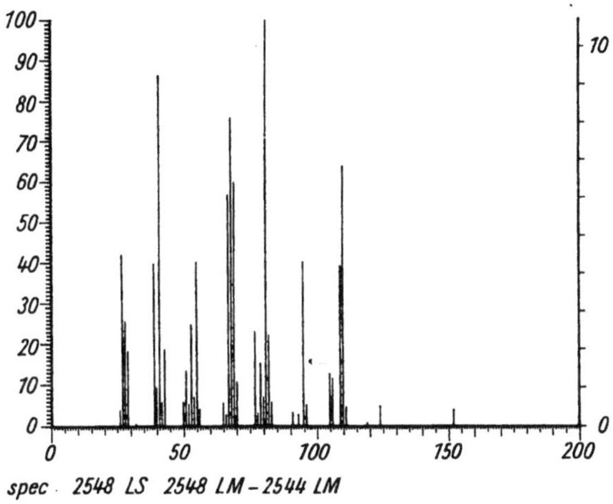

Fig. 4—12
By subtraction of spectrum No. 2544: corrected mass spectrum at location 2548 (see Fig. 4—10).

Table 4—3
Computer print-out for spectrum 2548

```
SEARCH6,57,2548,LS,20-160,
2548    3778  EPA/    314 C17.H31.BR     1-BROMOHEPTADEC-8-YNE
58
2548    5750  EPA/    138 C10.H18    CARANE
58
2548    6037  EPA/    138 C10.H18    CARANE
58
2548    6110  EPA/    152 C10.H16.O   (—)-CARANONE
58
2548    6111  EPA/    152 C10.H16.O   (—)-CIS-CARANONE
58
2548    9987  EPA/    152 C10.H16.O   THUJONE
70
READY
```

When the reference spectra are not present in the spectrum library, which in our own experience is often the case, the investigator can try to elucidate the structure of the compound from the fragmentation pattern of the mass spectrum. Additional information can be obtained by the application of high resolution mass spectrometry; exact mass measurements are thus obtained of the molecular and fragment ions. From these exact masses the elemental compositions can be calculated. These exact mass measurements can be performed manually with a so-called peak-match device (48). The information from the fragmentation pattern, together with the elemental composition of molecular and fragment ions, enables the investigator to make better structural proposals. These proposed compounds have to be synthesized and checked for their organoleptic, chromatographic and spectroscopic properties (49, 50).

Computerized MR (medium resolution) GCMS. With advancing computer and mass spectrometer techniques, exact mass determination in GCMS has become feasible even with capillary GC columns. The slow scanning required for high-resolution spectra used to be the restriction, but new fast-scanning medium-resolution equipment now permits sufficiently exact mass measurements even of sharp GC peaks of about 3 s width, as shown by the following example (51). A mixture of 40 ng each of anthracene, pyrene, p-terphenyl and triphenylene was chromatographed on a glass capillary column coupled to a Varian MAT 112S mass spectrometer and data system. The mass spectrometer was set to record mass range 30—350 at a speed of 2 s/decade with a fly-back time of 0.1 s and a resolution of 3,300. Figure 4—13 shows the total ion-current chromatogram as plotted from the data system. Figure 4—14 shows the element list of one of the compounds (triphenylene). The molecular mass (228 nominal) was measured with a deviation of 1.5 millimass units from the true value.
In many cases, retention times and mass spectra of the compounds to be determined are known and in general it will be sufficient only to confirm the presence of certain

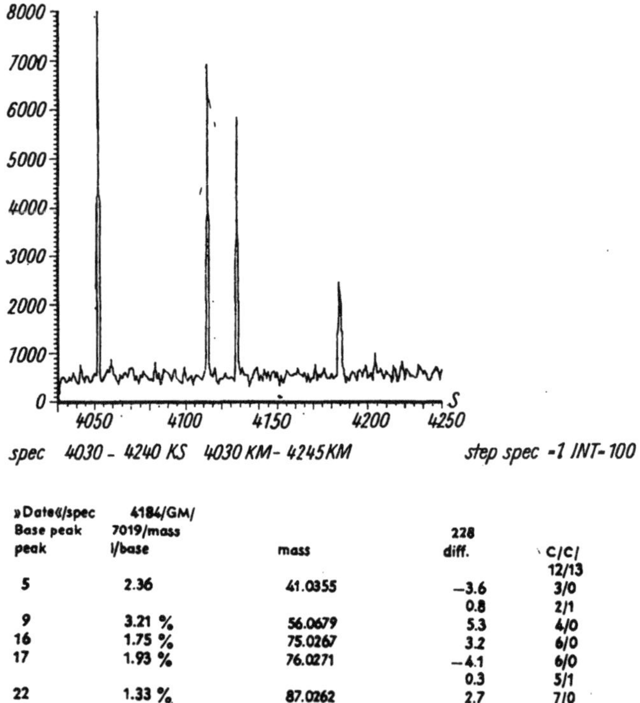

Fig. 4—13
Total ion-current chromatogram of a mixture of anthracene, pyrene, p-terphenyl and triphenylene (51).
GC conditions: 20 m × 0.15 mm (i.d.) glass capillary column, stationary phase SE-54. Temperature programmed from 150 to 280 °C, 6 °C/min. Injection port temperature 320 °C. Helium flow 2 ml/min.

»Date«/spec	4184/GM/				
Base peak	7019/mass		228		
peak	I/base	mass	diff.	C/C/	H
				12/13	1
5	2.36	41.0355	−3.6	3/0	5
			0.8	2/1	4
9	3.21 %	56.0679	5.3	4/0	8
16	1.75 %	75.0267	3.2	6/0	3
17	1.93 %	76.0271	−4.1	6/0	4
			0.3	5/1	3
22	1.33 %	87.0262	2.7	7/0	3
23	4.55 %	88.0266	−4.6	7/0	4
			−0.2	6/1	3
34	11.76 %	113.0313	−7.8	9/0	5
			−3.3	8/1	4
36	8.09 %	114.0527	5.7	9/0	6
62	1.86 %	188.0545	−3.6	14/1	7
78	6.16 %	225.0657	−4.7	18/0	9
79	26.49 %	226.0781	−0.1	18/0	10
			4.3	17/1	9
80	24.10 %	227.0872	1.1	18/0	11
			5.6	17/1	10
81	100.00 %	228.0924	−1.5	18/0	12
			3.0	17/1	11
83	16.69	229.0935	−3.6	17/1	12

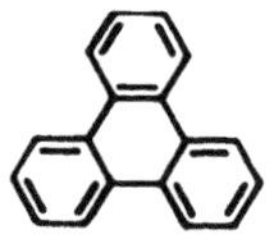

Fig. 4—14
Element list of triphenylene (51).

masses, at their respective retention times, which then results in the so-called mass fragmentogram. The different procedures are shown in Figure 4—6, route 2.

Computerized low resolution (LR) GCMS: This is performed in the same way as described before. Upon completion of the analyses, mass fragmentograms of masses can be recorded that are characteristic for the compound or group of compounds searched for.

LR and HR single-ion-detection (S.I.D.): The magnetic field of the mass spectrometer is at such a value that one specific pre-selected characteristic mass reaches the electron multiplier and is recorded. In the S.I.D. mode, the mass spectrometer acts as a highly specific detection system. However, only one mass of the compound or compounds searched for is used.

To diminish chances of misinterpretation when using S.I.D., other characteristic masses of the compound can be detected also. At present, with the use of multiple-ion-detection (M.I.D.), this can be performed simultaneously.

LR-M.I.D.: The mass spectrometer is equipped with a so-called peak-selector, which instructs the mass spectrometer to record a number of specified preselected masses characteristic of the compound(s) searched for. With magnetic sector mass spectrometers this can be performed by programming the magnetic field or the accelerating voltage. In the "fixed mass mode" the magnetic field is programmed by the peak selector, so that the sequence of the pre-selected masses successively reaches the electron-multiplier in a repeated cycle to yield a signal. In the case of very fast GC peaks, as produced by capillary columns, a very short cycle time is required (0.2 sec). However, this cannot be effected by changing the magnetic field because there is a limit to its rate of change. An alternative is found in changing the accelerating voltage; this can be done much faster. The only disadvantage of this electrical jumping is the limited mass range, which is about 40% of the lowest mass in the sequence. These limitations are inherent to magnetic field mass spectrometers only. With non-magnetic mass analysers (e.g. quadrupole), there are no problems in this respect. An example of the author's practice is illustrated in Figure 4-15; it shows the total ion-current chromatogram of a concentrate of the volatile compounds from white beans, together with mass fragmentograms of the characteristic masses of trichloroanisoles (210 and 212) and tetra chloroanisoles (244 and 246). These white beans, applied in confectionery industries as raw material, had a musty taint which, in line with previous off-flavour problems (50) could be caused by chloroanisoles. Naturally a direct search for these compounds was performed. The chloroanisoles (6 possible isomers of trichloroanisoles and 3 possible isomers of tetrachloroanisol) were searched for by computerized GCMS, S.I.D. and M.I.D. Spectra were recorded continuously, which resulted in about 500 spectra. The mass fragmentograms of a standard solution of tri- and tetrachloroanisoles are presented in Figure 4—16. Masses 210, 212 and masses 244, 246 are the molecular masses with their Cl^{37} isotopes of trichloroanisole and tetrachloroanisole, respectively. It is obvious that there is only one location in the fragmentogram of the white beans (see Fig. 4—15) where masses 210 and 212 of trichloroanisole occur, the retention time corresponding with that of 2,4,6-trichloroanisole and also that no tetrachloroanisoles could be detected. Next, S.I.D. and M.I.D. were performed. Figure 4—17 shows the mass fragmentogram of a standard solution of tri- and tetrachloroanisoles and Figure 4—18 the total ion-current chromatogram

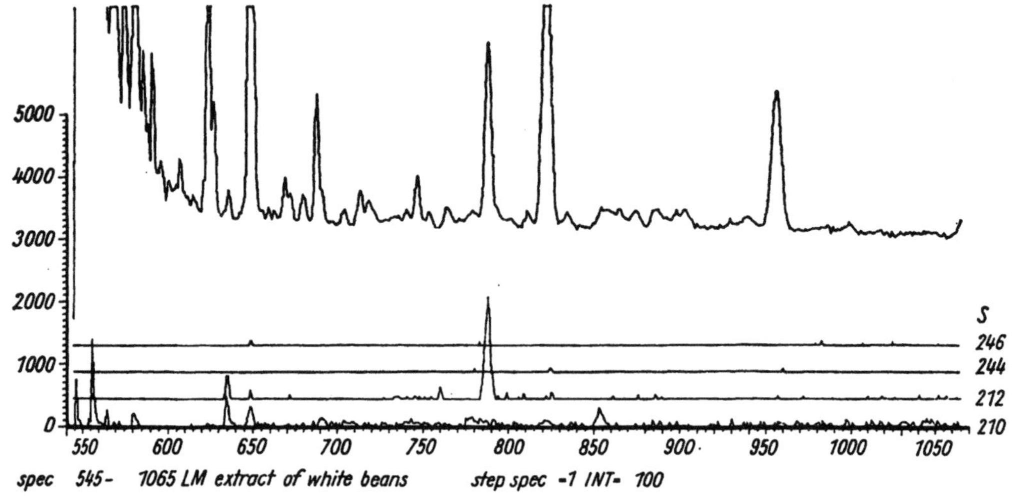

Fig. 4—15
Total ion-current chromatogram of an extract of white beans and mass fragmentograms of the m/e values 210, 212, 244 and 246.
GC conditions: 50 m × 0.5 mm (i.d.) glass capillary column, stationary phase SE-30. Temperature of column and injection port were 130 and 160 °C respectively. Helium flow 3 ml/min.

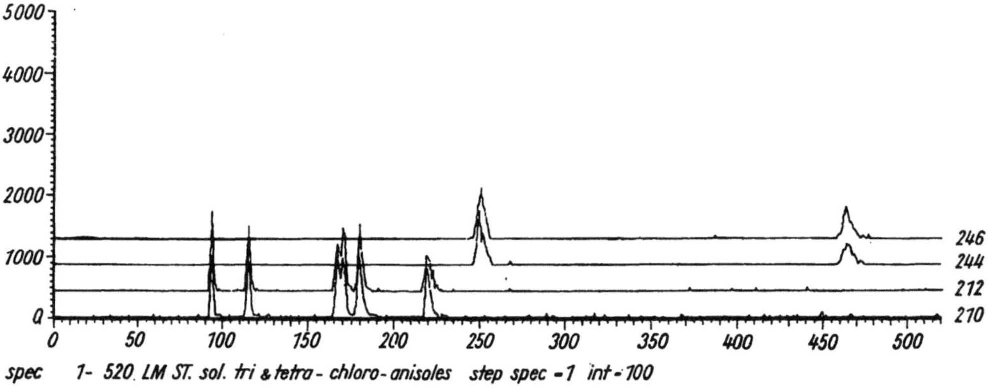

Fig. 4—16
Mass fragmentograms of a standard solution of tri- and tetrachloroanisoles. Conditions, see Fig. 4-15.

and mass fragmentograms of the extract of white beans obtained with the Varian MAT 112S equipped with a 4 channel multiple ion selector in the electrical jumping mode. Masses 210, 212, 244 and 246 were measured; the cycle time is 0.2 s and is made up of the integration time (0.03 s) for each individual mass, plus the jumping time (0.02 s). In the fragmentogram of the extract of white beans at 3 locations, masses 210 and 212 are present but only two are at retention times of trichloroanisoles. There are two locations where masses 244 and 246 are present at the retention time of a tetrachloroanisole. It is obvious that in the S.I.D. mode, where only one mass (e.g. 210) could be recorded, a false positive conclusion could be drawn for the presence of trichloroanisole. To diminish these chances of misinterpretation, more masses should be measured.

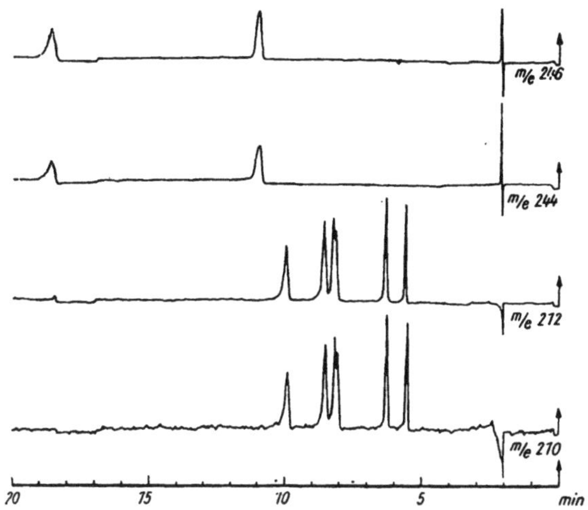

Fig. 4—17
Mass fragmentograms of a standard solution of tri- and tetrachloroanisoles obtained with a mass spectrometer equipped with a 4 channel ion selector.
GC conditions: see Fig. 4—15
Integration time 0.03 s per channel and jumping time accelerating voltage 0.02 s

Comparing the results of the three methods (computer GCMS, S.I.D. and M.I.D.), it will be clear that the advantage of the latter two lies in the increase in sensitivity, which results in the detection of a second tri- and two tetrachloroanisoles.
The minimum detectable amount applying computer GCMS is 100 pg — depending on the chosen mass or compound — whereas the detection limits for S.I.D. and M.I.D. are 10 pg. Computer GCMS, however, has an essential advantage over the S.I.D. and M.I.D. techniques. After the analyses, all the mass spectral information is available and the investigator is still able to record mass fragmentograms with other masses or even the complete spectrum belonging to the compound searched for.
With work in the computerized GCMS, S.I.D. or M.I.D. mode, chances do occur of interfering compounds with equal nominal mass. Determination of trace compounds in complex mixtures can often be falsified by these interferences. Increasing the resolving power of the mass spectrometer diminishes the risk of misinterpretation. The mass spectrometer, now, can distinguish between ions having equal nominal mass but different elemental composition (52—54). This technique can be performed in single-ion-detection mode and is described in Chapter 6.5.

4.3. Infra-red spectroscopy

R. Belz

4.3.1. Principles

The infra-red (IR) spectrum of a substance reflects the vibration modes of its molecule. These modes are determined by the structure of the molecule. Thus, an IR absorption spectrum yields information on the structure of a molecule and, in "favourable cases", it leads to complete identification of a substance.
As the name implies, infra-red absorption means absorption in the wavelength region that starts where the red region of the visible spectrum leaves off. This limit occurs at a wavelength of approx. 0.8 μm or 13,000 cm^{-1} (waves per centimetre, or wavenumber). Actually, the region of interest where the great majority of molecular vibrations give

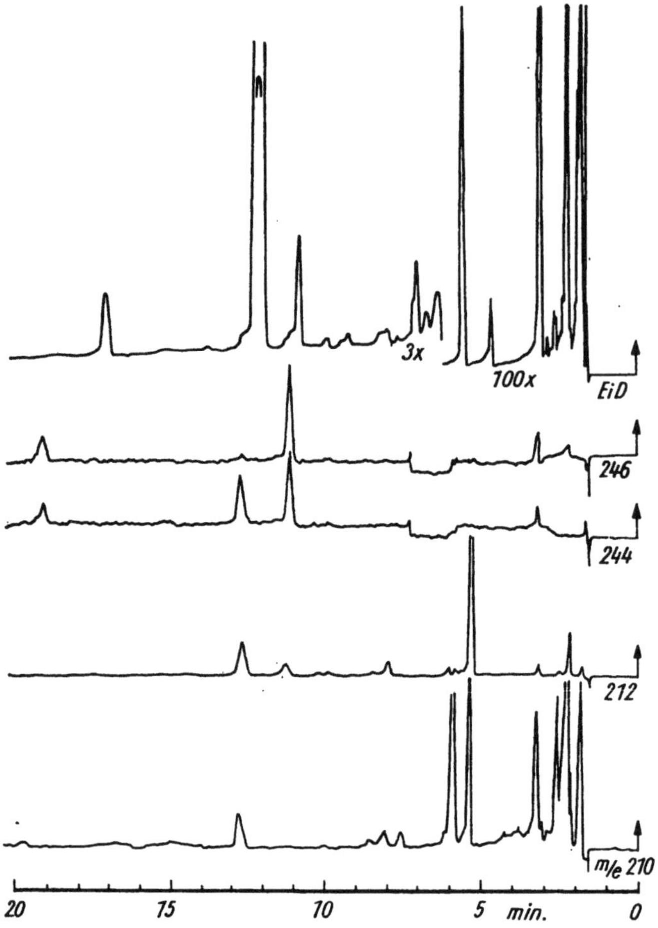

Fig. 4—18
Total ion-current chromatogram and mass fragmentograms of an extract of white beans obtained with a mass spectrometer equipped with a 4 channel ion selector.
GC conditions: see Fig. 4—15.
Integration time 0.03 sec. Electrical jumping time 0.02 sec.

rise to absorptions, begins at a somewhat lower wavenumber, viz. at about 2.5 μm or 4,000 cm^{-1}, and extends down to a few hundred wavenumber units. For purposes of chemical analysis, an IR absorption spectrometer should therefore cover the last-mentioned range; it is called the fundamental infra-red. For all practical purposes of aroma research, the so-called rock-salt region (down to 650 cm^{-1}) will suffice. This is the region that can be handled with sodium chloride optics, which are opaque at lower wavenumbers. The alkali halides are, in fact, about the only suitable transparent materials in the infra-red, and this means that all cells, windows, etc. are made from these materials; they should be protected from moisture while being handled.
The principle of an IR spectrometer is the same as that of any spectrometer; it incorporates a radiation source, a sample compartment, a monochromator for selecting wavelength, and a detector. A great variety of commercial IR spectrometers are available; practically all of them are of the ratio-recording type where the intensity

of a sample beam and that of a reference beam are compared, and have automatic scanning with pen-chart recording of absorption against either wavelength or wavenumber. The latter unit has become the more popular since the advent of spectrometers equipped with a grating monochromator, as such a device is most easily calibrated in wavenumber units. In the era of prism instruments, wavelength was favoured. Physically, the meaningful unit would be neither of these but, rather, the frequency, i.e. the number of vibrations per second. The wavenumber is equal to the frequency divided by the velocity of light, thus yielding numerical values of a convenient order of magnitude (several hundred to a few thousand, see above).

If the frequency of the radiation is equal to the frequency of a certain vibration mode of the molecule being irradiated, then radiation will be absorbed, giving rise to a band in the spectrum. The following picture may be given.

The atoms in a molecule can be thought of as balls held together by elastic springs representing the valence bonds. If brought into vibration, the atoms in such a model will all move simultaneously and perform complex motions relative to each other; this can be shown to be the resultant of $3n - 6$ independent vibrations, each with its own characteristic frequency (n = the number of atoms in the molecule).

In each of these vibrations, all the atoms of the molecule take part, so that the frequency of each vibration is determined by the masses of all the atoms, their configuration in the molecule, and the elasticities of the bonds connecting them. Each of these vibration modes will give rise to a specific absorption band in the IR spectrum. Such a spectrum is, therefore, a unique characteristic of the structure of the molecule; it is rightly called a finger-print of the molecule. Actually some vibrations may be expected to be more or less located at a certain site of the molecule. The frequencies of such vibrations are less characteristic of the molecule as a whole, and more so of specified groupings such as a hydroxyl or carbonyl group, paraffinic or aromatic structure, etc. These are, therefore, often recognized readily from the corresponding band(s) in the IR spectrum. Thus, the IR spectrum may perhaps be thought of as consisting of a "finger-print" part and a "functional group" part, and in fact most of the finger-print features can be found in the low-wavenumber range, roughly below 1,300 cm^{-1} whereas some of the most characteristic "functional group" features occur in the range above this wavenumber.

Besides the frequency, an IR band has a characteristic intensity which depends on the susceptibility of the corresponding vibration to excitation by the electromagnetic radiation. In general, the more polar a certain configuration, the more intense will be the IR absorption arising from it. For example, the intensity of an OH absorption will be much greater than that of an SH absorption, the fingerprint region of an ether will be more intense than that of a hydrocarbon, etc. An IR absorption spectrum, i.e. a plot of absorption intensity vs. frequency, is a truly twodimensional picture.

The art of interpreting an IR spectrum is revealed in several excellent textbooks; as aroma substances represent a fair cross section of all types of organic structures, any general textbook can be recommended for an introduction (e.g. 55—57).

Some further remarks should now be made. The interpretation of an IR spectrum will follow the peculiar nature of such spectra and will thus roughly consist of two main steps:

1. finding the functional groups in the molecule;
2. comparing the fingerprint picture against reference spectra of selected compounds having those functional groups.

In "favourable cases", this will lead to complete identification. If the substance is a hitherto unknown one, or if its spectrum is not available in the reference files or is not retrieved, e.g. owing to impurities in the sample, then only functional groups and sometimes certain structural elements can be assigned. This, however, is valuable information, and, together with other spectroscopical data, can lead to unambiguous identification.
From the above it follows that the IR spectroscopist should possess:

1. a working knowledge of correlations between functional groups and IR features;
2. a file of reference spectra, with entries to functional groups and wavenumber characteristics.

As to this second point, it might be presumed that an aroma specialist could do with a limited series of reference spectra covering only aroma compounds; in fact such sets do exist (e.g. 58—64). If, however, profound aroma research is done and new compounds are encountered, then it is preferable to have one of the general files (e.g. 65—68). Often, even if no complete match is found, valuable ideas crop up during the process of spectrum comparison. For a more detailed account of these and the following points, including an exhaustive bibliography, the reader may be referred to the review article by SHANKARANARAYANA et al. (69).
In aroma research, the crucial point more often than not is how to get good quality IR spectra at all. In order to yield an interpretable IR spectrum, a substance should be both reasonably pure and available in sufficient quantity. Unfortunately, from the nature of most separation and purification methods in aroma research, these two requirements are mutually contradictory. The aroma spectroscopist should therefore train himself, firstly, in recognizing contributions of impurities in a spectrum and, secondly, in working with micro amounts of sample. The first point is largely a matter of experience based on a knowledge of the frequencies (wavenumbers) of intrinsically intense bands of the more common contaminants such as grease, silicone oil (see Fig. 4-19), column bleeding, and the like. The second point is discussed in detail in the following section.

4.3.2. Micro sampling

In order to get the best possible IR spectrum from a limited amount of sample, the first thing to consider is a good match between the cross section of the sample cell and that of the radiation beam; there should be neither redundant cell volume nor excessive interception ("vignetting") of the radiation beam. The optimum cross section naturally depends on the type of instrument; it would typically be of the order of a centimetre by height and a millimetre by width. This area is determined by the monochromator slits; the actual illuminated area at the sample site may be much larger.
The radiation cross section can be diminished by inserting a beam condensor, which is available as an accessory to many instruments. A cross section of just a few mm^2 is thus feasible.

The total amount of sample in the cell is equal to the product of cross section × pathlength × concentration. This latter factor, i.e. pathlength × concentration, deter-

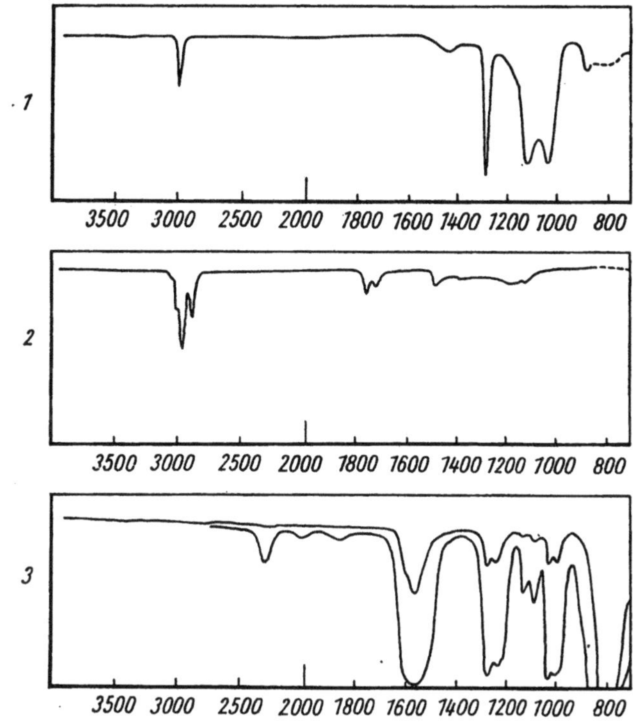

Fig. 4—19
Some infra-red absorption spectra.
1. Silicone grease.
2. Human skin fat.
3. Carbon tetrachloride.
1 and 2, approx. 5% solutions in carbon tetrachloride, cell thickness 0.1 mm, blank solvent in reference beam.
3, upper trace 0.1 mm, lower trace 1 mm, empty reference beam.

mines the intensity of the IR spectrum; for most spectrometers of a conventional type it can be said that a recognizable spectrum requires a (pathlength × concentration) value of at least a few mm· g/l. With a cross section of a few mm^2, this means that an amount of sample in the order of 10 μg is required for a recognizable IR spectrum. With intrinsically weak absorbers, such as most paraffinic type substances, this figure should be put somewhat higher.
Unless a direct GC-IR combination is effected, the sample will generally be available either collected in a GC cold trap, or as a non-volatile derivative separated by TLC or some other chromatographic method.
In the first case, if the sample is a liquid of rather low volatility, it may be washed onto a salt window using a volatile solvent, the solvent is then allowed to evaporate and the window is covered with another window so that the sample is held as a thin layer in between them (a "sandwich"). To keep the sample within the required boundaries, a groove may be cut in one window, tracing a rectangle congruent with the cross sec-

tion of the radiation beam (70). Otherwise, the sample may be rinsed into a special liquid cell by a suitable diluent, generally carbon tetrachloride, which dissolves almost all volatile substances except water. This latter point is especially important as water is often inadvertently condensed into the trap and would spoil the cell. Carbon tetrachloride has few IR absorptions of its own (see Figure 4—19) and these may be cancelled by placing a cell of matched thickness filled with pure CCl_4 in the reference beam. Still, the cell thickness should not be too great, otherwise too many blind regions will result owing to solvent absorption. A thickness of approx. 0.3 mm is about the maximum, and this restricts the volume of solvent to the order of one µl. This volume is so small that some sample is inevitably lost during the rinsing process, because of adhesion; but this loss can be kept small by choosing a narrow-bore GC trap, which can be "swept" rather effectively by dividing the solvent into a few consecutive loads.

In the second case, the sample is usually a solid and can be advantageously measured as a fine dispersion in a matrix transparent to IR, traditionally potassium bromide. By grinding the sample together with the appropriate amount of KBr to a fine powder of particle size smaller than the wavelengths of the IR radiation (< 2 µm), and pressing it into a pellet, one obtains a window transparent to IR except for the specific absorptions of the sample. The sample may be admixed to the KBr by dissolving it in a drop of any suitable volatile solvent and applying this to the KBr already present in the mortar. Most KBr presses have a piston and cylinder of circular shape; they produce round disks. The radiation beam has an oblong cross section, so that the two cannot be well matched. It is then best to opt for "vignetting"; it follows from theory that, in case the absorption intensities are small anyway, these will solely depend on the total amount of sample as long as the cross section of the window does not exceed the width of the radiation beam. A window diameter of choice would be 1 mm, if using a beam condenser, and 2 mm if using none. With a typical window thickness of 1 mm, this requires about 10 to 50 mg of KBr powder per sample, and the minimum amount of sample material, as with the liquid-cell method, is in the order of 10 µg.

The two "classical" sampling methods described above cannot easily be scaled-down much further. By using even stronger beam condensers, it would be possible to use correspondingly smaller microcells and KBr windows. In the first case, however, one will encounter increased difficulty from adhesion losses and from the heat effect of the radiation at the sample site. In the second case, the trouble arises from contamination accumulated from accidental sources such as solvent residues, grease, etc. In an IR spectrum of a derivative, contaminations are the more troublesome since the specific absorptions of the "volatile" moiety of the molecule are often already heavily overlaid by the unspecific absorptions of the "reagent" moiety. In the case of GC fractions, where accidental contaminations play only a minor role, essential improvement may be expected from more sensitive IR detectors or other devices to increase the signal-to-noise ratio, as this means that using the same cell the concentration of sample can be lowered because very weak absorptions will still stand out against the background noise. However, spurious bands caused by uncompensated solvent absorptions, surface contamination of cell windows, etc., become relatively more troublesome when going to lower and lower concentrations of dissolved sample.

The "top" in sensitivity, when one uses the sample collection technique, has probably been reached by the technique of collecting GC fractions into finely ground KBr powder. Even though the samples are liquid, there will be no "sweating out" during the subsequent pressing, precisely because the amount of sample is very small. Good spectra for GC fractions in the 1—10 μg range have been claimed (71, 72), the sensitivity limit being set by the declining trapping efficiency when going to smaller and smaller samples, and by contaminations occurring in the KBr powder.

4.3.3. Direct combination GC-IR

To circumvent transfer losses and contamination, direct combinations GC-IR have been devised. This generally implies that the GC fractions are passed to a gas cell placed in the IR measuring beam. Often the GC fraction must be held in the cell while its IR spectrum is being scanned (73, 74), but in "favourable cases" this scan can be completed while the carrier gas stream passes the gas cell unhindered (so-called "on-the-fly" spectra). This, however, requires good fast-scanning qualities of the spectrometer. In this respect, the following can be said. As stated in 4.3.1., the intensities of a sample beam and a reference beam are compared in any IR spectrometer; this serves to eliminate drift phenomena. The automatic alternation of sample and reference beam reaching the detector is mostly effected by a rotating "beam chopper"; its chopping frequency is limited by the response characteristics of the detector. The classical thermocouple by its very nature needs relatively low chopping frequencies, i.e. in the order of 10 per second, because it must be allowed sufficient time to heat up and cool down so as to develop its thermoelectric voltage. The time within which a complete spectrum can be recorded without band distortions is, in this case, at least a few minutes. The newer pyroelectric and photo-electric detectors permit much higher chopping frequencies, e.g. 400 per second (75), as they convert the radiation directly into an electrical signal. In such cases, the limit of scan speed is determined by the required signal-to-noise ratio, which, as follows from theory, is proportional to the square root of the acquisition time (i.e. the time span allowed for scanning). It is of course also proportional to the inherent sensitivity of the detector. Unfortunately some of the most advanced detectors carry the inconvenience of having to be cooled by liquid nitrogen (76) or liquid helium (75).
Sensitivity can be optimized by choosing a gas cell with a volume reasonably matched to the volumes of the GC peaks passing it or trapped in it, and with a pathlength as long as possible for maximal absorbance. These requirements have led to the employment either of "folded-path cells" (74) or of so-called "light pipes" with internally reflecting walls so that they can, in principle, be asnarrow as the slit image over their entire length. Light pipes are the cells of choice when working "on-the-fly", as such a cell is more or less automatically flushed between two consecutive peaks.
Sensitivity can be further enhanced by employing the so-called Fourier transform principle. This method obviates a monochromator with its narrow slits, which only pass such a small fraction of the radiation energy. Instead, the method uses an interferometer mirror as the scanning element. The interferograms thus obtained have to be converted to spectrograms using a digital computer. Moreover, the travel of the interferometer mirror is so critical that it has to be monitored by a laser beam. The gain in signal-to-noise ratio can be traded for even greater scan speed, the more so because double-beam operation can now be effected without using a mechanical chopper; the information of both beams is stored in the computer's memory. Much

pioneering work in this field has been done by Low and co-workers (77, 78). Using a triglycine sulphate pyro-electric detector, good IR spectra are claimed with one-second scan times on 20 μg fractions (79). Using a nitrogen-cooled detector, a two-second scan is claimed for 0.1 μg of sample (76).
An arrangement that could be considered as medium in cost and performance, is described by PENZIAS (80). It incorporates a heatable light pipe of 4 × 4 × 300 mm, a grating monochromator of a special design, a pyroelectric detector without cooling, and a low-inertia recorder, and yields good IR spectra in 6 seconds on-the-fly of GC fractions in the 100 μg range.
In some cases, the IR spectrum of a substance in the gas phase may show major differences from its spectrum in the condensed phase; most reference spectra on file have been obtained from liquid or solid sample spectra. This has been one of the considerations in undertaking the collection of IR spectra of gases on the one hand (82), and of devising liquid-phase GC-IR interfaces (81) on the other. However, this difficulty is mainly limited to:

1. very small molecules whose spectra show rotational fine structure in the vapour phase and
2. highly polar substances that have so much inter-molecular interaction in the condensed phase as to introduce major differences in the fingerprint region as compared with the spectrum in gaseous (or for that matter, dissolved) condition.

Fortunately, the great majority of compounds likely to be encountered in aroma research belongs to neither of these classes.

4.4. Other spectroscopic methods

R. BELZ

Of the other spectroscopic methods, we shall here briefly consider Raman spectrometry, nuclear magnetic resonance or, more strictly speaking, proton magnetic resonance, and ultra-violet spectroscopy.

4.4.1. Raman spectrometry

A Raman spectrum arises from coherent scattering of monochromatic incident light. The frequencies of the scattered radiation are equal to the frequency of the incident beam plus or minus the various vibration frequencies of the molecule. Accordingly, a Raman spectrum is identical to an IR spectrum as regards the relative positions of the bands, but the intensities are in this case determined by the polarizability characteristics of the various configurations. Often a less polar grouping is the more polarizable, so that a band that is weak in IR may be strong in the "Raman". This makes Raman spectroscopy especially useful in detecting features that escape direct IR detection owing to their intrinsically weak IR absorptions, e.g. some types of C=C bonds, sulphides, and ring systems (83).
The renaissance of Raman spectrometry occurred about 1965, when the continuous-wave laser Raman source was introduced. This allows focusing of the coherent radiation

on much smaller samples than was possible before. For example, the sample cell consists of an 0.1 mm i.d. glass capillary and contains a quantity of liquid sample in the order of 10 μg (84).
Because of experimental difficulties in the past, no files of reference data comparable in size with the IR files have as yet been accumulated; the gap is now beginning slowly to be filled, and some systematic work in this direction has meanwhile been done specifically in the aroma field (85, 86).

4.4.2. Nuclear magnetic resonance spectroscopy

In this context, we will restrict ourselves to proton magnetic resonance (PMR), i.e. the major form of application of nuclear magnetic resonance (NMR) in the aroma field. Protons (hydrogen nuclei) are the only nuclei which occur almost universally in organic compounds and give rise to magnetic resonance signals that are strong enough to be observable with sample sizes attainable in aroma research, using present-day instrumentation.

A proton has a spin momentum and, therefore, behaves as a tiny magnet. If a strong external magnetic field is applied to the sample by placing it between the poles of a powerful magnet, these little magnets may be visualized to "precess" in top-like fashion about the field direction with a frequency proportional to the field strength. When irradiated by radiation of that same frequency, resonance will occur and absorption is observed.

Present-day commercial instruments are equipped with a radio transmitter and receiver of fixed frequency, mostly in the order of 100 MHz, and have a magnet whose field can be adjusted to the resonance condition of protons at that frequency. Now, if a sample is placed between the magnet poles, and the magnetic field strength is swept past the expected resonance condition, one will generally encounter not just one absorption band but several; i.e. a spectrum. This is because each proton suffers a slight magnetic screening effect from the electron cloud surrounding it, and the magnitude of this effect differs slightly in accordance with the electronic configuration around the proton. Therefore, each type of hydrogen within the molecule will resonate at its own specific strength of external magnetic field. This effect is called the "chemical shift". Furthermore, a PMR band is doubled by coupling of the proton's spin to that of any immediately neighbouring proton of different chemical shift (see e.g. the doubling of the methyl signal in figure 4—1). If there are several identical neighbours, a corresponding binomial pattern will result. The separation between the lines, the so-called coupling constant, gives an extra clue to the local molecular configuration. Chemical shifts and coupling effects, together with some additional refinements, make PMR an invaluable tool for distinguishing the elementary segments of which a molecule may consist, such as: methyl, methylene, and methine groups. The chemical shifts of the protons within a specified grouping are but little influenced by atoms beyond its immediate environment, and this often makes PMR spectra gratifyingly simple to analyze. Several PMR data compilations exist (e.g. 87—89).

With conventional equipment, the necessary amount of sample is in the milligramme range; this generally limits the applicability of such equipment to fractions collected from preparative columns (90). The minimum amount of sample is dictated by the signal-to-noise ratio, which cannot be improved indefinitely by either sacrificing scan speed or using stronger radio-frequency power, as in both cases the protons will become "heated" to a high "spin temperature" during resonance and begin to absorb less; this is the so-called saturation effect. The other way round, this means that the signal-to-noise ratio need not deteriorate when using fast scanning, because radio-frequency power can be correspondingly raised; in this respect, "on-the-fly" scanning is possible with sample sizes of the same order as are usual with trapped samples, i.e. in the milligramme range (91).

The saturation problem may be overcome by using a great number of consecutive scans rather than a single slow scan (so that the protons may "cool down" between resonances) and digitally adding all the scans by a CAT (Computer Averaging Transients). These long acquisition times require long-term stability of the PMR instrument; an improvement of a factor of 10 is then easily reached. The technique, of course, is only applicable with trapped fractions.

A more fundamental improvement has been introduced by the so-called Pulse Fourier Transform method. Here the radio-frequency power is not given off continuously but in a fast sequence of equally spaced pulses; this results in a large number of side frequencies besides the fundamental frequency. This principle is analogous to the Fourier Transform technique in IR spectroscopy in that all frequencies are simultaneously active during scan. Here, too, the result is a much fuller utilization of the available radiation energy leading to an improved signal-to-noise ratio and, also, a need for a digital computer to convert the obtained data to a spectrogram, using the mathematical process known as Fourier transformation. With use of this technique, good PMR spectra have been claimed: 10 μg of sample dissolved in 5 μl of a proton-free solvent (deuterochloroform), the acquisition time being in the order of a few minutes (92). These developments open up new application areas in aroma research.

4.4.3. Ultra-violet spectrometry

An ultra-violet (UV) spectrometer is likely to be available in most laboratories. Although not primarily intended as an identification tool, it may be used to advantage when the presence of conjugated double bonds is presumed in the compound to be identified, as such structures give rise to characteristic UV absorptions. The intensities of such absorptions are generally such that using a conventional UV spectrometer, an adequate spectrum will be obtained with an amount of sample in the centimilligramme range; i.e. the same order of magnitude as with micro IR spectroscopy. It is always a simple matter, after an IR spectrum has been run, to rinse the contents of the IR cell into a UV cell using a UV grade solvent. It need hardly be recalled here that extensive literature exists on spectra-structure correlations in the UV. A special application of UV, i.e. class characterization of derivatives, is mentioned in Section 3.4.5.4. on TLC of 2,4-DNPH's. For the way in which UV may fit in identification schemes, the reader is referred to the textbooks (e.g. 93 or 94). In ref. 5 of this chapter, UV data are employed to a modest extent. The agencies listed in the section on IR spectrometry as issuing IR collections, actually sell UV collections, too.

4.5. Reaction gas chromatography

H. MAARSE

4.5.1. Introduction

Identification of components, separated by gas chromatographic analysis of mixtures, besides retention times, needs extra information which can be provided by various spectroscopic techniques, and reaction gas chromatography (RGC) can also yield useful information.

RGC (95) is a special form of gas chromatography; an injected compound is chemically altered within the GC system. The method was already applied in the early years of gas chromatography (96); it was introduced in the analysis of natural products by BEROZA (97) in 1962.

In this chapter, only those forms of RGC will be described that are of importance as an aid in the qualitative analysis of small amounts of compounds that are expected to be constituents of aroma mixtures.

As early as in 1966, a review on RGC by BEROZA and COAD (98) mentioned that 500 papers had appeared on the subject at that time. Many more have been published since. Here only those papers will be referred to that are important as illustrations of the techniques described.

4.5.2. Subtraction chromatography

One of the simplest forms of RGC is the removal of compounds with a certain functional group from complex mixtures by reaction with group reagents included *within* the GC system, to yield a volatile or non-volatile reaction product. This technique is called subtraction chromatography, and the compounds specifically removed by this technique are said to be abstracted. When one compares the chromatograms with and without the group reagent, the presence or absence of compounds can be shown, which is additional information for the identification of unknown compounds. However, comparison of both chromatograms might not give the right answer, since the retention times of a compound with and without a subtractor column usually differ. McKEAG and HOUGEN (99) recently reported these differences to be about 5%. To overcome this problem, NIGAM (100) built a special dual-channel micro-reactor. The reference channel is packed with support material, the reaction channel with support + reagent.

According to HAKEN et al. (101), a hypothetical, ideal subtractor column has the following properties:

- specific activity for a limited number of functional groups; i.e. preferably one, and inactivity towards other functional groups;
- complete abstraction;
- no retardation or peak spreading of compounds not abstracted;
- selected reagents must be suitable for use in a column such that it forms an integral part of the chromatographic equipment;
- reagents should be stable and reproducible in use;
- capacity of the reagent for a functional class being abstracted should be high.

None of the subtractor columns used meets all these requirements. Therefore, when mixtures of unknown constitution are analyzed, caution should always be exercised in the interpretation of results. Only a relatively small range of compounds have been tested, and much further study is needed.

4.5.2.1. Carbonyl compounds

Many reagents have been used for the removal of carbonyl compounds from mixtures. Table 4—4 reviews the reagents used; they are classified according to their efficiency and specificity.

Table 4—4
Reagents used for subtraction of carbonyl compounds.

reagent	subtraction**		
	efficiency	specificity	reference
F.F.A.P.	—	—	102, 106, 107
$NaBH_4$	±	+	109
$NaBH(OCH_3)_3$	+	—	109
$LiBH_4$	+	—	109
$LiAlH_4$	+	—	109
$Na_2S_2O_5$	+	+	101
benzidine*	+	+	99, 102, 107
o-dianisidine*	±	±	99, 101, 102, 107
semicarbazide	+	+	102
aminoalkylpolysiloxanes	+	+	103
polyethylene imine	+	+	104
polypropylene imine	±	±	104
3,4,5-trimethoxybenzylhydrazine	+	±	105

* benzidine and o-dianisidine both have carcinogenic properties; appropriate precautions should be taken whenever these compounds are used.
** + = sufficient
± = sufficient, with a few exceptions
— = insufficient

From the 12 reagents mentioned in the Table, only sodium metabisulphite, benzidine, semicarbazide, aminoalkylpolysiloxanes and polyethyleneimine are useful. It is hard, however, to compare these five reagents on the basis of the data in the literature, as not all investigators have used the same compounds to test their reaction columns. The number of carbonyl compounds used ranges from 11 to 50 (Table 4—5).

Semicarbazide gave good results (102) using a 25 mm × 3.25 mm i.d. precolumn containing 40% by weight of the reagent on 120—150 mesh Celite 545. Two factors limited the operating temperatures of the precolumn to the range of 95—115 °C. At higher temperatures, the reagent tends to self-condensation to the biurea derivative and, moreover, volatile byproducts (azines) are formed in the reaction of the reagent and the carbonyl compounds. Especially in temperature programmed runs, these byproducts may interfere with other minor components present in a sample.

Metabisulphite precolumns (10% on Celite 60—72 mesh), used by HAKEN et al. (101) are efficient only at low temperatures (65 °C) and, therefore, the precolumn must be separately heated outside the main oven. This is a disadvantage; high boiling (> 150 °C) non-carbonyl compounds pass the precolumn but the emerging peaks are very broad; this is almost certainly caused by adsorption on the surface of the support (Celite) in the subtractor column. A cold trap between precolumn and main column would solve this problem.

Benzidine precolumns (10% on Celite 545) showed at 125 °C subtractive properties similar to those of semicarbazide, but peaks of non-carbonyl compounds were generally broader than those obtained with semicarbazide (102).

Ashes and Haken (103) reported the use of **aminoalkyl polysiloxanes** as abstractors for ketones and aldehydes. Used in precolumns, these materials were only effective subtractors for aldehydes and lactones; ketones were not removed completely because of too short residence times. A disadvantage is, also, that reactions occur with alkyl halides.

Polyethyleneimine (20% on silanized, acid-washed Celatom 72–85 mesh) was evaluated as a specific carbonyl abstractor by Appleyard and Haken (104). This reagent was only tested on a 1.8 m column and from the results it was evident that use of this reagent in short precolumns would not give good results. One disadvantage could be that also epoxides and alkyl halides are subtracted.

Recently, Moore and Brown (105) introduced **3,4,5-trimethoxybenzylhydrazine** as a subtraction reagent for aldehydes and ketones. The top 15 cm of a 2 m × 3 mm i.d. glass column was packed with 10% of the reagent on Gas-Chrom P, 80–100 mesh. The main part of the column was packed with 5% Carbowax 20 M on Gas-Chrom Z, 80–100 mesh and preconditioned. The combined column was conditioned at 130 °C for about 30 min. and could then be used for 5–6 h at temperatures of 120 °C. Aldehydes and almost all ketones were completely removed. Only ketones, with the oxo group on any but the penultimate carbon in the chain, with alkyl groups in positions adjacent to the oxo group, or with α-β-unsaturation, showed less complete removal. The most striking example is 2,4-dimethyl-3-pentanone; only 15% of this compound was removed.

Table 4–5 summarizes the results obtained with seven carbonyl abstractors.

4.5.2.2. Alcohols

Two reagents have been used to subtract alcohols from mixtures: boric acid (see e.g. 99, 102, 107, 109, 111, 112) and 3-nitrophthalic anhydride (102). Cronin (102) compared both reagents in short precolumns in the injection heater of a gas chromatograph. He used these precolumns in conjunction with PLOT columns and came to the conclusion that either reagent has advantages and disadvantages. Boric acid forms borates which are rather volatile and disturb the chromatograms considerably when programmed temperature chromatography is applied. The 3-nitrophthalates have considerably higher boiling points than the borates, and no interference was observed. On the other hand, boric acid is reactive over a wide range of temperature (50 to 200 °C), while the reactivity of the anhydride is increasing with rise in temperature of the precolumn. The rather high optimum temperature (150–200 °C) might lead to dehydration of some sensitive compounds.

Table 4—5

Comparison of the various abstractors

Reagent	number of ketones				number of aldehydes				ref.
	examined	removed			examined	removed			
		completely	partially	not		completely	partially	not	
semicarbazide	13	11	2	—	11	11	—	—	102
metabisulphite	38	36	1	1	11	11	—	—	101
benzidine	11	5	6	—	14	14	—	—	107
o-dianisidine	6	—	—	6	6	6	—	—	99
polysiloxane	39	25	14	—	11	11	—	—	103
polyethyleneimine (column)	27	21	6	—	8	8	—	—	104
trimethoxybenzyl-hydrazine	30	26	4	—	6	6	—	—	105

REGNIER and HUANG (109) found that 1 and 5% boric acid coated supports caused more than 50% dehydration of tertiary alcohols at 150 °C. Lowering the percentage boric acid to 0.5% resulted in less dehydration, less reactivity to other functional groups and tertiary alcohols, but no change in reactivity with primary and secondary alcohols (Table 4—6). This low percentage precolumn, however, is rapidly saturated and can only be used with sample sizes $\leq 10\ \mu g$.

Table 4—6
Percentages for removal of various compounds with boric acid (109)

compound	removal (%)	
	0.5% boric acid	5% boric acid*
benzyl alcohol	100	100
1-octanol	100	100
2-octanol	97	100
tert-amyl alcohol	2	57
citronellal	26	75
1,2-epoxytridecane	0	77

* The reaction of 5% H_3BO_3 with tert-amyl alcohol was accompanied by dehydration and the formation of an alkane.

4.5.2.3. Carboxylic acids and phenols

Zinc oxide (ZnO) has been used as a subtractor for carboxylic acids (107). Straight chain carboxylic acids were quantitatively abstracted. Branched carboxylic acids, with α-substitution, were not quantitatively removed; after passing the ZnO column, peak heights of these compounds had been reduced considerably, the peaks tailed badly and the retention times increased.
Ethers, aldehydes, esters and ketones passed the ZnO column unaffected. Phenols and secondary and tertiary alcohols were abstracted to the extent of less than 20%, primary alcohols to 50%. ZnO powder, mixed with a coated support in the ratio 1 to 10 by weight, gave the best results (107).
Phenols can be quantitatively removed by a 5 cm column packed with a mixture of potassium carbonate and Chromosorb P (1 + 1). The behaviour of acids on this column was not investigated (108), but acids can be expected to be removed, too.

4.5.2.4. Sulphur compounds

Since the development of the sulphur-specific flame photometric detector, sulphur compounds can be detected in mixtures without any pre-separation. Therefore, there is no need for using RGC methods as regards this group of compounds (see Chapter 4.6.).

4.5.2.5. Other forms of subtraction gas chromatography

WALSH and MERRITT (113) used chemical group reactions to obtain, besides retention data, information on the identity of the compounds separated by GC. They bubbled

the effluent gas through specific reagent solutions and, from the resulting change in any one reagent, the functionality of the compounds was determined. CASU and CAVALOTTI (114) allowed the effluents to come in contact with filter strips wetted with reagent solutions. This form of chemical group reaction carried out directly in the effluent stream of a gas chromatograph, can by definition be called RGC.
Many other investigators (e.g. 113, 115—119) performed chemical group reactions in solution, or in the vapourphase, and compared gas chromatograms obtained before and after treatment. This procedure can give additional information on the identity of compounds in a simple way. Therefore, it is mentioned in this paragraph although, strictly speaking, it is not RGC (see also Chapter 2.). HOFF and FEIT (115) and KAMINSKY et al. (118) carried out the reactions in the vapour phase in syringes; most investigators, however, added a reagent to solutions of the unknown compounds. One disadvantage of this method is apparent; it is applicable only for rather low boiling compounds with vapour pressures that are high enough for direct analysis of the compounds in the vapour phase.
Table 4—7 presents a number of reagents used for different chemical groups.

Table 4—7
A number of chemical groups and reagents

chemical group	reagent	reaction conditions	reference
aldehydes	saturated $Na_2S_2O_5$ solution	stirring for 2 h	116
aldehydes and esters	2 n NaOH solution	15 h, 120 °C	116
C-C double bond	H_2/Pt	5—10 h; cooling	116
aldehydes, ketones and esters	$LiBH_4$ in ether	addition till gas formation stops	116
alcohols	dinitrobenzoyl chloride	vigorous stirring	117
acids and phenols	5% $NaHCO_3$ or NaOH	room temperature	120

4.5.3. Hydrogen reactions

4.5.3.1. General remarks

Three types of reactions catalyzed by e.g. palladium or platinum may occur in the vapour phase between organic compounds and hydrogen:

- hydrogenation: addition of hydrogen;
- dehydrogenation: abstraction of hydrogen;
- hydrogenolysis: cleavage, with hydrogen addition at the cleavage points.

Unless the reaction conditions are chosen very carefully, two or all three reactions take place simultaneously.
These reactions can be made to take place at different locations of the gas chromatograph concerned. Some investigators (20, 121, 122) placed the reactors between the end of a GC column and the inlet of a mass spectrometer. This procedure is useful only if compounds of interest are fully separated on a GC column. Otherwise mixed

spectra are obtained and their interpretation is cumbersome. During hydrogenolysis, a large number of reaction products is often formed even from a pure compound. Therefore, most investigators attach the reactor to the injection port (e.g. 123—128) or use the liner of an injection port as reactor tube (e.g. 125, 129). CRONIN and GILBERT (130), and STANLEY and KENNETT (131), adapted the method for use with capillary columns. CRONIN and GILBERT used 10 cm lengths of PLOT capillaries "coated" with neutral palladium catalyst to trap compounds of interest from capillary columns, using hydrogen as the carrier gas. Traps were immediately sealed off and stored at low temperature until required for analysis. STANLEY and KENNETT (131) also trapped samples from capillary columns in glass capillaries but they used a platinum catalyst spread over 3—4 mm on the wall of the capillary. After trapping the capillaries were filled with hydrogen and sealed off. Reaction products formed during heating for some minutes at temperatures of 150—200 °C were injected on a GC column, using a special crushing device (see Chapter 3.2.). Amounts of 0.01—0.1 μg could easily be trapped and hydrogenolyzed, when these methods were used.

4.5.3.2. Hydrogenation

Hydrogenation is a much used technique for determining the structure of unsaturated compounds. Nickel, platinum and palladium catalysts have been used in the analysis of hydrocarbons and methyl esters of fatty acids (132).
The choice of the hydrogenation temperature is very important; at too high temperatures the functional groups are attacked, too (hydrogenolysis). In this connection, a study by KUBECZKA (133) is of interest. He carried out the reaction in a closed system at room temperature. Even the tertiary alcohol, linalool, did not lose its OH group; it was saturated only (Fig. 4—20).

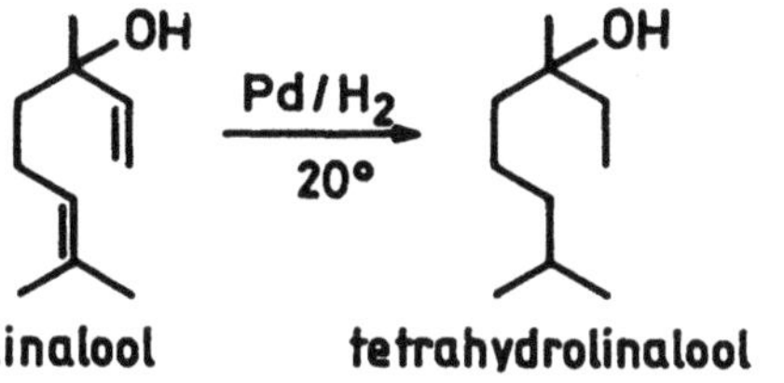

Fig. 4—20
Hydrogenation of linalool (133).

WIDÉN (134), using KUBECZKA's method, clearly demonstrated that opening of three- and four-membered rings of terpenes can also be largely avoided, saturation of multiple bonds still being accomplished.
SEVENANTS and JENNINGS (123) hydrogenated ethyl trans-2-cis-4-decadienoate, ethyl trans-2-hexenoate and trans-2-hexenyl acetate. The saturated compounds were obtained and no fission products formed, if the reaction temperature was taken just high enough to volatilize the unsaturated compound. These results agree with those of MOUNTS and DUTTON (127), who maintained the reaction temperature at 200 °C. On the other hand, BEROZA (97), using reaction temperatures at 280 °C, reported fission of even the shortest ester molecules.
ISSENBERG et al. (121) identified some phenols in wood smoke condensates with the aid of hydrogenation.

4.5.3.3. Micro-dehydrogenation (aromatization)

Micro-dehydrogenation has not reportedly been used in aroma research, but it has found some applications in the study of essential oils. As there is no reason why it may not come to be applied in aroma research, it is mentioned here. The technique yields information on the carbon skeleton of especially ring compounds, and can for example be applied in the qualitative analysis of terpenes.

As early as 1963, OKAMOTO and ONAKA (135) and some years later NIGAM et al. (136, 137) used micro-dehydrogenation in combination with GC in their respective studies on the identity of mono- and sesquiterpenes. For the study of a series of 42 monoterpenes, they used 5% platinum on alumina as a catalyst at temperatures ranging from 150° to 220 °C (136).

BUNGERT-HANSING and JORK (138), studying the composition of the essential oil of *Asarum europaeum* L., used dehydrogenation to obtain valuable information for the identification of four sesquiterpene alcohols. They used TLC to separate the reaction products.

STAHL and MÜLLER (139, 140) developed a system in which the reaction products were directly applied to a thin-layer plate using the so-called T.A.S. oven (thermo micro separation, transfer and application according to STAHL (141)). In it, solutions of the compound to be dehydrogenized are injected batch-wise into the T.A.S. oven. The highest conversion efficiencies were obtained with palladium-barium sulphate (10% Pd, Merck) catalyst. The solvent has a deactivating effect on the catalyst in that, after injection of 100 μl solvent, the catalyst has to be replaced. Some examples are presented in Table 4—8 (140).

Table 4—8
Catalytic dehydrogenation of sesquiterpene alcohols (140)

compound	temperature (°C)	reaction products
quajol	250	S-quazulene
globulol	300	S-quazulene
β-cadinene	250	cadalene
δ-cadinol	250	cadalene
(α + β)-selinene	250	eudalene
β-eudesmol	250	eudalene

According to STAHL and MÜLLER (140), fuller conversions are obtained with this Pd catalyst than with S or Se dehydrogenation procedures. Their technique, which they call carbon skeleton TLC, should have an advantage over the carbon skeleton gas chromatography. An advantage of the GC technique is that MS-analysis of the reaction products can very easily be accomplished.

Using TLC, a "sharp" injection is not necessary, while with GC "unsharp" injections result in broad peaks. This disadvantage, however, can be overcome by cooling either the oven or the first part of the GC column during the injection, and rapidly heating the oven to the end temperature.

For dehydrogenation in aroma research there is no obvious reason to prefer the TLC to the GC technique.

4.5.3.4. Hydrogenolysis (carbon skeleton chromatography)

Catalysts and reactors used for hydrogenation are also applicable in hydrogenolysis, but reactions are carried out at higher temperatures. THOMPSON et al. (142, 143) were the first to use hydrogenolysis; BEROZA et al. (97, 107, 144, 145) developed it further and called it carbon skeleton chromatography.

In addition to saturation of the multiple bonds (hydrogenation), heteroatoms are replaced by hydrogen, resulting in the parent hydrocarbon with the same carbon skeleton as the original compound, and/or the next lower homologue. Tables 4—9 and 4—10 summarize results obtained by BEROZA et al. (132) and by THOMPSON et al. (146), respectively.

Table 4—9
Carbon-skeleton chromatography products from different types of compounds (132)

compounds giving parent exclusively	reaction*
paraffin hydrocarbon	None
Unsaturated compound	Multiple bonds saturated
Halogen compound	C—X bond cleaved
Alcohol, *sec* or *tert*	C—O bond cleaved
Ester, alsohol part when *sec* or *tert*	C—O bond cleaved
Ether, *sec* or *tert*	C—O bond cleaved
Ketone	C=O bond cleaved
Amine, *sec* or *tert*	C—N bond cleaved
Amide, amine part on *sec* or *tert* carbon	C—N bond cleaved
compounds giving parent and/or next lower homolog	**reaction***
Aldehyde	$RCHO \rightarrow RH, RCH_3$**
Acid	$RCOOH \rightarrow RH, RCH_3$**
Anhydride	$(RCO)_2O \rightarrow RH, RCH^8$**
Alcohol, primary	$RCH_2OH \rightarrow RH, RCH_2$**
Ester, alcohol part (OCH_2R) when primary, acid part (R'CO)	$R'COOCH_2R \rightarrow RH, RCH_3$ $\rightarrow R'H, R'CH_3$**
Ether, primary	$RCH_2OCH_2R \rightarrow RH, RCH_3$
Amide, amine part ($NHCH_2R$) on primary carbon, acid part (R'CO)	$R'CONHCH_2R \rightarrow RH, RCH_3$ $\rightarrow R'H, R'CH_3$**

* catalyst temperature 300 °C, flow rate 20 ml/min, 1% Pd catalyst
** little or none of this parent hydrocabon is obtained.

Table 4—10
Carbon skeleton chromatography products from sulphur ring compounds (146)

compound	reaction* temperature (°C)	reaction product
2-methylthiacyclopentane	200	n-pentane
2,5-diethylthiophene	260	n-octane
3-methylbenzo [b] thiophene	170	(2-propyl)-cyclohexane
2-(4-thiadecyl)-thiophene	200	n-hexane + n-heptane
thiacycloheptane	200	n-hexane

* catalyst temperature as indiated, 1/8 inch diameter alumina pellets with 0.5% Pd

The so-called neutral palladium catalyst can be generally applied; it has indeed been used by many investigators. Its preparation is as follows (129).

A 330 mg sample of palladium chloride is dissolved in 100 ml of 5% aqueous acetic acid by heating on a hot plate for about 1 h with occasional swirling by hand. The solution is cooled, 200 mg of anhydrous sodium carbonate are added, and the solution is evaporated to dryness in contact with 10 g of Chromosorb W, 60—80 mesh, in a rotary film evaporator. The catalyst preparation is then heated in an oven at 110 °C for 1 h.

The catalyst is activated in the reactor tube (so-called "liner") with a hydrogen flow of 25 ml/min for $^1/_2$ h at 150 °C and for 1—3 h at 280 °C.

Reaction products can be identified by measuring retention times on one or two GC columns with stationary phases of different polarity, and by spectroscopic means. GC-MS is most appropriate, especially if a very small amount of material is available because MS reference files are especially well provided with hydrocarbon spectra.

Cronin and Gilbert (147) developed a special method for the detection of nitrogen-containing compounds in eluates of gas chromatographs. By hydrogenolysis with a Pt catalyst, nitrogen is converted into NH_3, which is detected either by phenolphthalein indicator in a PLOT capillary or by means of ninhydrin reagent coated on a PLOT capillary containing activated alumina. The detection limit is 0.2—0.4 μg. Grúz (148, 149) used a Zn catalyst, prepared by mixing Zn powder with a GC support material. Advantages of this catalyst should be its simple preparation; no activation in hydrogen stream is needed, and there is no formation of saturated and partially separated compounds from aromatic compounds and trapping of low molecular tacds (HCl, H_2S) by the catalyst material. A disadvantage is the high reaction temperature (400 °C), which will result in changes in the skeleton of heat labile compounds.

Although carbon skeleton chromatography could have been used in flavour chemistry for the identification of small amounts of compounds, the examples may well be hidden in the experimental parts of many papers. Two applications of the technique at our institute have been published by Kleipool et al. (150, 151). They identified an interesting constituent of rum and cognac, 1-(2,3,6-trimethylphenyl)-3-ethoxy-1-butene, using this technique in combination with IR, MS and NMR (151). They also used it for the identification of 3,4-dihydro-4-methyl-3-trans-propenyl-1,2-dithiin, one of the reaction products of 2-butenal and H_2S; hydrogenolysis of this compound gave 3-methylheptane (150).

Examples of applications of the technique for identification of terpenes have been reported (129, 134, 152, 153). Kepner and Maarse (129) applied hydrogenolysis to 15 representative monoterpenes, using a 2% neutral palladium on Chromosorb W catalyst in a GC injection port. The products were identified by mass and infra-red spectroscopy, and their Kovats Indices on Carbowax 20 M and SF 96 (50) were determined. Widén (134) studied the influence of the type of catalyst and the reaction conditions on the composition of the reaction mixture obtained by hydrogenolysis of some monoterpenes. He points out that the amount of information can be greatly increased by subjecting the same compound to analysis with different catalysts and under different experimental conditions. Results of Galfré and Guiochon (152), and those of Kepner and Maarse (129), indicated that cis-trans-isomerization takes place during hydrogenation and hydrogenolysis. The former authors made a special reaction tube by coating palladium metal directly on the wall of a glass capillary (12 m ×

× 0.3 mm i.d.). Reaction with α-pinene in this tube was 90% stereospecific (Table 4—11).

Table 4—11
Catalytic reaction of H_2 with α-pinene under different conditions

method	reaction product		reference
	cis-pinane	trans-pinane	
2% Pd on Chromosorb W	55	40	129
5% Pd on Chromosorb P	36	47	152
Pd coated on glass capillary	90	10	152

Another interesting example of the influence of reaction conditions on the composition of the reaction mixture was given by MAARSE (153). Germacrone was hydrogenolyzed in one and in two steps, using a 2% neutral palladium on Chromosorb W catalyst.

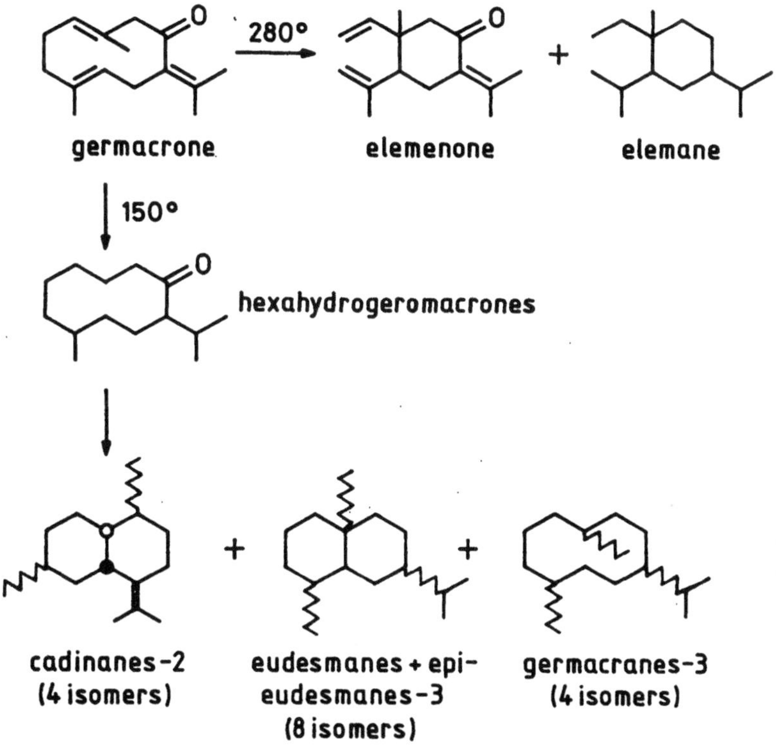

Fig. 4—21
Hydrogenolysis and hydrogenation products from germacrone (153). The figures behind the names indicate the number of isomers actually formed, and the figures in parenthesis indicate the number of possible isomers.

Hydrogenolysis in one step at 280 °C resulted in rearranged products only; the carbon skeleton of the original compound was not obtained. The two-step reaction gave a large number of products; three of these certainly were germacranes (Fig. 4-21).
Speaking generally, it can be said that during vapour-phase hydrogenation and hydrogenolysis other products can be formed than those obtained in solution at room tem-

perature. Therefore, care has to be taken in evaluating the results of catalytic reduction in the vapourphase. Hydrogenation at low temperature, before hydrogenolysis at high temperature, can reduce the chances of formation of "artifact skeletons".

4.5.4. Ozonolysis

Micro-methods for the location of double bonds in unsaturated compounds *via* ozonolysis have been used by many investigators (e.g. 130, 131, 154—160). Like other forms of reaction chromatography, the development of micro-ozonolysis owes much to the work of BEROZA et al.

BEROZA and BIERL (157) constructed a flow-through ozonizer, which is now commercially available, and used triphenylphosphine ($(C_6H_5)_3P$) to cleave the ozonide. The reaction proceeds as follows:

$$\begin{array}{c} R_1-C=CH \\ \;\;| \quad\; | \\ \;\;R_2 \;\; R_3 \end{array} \xrightarrow[2.\,(C_6H_5)_3P]{1.\,O_3} \begin{array}{c} R_1-C=O \\ | \\ R_2 \end{array} + \begin{array}{c} O=CH \\ | \\ R_3 \end{array} \qquad \text{eq. 4.1.}$$

Compounds purified by TLC or GC are dissolved in carbon disulphide (CS_2) or pentyl acetate, the solution is cooled at —70 °C and, for 15—90 sec., ozone is led into the solution; oxygen and ozone are flushed out with nitrogen, and triphenylphosphine is added. The purpose of this reagent is to avoid peroxide formation in the cleavage of the ozonides, and to thereby minimize side-reactions (130). About 10 minutes after the addition, an aliquot of the solution is injected into the gas chromatograph.

Pentyl acetate is the solvent if fragments up to C_5 are to be analyzed. CS_2 for fragments from C_5 upwards. Amounts of 1 μg of an organic compound are sufficient to identify the resulting products and so to locate the double bond.

An improvement of BEROZA's procedure was published by CRONIN and GILBERT (130). They made the method more sensitive (0.05 μg) by carrying out the reaction *in situ* without any solvent, and detected the presence of formaldehyde, resulting from oxidation of terminal methylene groups, after reaction with chromotropic acid. Compounds were trapped from capillary columns in short lengths of PLOT capillaries which next were sealed and stored, or ozonolysis was performed directly after trapping. Ozone was passed through the cooled (—70 °C) traps at a flow rate of about 4—6 ml/min up to 1 min, depending on the amount of compound. Excess of ozone was removed by purging with N_2. Depending on the nature of the fragments to be studied, one of the following procedures was applied to the ozonides retained in the trap:

- pyrolytic cleavage (at 100 °C) for examination of simple monofunctional carbonyl fragments (formaldehyde excepted) and G.C. analysis of the fragments;
- addition of o-phenylenediamine in ethyl acetate to the trap, if glyoxal or its homologues are to be expected. This reagent was also used by MOORE and BROWN (155), but they performed the reaction in solution and needed 10 μg of material. After sealing, the trap is heated in boiling water for a few minutes. The formed quinoxalines (Fig. 4—22) are analyzed by GC;
- addition of chromotropic acid solution as a test for terminal methylene groups. After sealing, the trap is heated at 100 °C for 2 minutes; purple coloration, stable for several hours, indicates a positive test.

An interesting result of the work of CRONIN and GILBERT is that they found that addition of triphenylphosphine was not necessary; it only changed the yield of carbonyl fragments to some minor extent. This confirmed similar results of WHITE et al. (158), who suggested that traces of reducing agents present as impurities sufficed to cleave the ozonides.

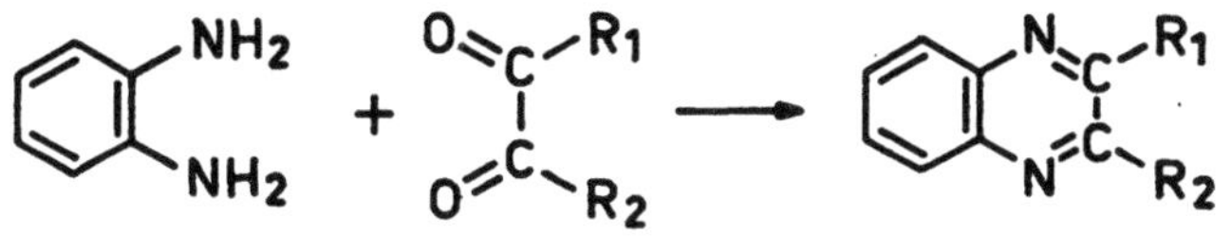

Fig. 4—22
Formation of quinoxalines from vicinal dicarbonyl compounds (130).

MOORE and BROWN (155) applied micro-ozonolysis to the identification of terpenes (Fig. 4—23). They characterized formaldehyde (I) as the cyclohexane-1,3-dione (II) derivative. The pyran compound (IV), which is formed from the primary reaction product (III) by heating to 90 °C in aqueous solution, can be analyzed by GC

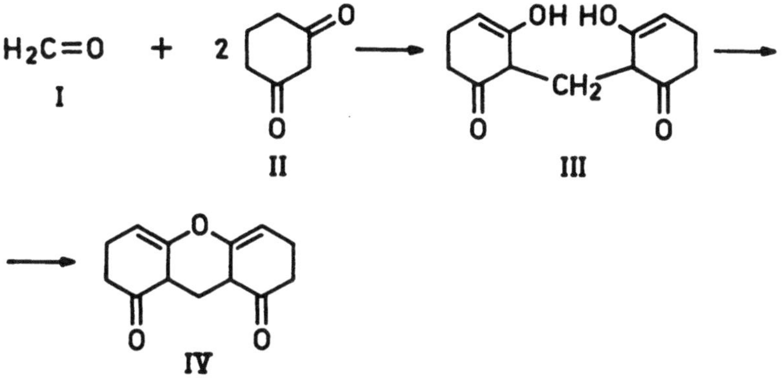

Fig. 4—23
Reaction between formaldehyde and cyclohexane-1,3 dione (155).

4.5.5. Other forms of reaction chromatography

Pyrolysis (thermal degradation of materials). This technique is used in the study of non-volatile material such as polymers and co-polymers, but it can also be applied in the identification of volatile compounds (e.g. 161—165). It has sometimes been called "the poor man's mass spectrometer".
As the GC-MS combination (see Chapter 4.2.) became available to many research teams over the last 5—10 years, the need for pyrolysis as a means of identification for volatile compounds has become less pronounced. Pyrolysis can, however, still give a positive contribution; for example in the identification of structural isomers. From the mass spectra it is often impossible to arrive at definitive identification, whereas the isomers can be readily distinguished by their pyrograms (chromatograms of the fragments obtained by pyrolysis). An example is given in Figure 4—24 (165). The mass

spectrum of six C_6H_{12} isomers and the normalized bar graph pyrograms of the same compounds are shown. It is evident that unambiguous identification can only be done from the pyrograms.

The method did not find much application, because of its complexity and owing to non-reproducible results. MERRITT and DI PIETRO (163) used the technique to determine the type of functional groups in a compound. This result can much more easily be obtained by determining the KOVATS Index (I) of a compound on a polar and an apolar column and calculating the ΔI value, which is a characteristic for a given homologous series of compounds (see Chapter 3.2.6.).

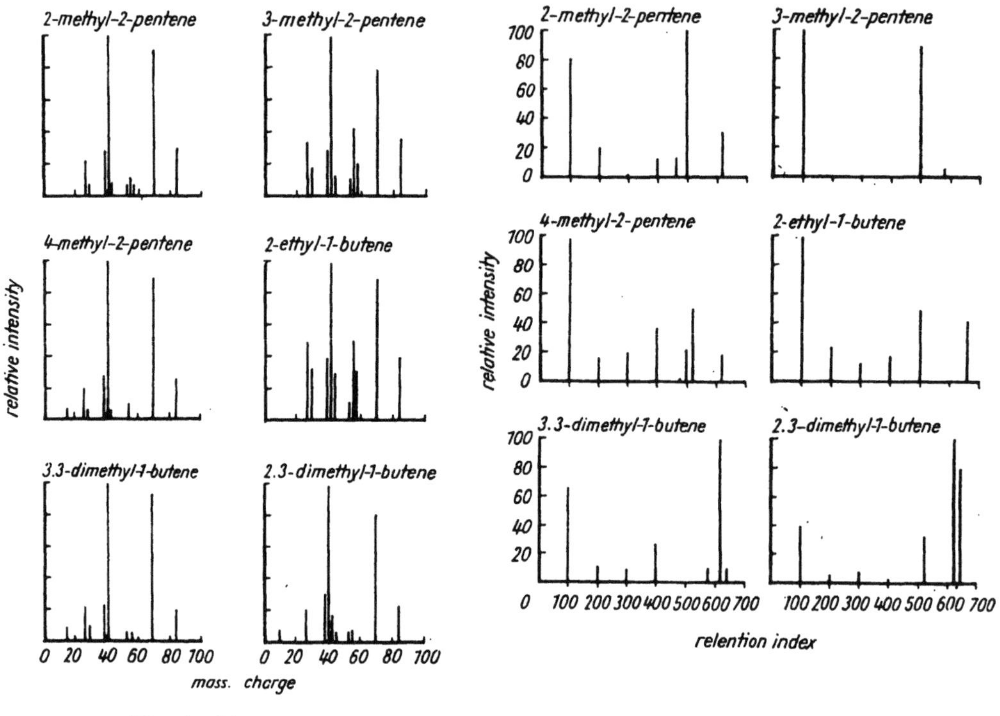

Fig. 4—24
Mass spectra (left) and pyrograms (right) of six different C_6H_{12} isomers (165)

On-column esterification. The separation and analysis of fatty acids is difficult, because they are polar compounds. Conversion to less polar, more volatile materials is commonly brought about by methyl esterification external to the gas chromatographic process (see 3.2.5.). Especially where only very small amounts of material are available, R.G.C. should be a useful alternative if a quantitative conversion to a more volatile derivative could be accomplished on a reactor column.

Studies of on-column esterification procedures have been made (166) but a reliable method to replace external esterification has not yet been found. Nor is it so much needed any more; the external method has been scaled down and speeded up so that $^1/_2$ mg of an edible fat or oil can be analyzed in about 10 min. (see refs. in 3.2.5.5.).

4.5.6. Conclusion

RGC is a valuable aid in the identification of volatile compounds, especially if such small amounts are available that they are sufficient only to obtain a mass spectrum but do not suffice for IR and NMR spectra. Often a large number of possible structures can be ruled out if additional information is available from hydrogen reactions, ozonolysis and subtraction chromatography. Application of these techniques seems too infrequent (e.g. 150, 151, 167). Possibly, many investigators have also used one of the RGC methods, and also published their results. However, we may well have missed those since they are hidden in the experimental parts of pertinent papers. The recent developments of these techniques in combination with glass capillary columns (e.g. 102, 130, 131) will undoubtedly lead to an increase in their applications.

4.6. Selective detection

P. J. Groenen

4.6.1. Introduction

Several selective detector systems for gas chromatographic effluents have been available for at least a decade now. Major areas of application have been the analysis of pesticides, pharmaceutical products, human body fluids and excretions, and nitrosamines. This section describes the application of selective GC detectors in aroma research, which is of a relatively recent date, and discusses some of the pros and cons of these detector systems. Selective detectors for separation techniques other than gas chromatography are available. They remain undiscussed, in view of their limited applicability in aroma research.

A detector can be termed selective, in a practical sense, if its response per mole for one group of compounds in a mixture is higher by several orders of magnitude than its response per mole for all other compounds in that mixture. Thus, selectivity may mean that the detector responds to all the compounds in the mixture, but with an enhanced response for one particular group of them; it may also mean, — and this is the more attractive situation —, that the response for one group of compounds is for instance as in the FID while the response for all other compounds is sharply reduced or fully suppressed (proximate or full element-specificity or group-specificity). These situations are extremes, and intermediate situations, corresponding to different degrees of specificity, can be thought of. Of course, a comparison with the properties of the FID is only a matter of convenience. It has no direct meaning, except for those detectors (in fact the two most prominent selective detectors) that are a modification of the standard FID. The human nose, or more precisely the human olfactory organ, is by definition a group-specific detector system, as it essentially responds to odorous substances only.

The FID is the GC detector that has been used most frequently over the years. Its principle has been discussed in Chapter 3. It is generally looked at as a universal, non-selective detector, but, in a way, even the FID should be termed a selective detector, since under standard conditions it responds to organic carbon compounds only. This potential element-specificity of the FID does not come to the fore and is seldom made use of, however, as most GC work is concerned with organic compounds only. Still, the insensitivity of the FID to water is an aspect of its element-specificity which is often used to advantage.

However, detectors more selective than the FID were needed and analysts soon began to look for detectors with the same high sensitivity but with specificity towards certain elements or certain classes of compounds. Obvious target elements were nitrogen, sulphur, halogens, phosphorus and metals. Of these, phosphorus and metal compounds lie outside the field covered in this monograph. Still higher specificity than just one-element specificity was sought, but in aroma research it has so far been obtained only exceptionally (*viz.* for di-carbonyl compounds, compounds with nitro groups, and infra-red detectable compounds; see paragraph 4.6.3.4.). Outside the aroma field, a remarkable detector system (168—170), which selectively detects N-nitroso compounds, should be mentioned. Some other selective detection methods for N-nitroso compounds will be mentioned under "Alternative nitrogen-detectors". An older example of group-specificity, which dates back to the earliest days of gas chromatography, even before the advent of the FID, and which involved the acid-base titration of aliphatic acids eluting from a GC column (171, 172), is only of historical importance. Preferential enrichment of groups of compounds can often be obtained during the isolation procedures. Examples of this have been given in Chapters 2 and 3. Methods for the selective elimination of groups of compounds have been discussed in the section on reaction chromatography (section 4.5.).

Table 4—12
Selective GC detectors (other than MS and IR) currently available for potential use in aroma research[a)]

detector system	compounds selectively detected [b)c)]				
	N	S	hal.	$^3H/^{14}C$	other
Human nose					odorants
Flame ionization detector (FID)			(+)		carbon compounds
Alkali flame ionization detector (AFID)	+	(+)	(+)		
Flame photometric detector (FPD)	(der.)	+			(der.)
Electrolytic conductivity detector (CECD/HECD)	+	(+)	(+)		
Microcoulometric detector (MCD)	(+)	(+)	+		
Microwave plasma detector (MPD)	(+)	(+)	(+)	(+)	(carbon compounds)
Radioactivity detector				+	
Electron-capture detector (ECD)	(der.)	(+)	+		diketones; (der.)
Selective electrode		(+)			

a) Some of these detectors have additional (sometimes major) applications outside the field of aroma research, such as: analysis of pesticides and drigs (ECD, FPD, AFID, CECD/HECD, MCD), N-nitrosamines (AFID, CECD/HECD), and Hg, Cr, Se, or As compounds (MPD)

b) hal. = halogen; der. = after introduction of a(nother) hetero-element by derivatization; (+) = less important than the other detectors available for this element.

c) Not necessarily with 100% specifity.

The most prominent selective GC detectors available to-day are listed in Table 4—12. They will be briefly discussed, as far as they have been used, or might be used, in

aroma research. An in-depth discussion of selective detection is not within the scope of this monograph, and would be inappropriate in view of the existence of some recent text-books and reviews devoted to selective detectors (173—186), or to GC detectors in general (187—194). The interested reader should consult these for detailed information. Here, emphasis will lie on the practical application of selective detection in aroma research and related fields.
Selective detectors may serve two purposes:

a) they may contribute to the identification of unknown compounds;
b) they may enable the quantitative determination of specified compounds, hidden in complex mixtures.

Though this chapter deals with the former subject, we shall at times address ourselves to the latter aspect, too.
As Table 4—12 shows, several selective detectors can be used for more than one element. A treatise on selective detection might therefore start from the detectors available, and successively discuss the potentials of each of them for various elements. Laboratory practice, however, goes the other way round, by asking which detectors are available for a particular element. The latter, practical approach has been chosen for this section.

4.6.2. Selective detectors available to-day for use in aroma research

4.6.2.1. Nitrogen-specific detectors

Selective detection of nitrogen compounds has been used mainly in the analysis of pesticides, in pharmaceutical and clinical chemistry and in the determination of N-nitrosamines. Reports on the use of nitrogen-specific detection in aroma research could until recently be counted on the fingers of one hand. The detector used exclusively is the alkali ionization detector (AFID). Other N detectors, such as the electrolytic conductivity detector, the microwave-excited emission detector or the micro coulometer, appear not to have been used for aroma studies.

The alkali flame ionization or thermionic detector. The alkali flame ionization detector is a modified FID. In its earliest forms (195—198), a sodium or potassium salt was positioned in the flame. In the atmosphere of the heated salt, compounds containing phosphorus elicited a much higher ionization current than in the normal FID, whereas for other compounds the signal was enhanced to a much lesser extent (halogenated compounds) or not enhanced at all. Later it was found that if potassium, rubidium or cesium salts were used, a selective response could be obtained for phosphorus and nitrogen (199—201), and the response to other compounds could be largely suppressed (201). Until recently, commercial versions of the AFID were all based on this principle. They contain a piece of crystalline Rb_2SO_4, RbBr, RbCl, CsBr or KCl, positioned above the flame, around the flame, or on top of the jet proper. The potentials of these respective AFID versions have been amply described (184, 202—210). A selectivity factor of about 5,000 for nitrogen vs. non-nitrogen can generally be obtained. The lower detection limit for nitrogen is typically of the order of 10^{-12}—10^{-13} g N/s, or

200−20 pg per injection for a compound containing 10% N with a peak half-width of 20 s.

The performance of the AFID is, however, highly susceptible to variations in the hydrogen flow and, to a lesser degree, the carrier gas and air flows and the temperature of the salt. This susceptibility considerably diminishes its usefulness for quantitative measurements and especially necessitates high-quality hydrogen flow control.

In some AFID detectors, the position of the salt above the flame has to be optimized daily; some AFID detectors require the additional adjustment of the position and the voltage of an auxiliary electrode. All in all, this results in a fairly complicated multi-variate system.

Gough and Sugden (209) have systematically studied the stability of one version of the AFID in comparison with an FID. They concluded that the AFID was far less stable than the FID. Coefficients of variation of 20% and 3.5%, respectively, were observed with a standard solution, injected 126 times in a period of 5 weeks. Hoodless et al. (210) also have made an in-depth study of the performance of this version of the AFID. They propose no less than seven sets of usable operating conditions, each with its own advantages and disadvantages. An AFID optimization detail has been described by Chamberlain (211). He stresses the tediousness and time-consuming character of the standard optimization method. A needle filled with 2,6-dinitrotoluene is inserted through the septum of the injection port and left there to maintain a steady stream of material through the column to the detector until optimization has been achieved.

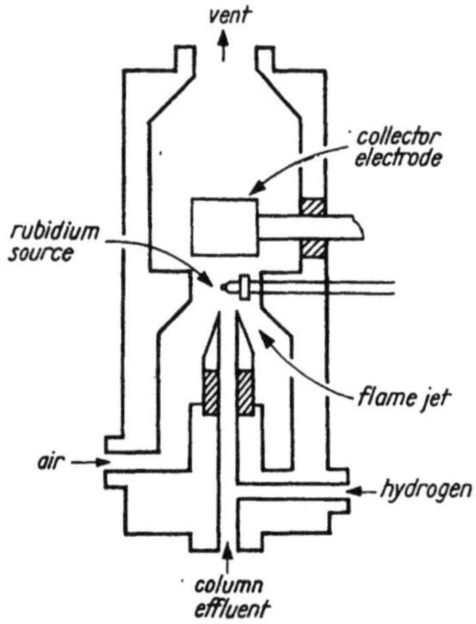

Fig. 4−25
Schematic diagram of the phosphorus-nitrogen detector (PND) (214).

A novel type of thermionic detector specific for compounds containing nitrogen and phosphorus has been described by Kolb and Bischoff (212) and Kolb et al. (213). Its stability, sensitivity, and selectivity, and its optimization characteristics are all claimed to be superior to earlier thermionic detectors, and to other nitrogen-specific detector systems. In this new detector, called phosphorus-nitrogen detector or PND, the alkali source is a non-volatile rubidium silicate bead, which can be heated electrically (Fig. 4−25). The PND can be set to detect both nitrogen and phosphorus (NP mode), or phosphorus alone (P mode) (Fig. 4−26).

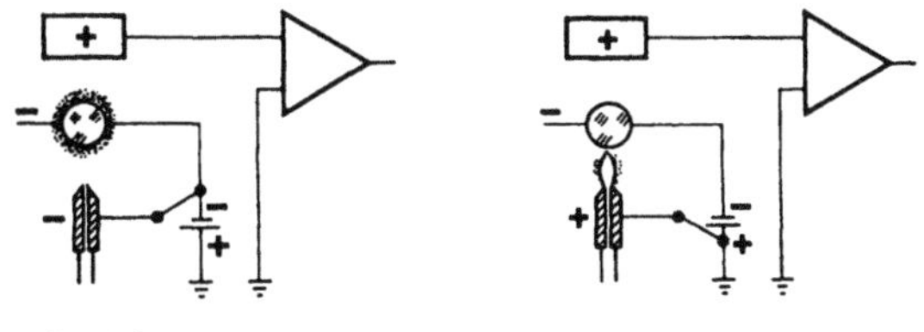

Fig. 4—26
Construction and principle of the phosphorus-nitrogen detector (PND) (212). For the sake of clarity, the electrical heating of the bead is not included.
(Reproduced by permission of Preston Publications, Inc.).

In the NP mode, the flame no longer has to perform the two functions it has in other thermionic detectors, viz. ionization of the compounds eluted from the column, and heating of the alkali salt. Thus, the conditions of the flame and bead can be separately optimized, which eliminates many of the awkward adjustments required with other thermionic detectors. Only 1—3 ml H_2/min are fed to the burner; this results in a relatively cold flame, adjusting itself around the heated bead. In the NP mode, cyan radicals are produced, which take up an electron from the alkali, resulting in a cyanide ion and an Rb^+ ion. The latter is recaptured by the bead, while the cyanide migrates to the collector electrode and generates the response. A detection limit of 1.2×10^{-13} g N/s is reported for pyridine. Like many other reported innovations in selective GC detection, these results originate from a GC manufacturer's laboratory and, understandably, will underline positive aspects of performance. Independent investigators, however, have confirmed the stability and precision of this detector (214 to 218). This novel type of nitrogen detector obviously has advantages over other thermionic detectors. Yet, some criticism has also been expressed. LUBKOWITZ et al. (219) have reported, that as the rubidium silicate bead ages, frequent increases in heating current are needed to maintain the sensitivity of the detector; this again accelerates bead aging, and so on. Bead life may be as short as two months, according to those authors.

BURGETT et al. (220) have recently described a similar nitrogen-phosphorus detector with an electrically heated alkali salt, but with a low-temperature plasma instead of a flame for ionization of the eluted compounds.

Though the AFID detectors commercially available are simple and relatively inexpensive, several investigators have constructed their own AFID, e.g. by shaping and boring an Rb_2SO_4-KBr (1:1) disc to fit on top of an FID jet (221), or by simply leaning a piece of such a disc against the flame tip (222). Similar techniques have been reported by other investigators (223, 224). These simple detector systems are useful for qualitative screening purposes.

Several investigators have reported that the AFID not only responds to phosphorus and/or nitrogen, but can be made to detect sulphur compounds as well. This will be dealt with in the relevant part of this section on selective detection.

Alternative nitrogen detectors. Other nitrogen-specific detectors available to-day are the electrolytic conductivity detector, the microcoulometric detector and the microwave-excited emission detector. These detectors seem not to have been used in aroma research and will here be dealt with briefly. They can be set to detect other elements as well (see Table 4—12).

The electrolytic conductivity detector, as a selective detector system, was introduced by COULSON (225, 226). Its principle of operation is entirely different from that of the AFID. When it is operated in the so-called reductive mode, the GC effluent is introduced into a high-temperature hydrogenation furnace where nitrogen compounds are converted to ammonia over a nickel catalyst. The vapours are led into a stream of deionized water and the resulting change in electrolytic conductivity is measured (Fig. 4—27). H_2S and HCl, the hydrogenation products of sulphur and chlorine compounds, are removed with an $Sr(OH)_2$ absorbent. The response of the detector is proportional to the amount of nitrogen in each compound and 0.1—1 ng of organic nitrogen is detectable (183, 226, 227). In this mode, the detector is completely insensitive to chlorine, oxygen, hydrogen and sulphur up to the milligramme level, and, unlike most versions of the AFID, is also insensitive to phosphorus. In other words, the CECD in the reductive mode is fully nitrogen-specific.

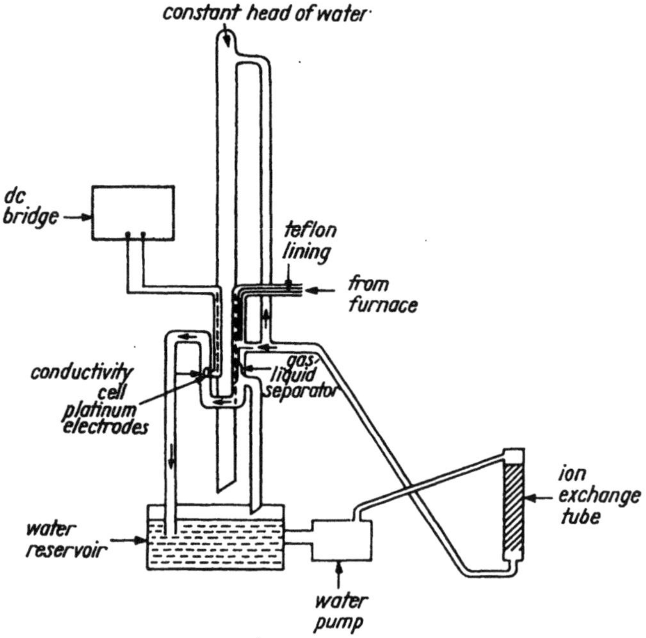

Fig. 4—27
Schematic diagram of the Coulson electrolytic conductivity detector (CECD) (176).

RHOADES and JOHNSON (228, 229) used the CECD in a pyrolytic mode without a catalyst, at a lower temperature with an argon stream. In this way, the detector was said to be specific for amines and nitrosamines and, after removal of amines, specific for nitrosamines. Confirmation of positive results with GC-MS is nevertheless recommended (230, 231).

Under oxidative conversion conditions, halogen or sulphur compounds or both can be detected selectively (see 4.6.2.2. and 4.6.2.3.).

GREENHALGH and COCHRANE (232), and PALFRAMAN et al. (233) have compared performance characteristics of the CECD and the AFID for the determination of nitrogen compounds. The sensitivities of the two detectors were found to be similar, but the CECD was less disturbed by small changes in operating parameters. It was considered more convenient and reliable, and more suitable for routine use.

The Coulson detector — in spite of its sensitivity, selectivity and reliability — has not received the attention it deserves. HALL (234) has ascribed this to the rather large size of the detector system, which has to be mounted as a separate unit, external to the chromatograph, *via* heated transfer tubing, which is a site for contamination build-up and sample decomposition. Another explanation, according to HALL, may have been the fact that initially the detector could only be used with Coulson Instruments Company gas chromatographs. Moreover, after every injection the solvent peak had to be vented, and the water flow in the detector cell was subject to irregularities and stagnation. HALL (234) therefore developed a microversion of the CECD, which is said to have 20—50 times the original sensitivity, a linear range of 10^5, a selectivity of 10^5—10^6 for N, S, or Cl, and none of the drawbacks mentioned. The micro-conductivity cell also serves as a gas-liquid separator cell; it contains concentric cylindrical electrodes and is connected to an a.c. instead of a d.c. Wheatstone bridge. Ethanol is the conductivity solvent, instead of water. The reaction tube in the reactor furnace and all connective tubing have an internal diameter of 1 mm or less. Hitherto, applications of this detector have been surprisingly scant.
Von RAPPAR et al. (235) used a modified HALL detector in the pyrolytic mode for the highly selective detection of nitrosamines. For simple nitrosamines, the detection limit was 50 pg per injection. A chromatogram of 100 pg of dimethylnitrosamine and 10 µg of two pyrazines showed only the nitrosamine peak, so that a chemical-group specificity is evident. The general validity of this needs further study, but the results are consistent with those for the CECD mentioned above. In this mode, the detector would of course be unsuitable for aroma studies. Experiences with the determination of nitrogen compounds with this detector in its reductive mode are therefore badly needed. One step in this direction has recently been set (236).

The microcoulometric detector (MCD) was designed by COULSON and CAVANAGH (237) and introduced as a selective GC detector for halogen and sulphur compounds by COULSON et al. (238). A nitrogen-specific version was developed by MARTIN (239). Nitrogen compounds are converted into NH_3, which is continually neutralized by coulometric titration with H^+-ions. The method bears some resemblance to the determination of aliphatic acids, eluting from a GC column, by acid/base titration (171, 172), which was mentioned in paragraph 4.6.1. as a historical example of group-specific detection. The MCD can detect about 1 ng of nitrogen (239, 240) and has a very high selectivity (about 10^6). In his review of selective detectors, ADLARD (176), however, states that it is unlikely that the present design of this detector will continue to be used as a GC detector, in view of the comparatively long response time of about 20 s of the cell and its associated electrical circuitry. The use of reactor tubes of narrower bore than those originally present would offer no advantages. A modern version of the MCD, however, has recently become available.

The microwave-excited emission spectrometric detector or microwave plasma detector (MPD), which was introduced for use with GC by MCCORMACK et al. (241), is still another, quite different kind of detector. It can, in principle, be tuned to any particular element; it is based on the highly specific emission of spectral lines and bands by excited atoms or molecule fragments. The spectra are excited in the plasma of a 2,450 MHz electrode-less discharge. Such an excitation system was found to be superior to a flame emission system or to high-voltage d.c. and a.c. glow discharges. The detector has many attractive properties, such as high sensitivity and selectivity, a small volume, a rapid response, and a wide dynamic range. The hetero-elements, relevant to aroma research (nitrogen, sulphur and halogens), can be selectively detected; in the case of nitrogen, leakage of traces of N_2 from ambient air must be rigorously precluded. Many other elements, such as carbon, hydrogen, oxygen, phosphorus, mercury, selenium or chromium can also be monitored, and the over-all potentials in this field are so impressive that an example seems appropriate here (Fig. 4-28). It has been taken from the work of McLean et al. (242).

The MPD is a rather expensive instrument, but its potentials exceed by far those of other selective detectors, and are said to approach those of the mass spectrometer when it comes to detection of elements and determination of inter-element ratios (242—244), or the identification of unknown compounds from their molecule spectra (245). Recently, a thorough study by Van Dalen et al. (246) has provided detailed information on the optimum operating conditions of the MPD. The linear dynamic range of the detector spans three to four decades, and its detection limits are of the order of 0.1 ng/s. Van Dalen et al. are sceptical of the claim that the elemental composition of unknown compounds can be established with the MPD. The variation of the sensitivity between different compounds — for carbon, and the more so for other elements — was much too large.

Selective detectors for derivatives, prepared from nitrogen. At the end of this discussion on the detection of nitrogen compounds, it is appropriate to briefly mention a quite different approach that has been adopted, namely: conversion of these compounds into e.g. electron capturing or sulphur-containing derivatives, so that other detectors can be called upon. Especially for amines have such procedures been devised. Examples will be given in paragraph 4.6.3.1.

4.6.2.2. Sulphur-specific detectors

Sulphur-specific detection has been extensively used in pesticide analysis and in air pollution studies. Applications in aroma research are also numerous, quite unlike the situation which exists with regard to nitrogen detection in aroma research. This is presumably due to the early availability of one highly convenient sulphur detector, the flame photometric detector or FPD. Other detectors currently available for the selective detection of sulphur compounds include the microcoulometric detector (MCD), the electrolytic conductivity detectors (CECD and HECD), the microwave plasma detector (MPD), the sulphide-ion specific electrode, and even the alkali flame ionization detector (AFID) and the electron capture detector (ECD), but examples of actual applications in this field are rare. All these various sulphur detectors will be described hereafter, with emphasis on the FPD. Examples of their application in aroma research will be discussed separately in paragraph 4.6.3.

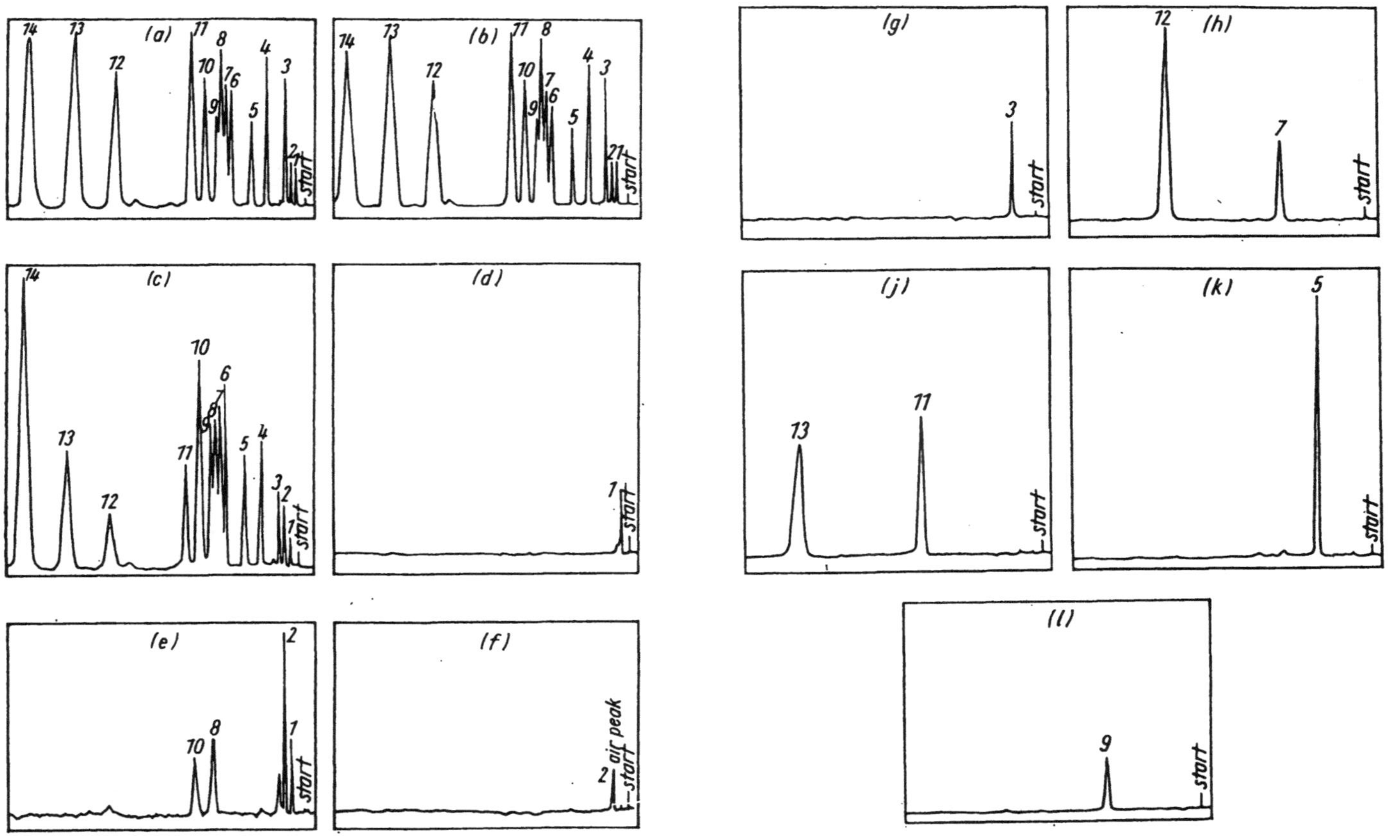

Fig. 4—28
Element-specific detection with the microwave plasma detector (MPD) (242). The test sample contained: 1) deuteroacetone; 2) nitroethane; 3) fluorobenzene; 4) toluene; 5) butyl iodide; 6) nonane; 7) chlorocyclohexane; 8) anisole; 9) diethyl disulphide; 10) 2-octanone; 11) bromobenzene; 12) o-dichlorobenzene; 13) o-bromotoluene; 14) undecane. An FID-chromatogram of the mixture is shown under (a). The MPD successively was tuned to carbon (b), hydrogen (c), deuterium (d), oxygen (e), nitrogen (f), fluorine (g), chlorine (h), bromine (j), iodine (k) and sulphur (l).

The flame photometric detector. The FPD is based on a principle developed by DRAEGER and DRAEGER (247), and CRIDER (248), and was introduced as a selective GC detector by BRODY and CHANEY (249).
Spectra emitted by a hydrogen-rich flame were shown to exhibit characteristic lines for sulphur and phosphorus compounds, and with comparatively simple interference filters (half-band width 5 and 7 nm) a typical sulphur emission line at 394 nm and a phosphorus line at 526 nm can be monitored. A parallel FID signal with about one-fifth the intensity of a normal FID signal is available. The optical filters can be exchanged for successive detection of sulphur and phosphorus. A dual-channel FPD, which enables their simultaneous detection, is also available. The detector is compact and robust, and — though other versions of the FPD have been developed — the initial design has proven itself over the years, without any essential changes being required. The construction of a single-channel FPD is shown in Figure 4—29. The flame is shielded from direct view of the multiplier and mirror; only the area above the tip

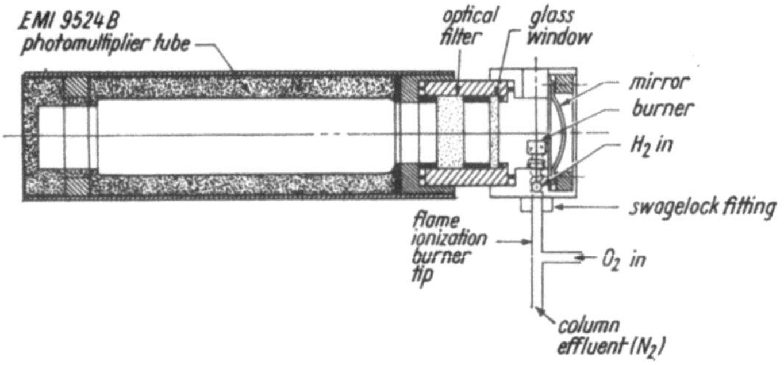

Fig. 4—29
Construction of a single-channel flame photometric detector (FPD) (249).
(Reproduced by permission of Preston Publications, Inc.).

of the flame envelope, where sulphur and phosphorus emission occurs, is viewed. The response to sulphur varies approximately with the square of the sulphur concentration in the 1—100 mg/l range. The application to phosphorus compounds will remain undiscussed here, as not a single phosphorus compound has ever been found as an aroma constituent in a food product (250). Characteristic gas flow rates for the FPD are 170 ml H_2/min, 20 ml O_2/min, and 100 ml air/min, with a carrier gas flow of 80 ml N_2/min (251), but quite different flow rate regimens have also been used. Amounts of 0.2 ng of S can be detected. The characteristics of the FPD in the sulphur mode have been investigated by many authors (see for instance 252—256). MIZANY (252) studied the influence of oxygen-hydrogen flow ratios and total gas flow; both affect the temperature of the flame and hence the degree of formation of the emissive S_2 species. He found that the optimal gas flows for sulphur detection were not the same as those for phosphorus detection. For sulphur, an O_2/H_2 ratio of 0.4—0.5 gave the highest response, with hydrogen set at 50—100 ml/min.
These conditions are quite different from those recommended by the supplier, Tracor Inc., and have not come into general use. According to MIZANY, the response to sulphur varied almost linearly with the sulphur mass at some gas flow rates and with its square under other conditions. S_2 formation depended on the degree of oxidation

of sulphur in the molecules (disulphide, sulphide, sulphone, sulphite, sulphonate, sulphate).

An FPD of slightly different design is commercially available. The burner here consists of two concentric tubes: the inner tube for air, and the outer ring for column effluent plus hydrogen. Optimization and response of this FPD have been studied by GREENHALGH and WILSON (257).

The presence of background sulphur in the FPD, either from column bleeding or from impurities in the gases used, can cause significant errors in quantitative measurements because of the non-linearity of the detector (254). Similarly, a small sulphur compound on the tail of a larger one may suffer the effect of non-additivity. Obviously, frequent calibration of the detector, and correction for background wherever possible are indicated. That the non-linearity of the FPD can also be turned to advantage was shown by CRIDER and SLATER (258) and by ZEHNER and SIMONAITIS (259); those authors added SO_2 to the hydrogen supply of the flame and observed a five-to-tenfold increase in sensitivity. STUBBS (254) also cautioned for a suppression of the sulphur response by high amounts of co-eluting hydrocarbons; see also BLOMBERG (260) and references given by that author. Suppression of the sulphur signal seems to be absent in a double-burner FPD, developed by RUPPRECHT and PHILLIPS (261). In this design, the decomposition of the compounds in the GC effluent is brought about in flame number one; the second flame gives the emission, characteristic for sulphur compounds. Such a double-burner FPD is commercially available (262, 263).

DEVONALD et al. (255) emphasized the limited dynamic range, of about 0.3—30 ng, of the FPD. They found that a series of dilutions may be required when sulphur compounds are present at widely different concentrations.

SUGIYAMA et al. (256) reported that the response of the FPD varied enormously with the type of sulphur compound. As an extreme example, butyl sulphide was about 100 times more responsive than pentyl disulphide. In view of this, and of the non-linearity of the detector, they anticipated that analysts would have great difficulties in determining sulphur-containing compounds with the FPD.

Another drawback of the FPD consists in the fact that the hydrogen-rich flame is extinguished by injections of more than one μl of organic solvent and has to be re-ignited at low hydrogen flow rate, but this can be prevented by a solvent-venting device behind the column. Such solvent problems, of course, are not restricted to the FPD; solvent venting will benefit any GC detector's performance. According to BURGETT and GREEN (264), solvent flame-out does not occur even with 25 μl injections if the hydrogen and air/oxygen inlets are interchanged.

The limitations of the FPD have been reviewed in more detail by FARWELL and RASMUSSEN (265). It is remarkable that up to 1978, i.e. twelve years after the original paper of BRODY and CHANEY (249), contradictory reports keep appearing as to the response of the FPD to different sulphur compounds and on the optimum gas flow ratios to be used (266—268).

MARUYAMA and KAKEMOTO (267) hold the extreme view that the response of the FPD is determined by the number of sulphur atoms in the molecule, and does not depend on the structure or the oxidation state of the compound, in contrast to the reports of MIZANY (252) and SUGIYAMA et al. (256). They acknowledge, however, the influence of the experimental conditions and of the geometry of the detector.

A selectivity of 10^4 or better for sulphur versus carbon can easily be obtained with the FPD.

Alternative sulphur detectors. In spite of its limitations, the FPD has become the S-detector of choice; it dominates the field. For that reason, other sulphur detectors have not gained much importance, and a detailed discussion would be inappropriate here. Only short descriptions will therefore follow. The principles of several of these detectors have been outlined under "Alternative nitrogen detectors" and need not be repeated.

The microcoulometric detector (MCD) in the reductive mode can be made to selectively detect HCl plus H_2S by titration with electrolytically generated Ag^+-ions. The current needed to restore the original concentration of Ag^+-ions is recorded. An analysis in the oxidative mode will measure only the Cl-compounds, again as HCl, and the S-compounds are found by difference (238).

BURCHFIELD et al. (269) determined sulphur compounds with the MCD in the oxidative mode, with an iodine cell instead of a silver cell; the SO_2 formed is titrated with electrolytically generated iodine. An improved titration cell with a faster response was designed by ADAMS et al. (270); those authors determined sulphur compounds as such, without a conversion furnace between column and detector, by their iodine or bromine consumption. Olefins, however, would interfere in this direct approach, by also consuming titrant.

GIUFFRIDA and IVES (271) demonstrated that the initial MCD design could considerably be improved and simplified by using a shorter oven, and by decreasing dead volumes. The exact position of the electrodes was also capable of improvement. Amounts of 10—20 ng of some S-containing compounds were detectable.

ŠEVČÍK (272) used a GC-FID-MCD sequence, by coupling the MCD to the exit side of an FID.

The Coulson electrolytic conductivity detector (CECD) (225, 226) and its modified version according to HALL (HECD) (234), when operated in the oxidative mode, can selectively detect sulphur compounds after conversion to SO_2 or SO_3. Selectivity for sulphur versus carbon is 10^4—10^5. The response to sulphur is linear, and 0.5 ng of S can be detected (273). Hydrochloric acid, formed from chlorine compounds, can be removed with a heated $AgNO_3$ scrubber. Carbon dioxide allegedly does not interfere due to its low dissolution and ionization rate (225). By contrast, ADLARD (176) has found the CECD "virtually unusable" in the oxidative mode, because of the large response produced by CO_2. According to HALL (234) and PAPE et al. (236), interference of CO_2 may well be a problem. It can, however, be solved by using a non-aqueous solvent such as pure ethanol or n-butanol. The response of the HECD to nanogramme amounts of sulphur compounds is twenty times that of the FPD, and its baseline noise level is lower (234).

The alkali flame ionization detector (AFID), which is used basically as a nitrogen and phosphorus detector, can be used to detect sulphur compounds as well. Many authors have reported a response, — negative and/or positive —, to sulphur (224, 274—277), but in this mode the detector is not considered competitive (177). Yet, HOODLESS et al. (278) obtained an analytically usable, positive response to low levels of sulphur, with a limit of detection of a few pg per second, similar to that obtained for nitrogen compounds. Quantitative determinations of sulphur compounds with the AFID were considered much more convenient than determinations with the FPD, because of the latter's non-linearity. MELLOR (175) has shown that the AFID can be tuned as either a phosphorus/nitrogen or a phosphorus/sulphur detector. The latest results of HOODLESS et al. (210) suggest no less than seven sets of usable operating conditions for the AFID, and show that specificity for phosphorus, or nitrogen, or sulphur can be achieved.

The highly selective microwave-excited emission detector or microwave plasma detector (MPD), described under "Alternative nitrogen detectors", which is a multipurpose detector, can be tuned to detect S compounds (243, 279—281) (see Fig. 4—28).

DAGNALL et al. (243) tuned one monochromator to carbon and another to sulphur or other hetero-elements. The minimum amount of sulphur detectable was $4.0 \times \times 10^{-11}$ g/s; selectivity for sulphur versus carbon was 460. Hetero-element/carbon ratios in a molecule could be assessed using this combination (242—244).

A sulphide-ion-specific membrane electrode has been used as a GC detector by KOJIMA et al. (282). With hydrogen as the carrier gas, sulphur compounds were converted to H_2S in a post-column reactor furnace, after which the vapours were contacted with a streaming aqueous solution of NaOH, $NaNO_3$, Na_2SO_3 and a trace of Na_2S. The resulting solution was led into a microcell with a sulphide-ion-specific electrode, which detected changes in S^{2-}-concentration. Sensitivity and response time are not specified in the abstract available, but sulphur compounds could be easily distinguished from non-sulphur compounds.

Finally, even the electron capture detector (ECD), in combination with an FID, has been used as a sulphur detector (283). The ratio of ECD response to FID response was found to be a highly informative parameter, which enabled the authors concerned to discriminate between different types of S compounds.

4.6.2.3. Selective detectors for other compounds

The preceding paragraphs on detectors for nitrogen and sulphur compounds discussed the most important areas for selective detection in the aroma field. Other groups of compounds, however, can also be detected selectively. Examples are electron capturing compounds, and radioactive compounds.

The electron capture detector has repeatedly been mentioned, earlier in this section on selective detection. A description of the construction and the operation of this detector will be omitted. It was felt that this would require a disproportionate amount of space, not in keeping with the limited importance of the ECD for the general aroma worker. The interested reader should consult LOVELOCK's early papers (e.g. 284 and 285). The review by AUE and KAPILA (286) contains useful information on the many controversies around the ECD. In their words, the detector is inexpensive, highly versatile and supersensitive; easy to operate but difficult to understand; yet, for better of for worse, indispensable in to-day's trace analysis (286). Prospective users of the ECD must be aware of the unpredictability of its response, as the electron absorption coefficients of various compounds and classes of compounds differ enormously (285). It is often impossible to decide whether an unknown peak represents a small amount of a compound of high electron affinity, or the reverse (176). Recently, many interesting studies have appeared outside the aroma field, in which the ECD was

coupled with capillary GC columns (e.g. 287–293). Increased confidence can be put in identifications or quantitative determinations with such a combination. As early as 1961, LOVELOCK also coupled his ECD with a capillary GC column (285).
The ECD is especially suited for determination of halogenated compounds, of compounds containing nitro groups, and of diketones and quinones. As indicated in Table 12, halogen compounds may be determined with the Coulson electrolytic conductivity detector, too. This detector is more selective, but less sensitive, than the ECD. COCHRANE and GREENHALGH (174) have predicted that the electrolytic conductivity detector as modified by HALL (234) will eventually replace the electron capture detector for the determination of Cl-containing compounds. Other halogen-specific detectors are the microcoulometric detector (238, 269, 294), the alkali flame ionization detector (195, 197, 295), and the microwave-excited emission detector (242). KARAGÖZLER and SIMPSON (296) and GOUGH et al. (297) have recently shown that even the normal FID, if operated at a very high hydrogen flow, will respond relatively selectively to chlorinated hydrocarbons.
A new microcoulometric GC detector from DOHRMANN is being advocated as the most sensitive, linear, specific halogen detector available to-day.
The selective detection of radio-active compounds in GC effluents, though quite different in nature from the detection principles discussed so far, certainly merits attention in this survey.
Methods based on radio-activity have been developed as early as 1959. Radio-active compounds (^{14}C- or ^{3}H-labelled) leaving the GC column can be measured with an on-line liquid-scintillation counter after absorption in a circulatory gas scrubber (298), or can be measured in the vapour state in an ionization chamber (299). The radio-active compounds in the eluate can also be converted to $^{14}CO_2$ (300–302), to $^{14}CH_4$ (302), or to $^{3}H_2$ (301, 303). The gases produced can be monitored in an ionization chamber, a Geiger-Müller counter, or a proportional flow counter. Besides these on-line detection techniques, methods are available for off-line measurement of the activities of GC fractions. According to DRAWERT et al. (304), the on-line and off-line methods have about the same sensitivity, but the on-line combustion methods are more attractive, as they are less laborious and time-consuming, and require less costly apparatus than does liquid-scintillation counting.
A review of these radio-GC detection methods has been given by DRAWERT and BACHMANN (302). Applications of the detectors mentioned will be discussed in paragaph 4.6.3.4.

4.6.3. Application of selective detection in aroma research

4.6.3.1. Nitrogen compounds

Applications of the alkali flame ionization detector

Reports on the use of nitrogen-specific detection in aroma research could — until recently — be counted on the fingers of one hand. The detector used exclusively is the alkali flame ionization detector.
MILLER et al. (305–307) used the AFID for the quantitative and selective determination of dimethylamine and trimethylamine in fish and fish protein concentrate. The

concentrations of these compounds reflect the quality and the freshness of the product; the amines are also of interest as potential precursors of nitrosamines. In one of these studies (307), the AFID was coupled to a wide-bore capillary GC column (153 m × × 0.75 mm i.d.). The correctness of the AFID identifications in these studies was confirmed with mass spectrometry.

DUMONT et al. (308) studied an odour defect in Gruyère cheese, which had developed during ripening of the cheese. The odour was described as "raw-potato-like" or "cellar-like". Aqueous distillates of the cheese also contained the odour defect, which disappeared on acidification. Isolation of the bases and analysis by GC-AFID enabled detection of fourteen nitrogen compounds, most of which were absent or much less prominent in extracts of good-quality cheese. Subsequent analysis by GC-MS led to the identification of most of these compounds, including pyrazines, pyridines and indole. The compound responsible for the off-flavour, however, was only a minor peak in the chromtogram obtained with the nitrogen detector. It was tentatively identified by mass spectrometry as 3-methoxy-2-propylpyridine.

CRONIN and STANTON (309) identified an important contributor to carrot aroma, *viz.*, 2-methoxy-3-*sec.* butylpyrazine. GC-MS enabled the eventual identification of the compound. Here, AFID detection in combination with a PLOT GC column was helpful in localizing the pyrazine in a fraction obtained from a preparative GC column but still of considerable complexity.

MUSSINAN et al. (310) identified 33 pyrazines in pressure-cooked beef with capillary-GC/MS. A capillary-GC/AFID combination provided additional confirmation. An Rb_2SO_4-KBr pellet, shaped and bored to fit the flame tip, as described by CRAVEN (221), was used in this work.

More applications of the thermionic detector can be found in areas outside, but more or less related to, the food aroma field, such as the determination of volatile N compounds in tobacco (214), in tobacco smoke (41, 311, 312), in polluted air (214), in plastics (313), and in urine (214). Chapter 6 of this monograph deals with such subjects. Several authors have used a combination of nitrogen and sulphur detection. Examples will be given further on, after the discussion on the selective detection of sulphur compounds in aroma research.

Application of alternative nitrogen detectors

Other N detectors, such as the electrolytic conductivity detector, the microwave plasma detector or the microcoulometer, appear not to have been used for aroma studies. In a related field, *viz.* that of tobacco smoke analysis, GUERIN et al. (314) have used the Coulson electrolytic conductivity detector to obtain nitrogen-specific profiles of the smoke from different types of tobacco.

Selective detection of nitrogen compounds via *derivatization*

A quite different approach that has been adopted for the selective detection of certain nitrogen compounds is conversion of these compounds into electron-capturing or sulphur-containing derivatives, so that other detectors can be called upon. This has enabled the highly sensitive and quite selective detection of amines. HOSHIKA (315) recommends conversion into fluor-containing Schiff bases with pentafluorobenzaldehyde, which is said to be a facile procedure with a capacity to discriminate primary amines from NH_3 and secondary amines. Twenty pg of the Schiff base can be detected with an ECD. Amines have also been converted into dinitrophenylamines with 2,4-dinitrofluorobenzene (316, 317), allowing detection of amounts as low as 2—20 pg. These techniques might be useful in aroma research. Applications in this area, however, have not been reported so far. In the field of odour nuisance, Mosier et al. (318) determined aliphatic amines volatilized from a cattle feed yard after conversion into pentafluorobenzamides.

Other useful electron-capturing derivatives are the perfluoroalkylamides, especially the trifluoroacetyl-, pentafluoropropionyl- or heptafluorobutyryl derivatives of the amines (see e.g. 319—321).

Finally, Hoshika (322) has even devised a flame photometric detection method for amines. Thirteen C_1—C_7 aliphatic primary amines were converted to S-containing Schiff bases by reaction with 2-thiophene aldehyde; they were determined selectively with capillary GC/FPD. This remarkable example of detection of N compounds with an S detector brings us quite smoothly to the next section, which is on the detection of sulphur compounds.

4.6.3.2. Sulphur compounds

In spite of some limitations, already discussed, the FPD has met with great enthusiasm and confidence, right from its appearance. Its use for qualitative purposes has certainly stood in the first place, but much successful quantitative work has also been reported.

An important area for sulphur-specific detection is the quality control of beer (323 to 335). Bärwald and Miglio (325) found dimethyl disulphide to be the major sulphur compound in the head space (50 °C, nitrogen atmosphere) of natural hop. Traces of ethanethiol, dimethyl sulphide and diethyl disulphide were also present. Drews et al. (323) developed a method for the qualitative and quantitative analysis of volatile sulphur compounds, and compared good-quality and off-flavoured beers. The latter contained higher amounts of thiols and sulphides. In British beer, dimethyl sulphide, in amounts of 2—6 μg/l, was the only sulphur compound detected by Harrison and Coyne (324); thiols were not found, even when a nylon capillary column was used. Adsorption at the metal surfaces between column and FPD was considered a possible explanation. Richardson and Mocek (327) studied the formation of artifacts in the GC determination of sulphur compounds in beer. Important factors were the time and temperature of sample conditioning, and the type of column used. The same workers (326) detected dimethyl sulphide and, occasionally, traces of methyl-n-butyl sulphide and dimethyl disulphide in beer (Fig. 4—30). The early peak in the FPD beer chromatograms was not due to H_2S but to CO_2, which confirms an earlier conclusion of Drews et al. (323). The FPD also gave a peak for another very abundant non-sulphur component, *viz.* ethanol.

McGowen et al. (328) detected dimethyl sulphide and H_2S in beer, and demonstrated the presence of distinct differences between hop varieties as to sulphur compounds. Jansen et al. (329) detected dimethyl sulphide and traces of H_2S, methanethiol,

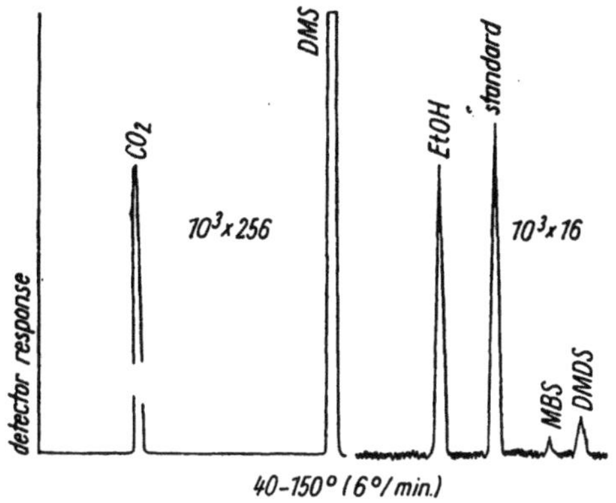

Fig. 4—30
Flame photometric detector response, obtained for a 4 ml injection of beer headspace vapour (326). DMS = dimethyl sulphide; EtOH = ethanol; MBS = methyl butyl sulphide; DMDS = dimethyl disulphide. Internal standard: diisopropyl sulphide. (Note the absence of base-line drift during temperature programming).

ethanethiol and dimethyl disulphide in beer; their results suggested a negative correlation between levels of sulphur compounds and organoleptic preference.

SCHREIER et al. (330—332) and SCHREIER and DRAWERT (168) used both GC-FPD and GC-MS for the analysis of beer and wine, and identified a number of new thioether compounds. Especially 3-methylthio-1-propanol (methionol) and the corresponding acetate were found to occur in amounts 100—1000 times higher than those of other volatile S-compounds. Their sensory thresholds were estimated to be higher to a similar extent. In fact, all known sulphur compounds in beer were thought to be present at about their threshold concentration. Model experiments (334) demonstrated that methionine was the main precursor of the thio-ether compounds. In wine, the N-(3-methylthio-propyl) derivative of acetamide was another new sulphur compound, together with some other amides (333). The FPD response was approximately quadratic, the exponent being 1.89.

Sulphur compounds in raw grain spirit were studied by RONKAINEN et al. (336). They observed the disappearance of several volatile S-compounds during storage; dimethyl disulphide persisted. The FPD response was found to be quadratic at low concentrations but it became linear at higher concentrations. MIGLIO (337) investigated the importance of S-compounds for the aroma of brandies and elucidated reaction mechanisms.

FPD applications in research on aroma of food products are numerous. WILSON et al. (338) studied pressure-cooked beef with GC-FPD and GC-MS. They report that the FPD information greatly facilitated the interpretation of mass spectra, especially where mixtures were involved. Volatiles in roasted filberts were investigated by KINLIN et al. (338). HEIDEMANN and WISMER-PEDERSEN (340) studied the influence of curing and storage on the volatiles in pork; they saw a decrease of hydrogen sulphide and methanethiol, and an increase of sulphides, disulphides and some sulphur-containing ring compounds. PERSSON and VON SYDOW (341, 342) identified 21 sulphur compounds in the aroma of canned beef with GC-FID, GC-FPD and GC-MS. Volatiles formed from unconventional proteins by heat treatment were investigated in the same way by QVIST and VON SYDOW (343, 344) and by QVIST et al. (345). MANNING (346) studied Cheddar cheese aroma, and obtained strong indications that the presence of methanethiol, though not methanethiol alone, is essential. FREEMAN (347) and FREEMAN et al. (348) studied S-compounds in parsley, onion, leek and garlic. The formation of odorous compounds from hydrogen sulphide and 2-butenal was followed by BADINGS et al. (349) with GC-MS, GC-FPD, IR and NMR. Apart from the simple addition product,

3-mercaptobutanal, which had an omelette-like odour, they identified several thiane, thiin, and dithiin derivatives, which had odours reminiscent of the odours of such products as onion, asparagus, leek and boiled potatoes. TRESSL et al. (350) ideritified over 100 compounds in cooked asparagus with GC-MS; the sulphur compounds present were thiophenes and thiazoles. GC-FPD analyses contributed to these identifications. Sulphur-containing acids, present in uncooked asparagus, e.g. 1,2-dithiolane-4-carboxylic acid (asparagusic acid), were identified in the same way (351). The determination of dimethylsulphide in the head space vapour of green tea with FPD and FID was shown by KAWABATA et al. (352) to be an effective and simple method for the evaluation of tea quality.

Many applications of the FPD as a sulphur detector can be found in research areas more or less related to the aroma field, such as tobacco smoke or air pollution. These subjects lie outside the scope of this monograph, and will be reviewed briefly in Chapter 6; for particulars the reader is referred to the original literature, as summarized in the following.

Applications to tobacco smoke have been reported by GROENEN and VAN GEMERT (353), GUERIN (354), HORTON and GUERIN (355) and GUERIN et al. (356), who studied the differences in sulphur profiles between the smoke of various tobacco varieties and the effect of filter types; also by BERTSCH et al. (357, 358), and BLOMBERG (260), who advocated the use of nickel or glass capillary columns respectively for the analysis of S-compounds in tobacco smoke.

Volatile sulphur compounds in polluted air have been studied intensively with FPD (255, 359—370). BARYNIN and WILSON (364), in a field study, reported a remarkable selectivity aspect of the FPD, which is worth being preserved and passed on to the readers of this monograph, *viz.* the ability to distinguish cars from lorries, passing by on a neighbouring road.

Other FPD applications of use to the aroma investigator, include the determination of volatile S-compounds in natural gas (371), coal gas (372), animal manure (373, 374) and mouth air (375).

Application of alternative sulphur detectors

A lone application of the microcoulometric detector (MCD) in the aroma field is a study by SCHAIT and CUZNER (376). They determined thiols and sulphides in beer. Thiols were to be measured directly by titration with Ag^+, without reduction to H_2S or oxidation to SO_2, but were not detected, at a sensitivity level of 1 μg/l. Dimethyl sulphide and diethyl sulphide were detected and quantitatively determined after conversion to hydrogen sulphide.

Two studies by MARTIN and GRANT (377, 378) were aimed at discriminating sulphur compounds by type, such as non-thiophenic, benzothiophenic, dibenzothiophenic and naphthothiophenic, with the microcoulometric iodine cell. Some of these compounds might have sensory importance, which is the reason for here referring to this work; the samples investigated were petroleum samples and the like. A GC column, separating by boiling point, was used to obtain qualitative and semi-quantitative information on groups of compounds, but identification when needed was by mass spectrometry. Simulated stack effluent from a kraft pulp mill, containing H_2S, SO_2 and CH_3SH, was

analyzed with automated GC-MCD by APPLEBURY and SCHAER (379). Simulated liquid effluents from a kraft mill were analyzed with GC-MCD via the headspace technique by VITENBERG et al. (380).
Even the electron capture detector (ECD), in combination with an FID, has been used as a sulphur detector (283). The ratio of ECD response to FID response was found to be a highly informative parameter, which enabled the authors to discriminate between different types of S-compounds. It was 180—350 for trisulphides, 8—20 for unsaturated disulphides, 0.6—3.3 for saturated disulphides, 0.13—0.48 for thiols, and ≤ 0.04 for monosulphides. The method was used for the analysis of garlic headspace and extracts. Calculation of the ECD/FID response ratio for each peak, plus determination of relative retention times on two columns of different polarity, allowed tentative identification of ten S-compounds. The possible interference of other electron-capturing compounds was not studied.
Other potential S detectors, the Coulson or Hall electrolytic conductivity detector, the alkali flame ionization detector and the microwave plasma detector, have not hitherto been applied by aroma workers. In a related field, *viz.* the analysis of odorized natural gas, the Hall electrolytic conductivity detector in its oxidative mode has been successfully used as a sulphur detector. Advantages as compared with the FPD were its linearity and its negligible susceptibility to hydrocarbon interference (381). The alkali flame ionization detector has been used as a sulphur detector to study the gasification products of brown coal (H_2S, CH_3SH, COS, CS_2) (382).

4.6.3.3. Nitrogen and sulphur compounds

Several authors have used a combination of sulphur and nitrogen detection, mostly the flame photometric and the thermionic detector. HŘIVNÁČ et al. (383) developed a gas chromatographic multi-detector coupled to a glass capillary column. The arrangement consists of an FID, serving as a general monitoring device, an FPD, a nitrogen-phosphorus-selective detector (NPSD) plus a sniffing port for the sensory evaluation of column effluent. A two-step four-way splitter with shrink-tubing connections and some purge gas inlets as required, enable the simultaneous utilization of the detectors. As a demonstration of the potentials of this multi-detector system, the authors show chromatograms of volatiles from roasted meat flavour, monitored with FID, FPD and NPSD. The report does not reveal the identities and odour properties of the many compounds in the mixture. The authors express a regret concerning the poor selectivity for nitrogen compounds over hydrocarbons (400:1) of their NPSD, which seems to act primarily as a phosphorus-specific detector. The other detectors of the AFID type are said to behave similarly. The authors suggest that for N-compounds the electrolytic conductivity detector (225) in the refinement of HALL (234) is more promising, if a selectivity comparable to that of the FPD for S-compounds is called for.
TRESSL et al. (384) used both nitrogen-specific and sulphur-specific detectors in their studies of the biogenesis of volatiles in fruit and vegetables (Fig. 4—31). They identified several methylthio-alkyl and methylthio-alkenyl isothiocyanates and ditto nitriles in extracts of radish and cabbage, in which enzyme activity had been inhibited. The selective detectors facilitated the analyses, but the final identifications were provided by mass spectrometry.

Investigators at the laboratories of International Flavours and Fragrances Inc. describe how they routinely survey natural extracts with 0.5 or 0.75 mm i.d. open tubular

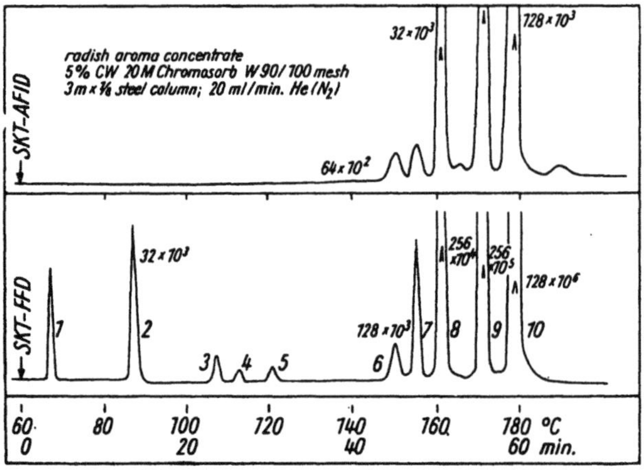

Fig. 4—31
Gas chromatograms of aroma extract of radish (384). Upper chromatogram: nitrogen-specific detection; lower chromatogram: sulphur-specific detection. (Note the absence of base-line drift during temperature programming).

GC columns coupled to an AFID-FPD-FID three-detector system before they proceed to GC-MS analyses (222, 338, 339, 385). In one study (339) they investigated the volatile components of roasted filberts. One hundred and eighty seven compounds were newly identified, including sulphur- and/or nitrogen-containing compounds. They established that mass spectral interpretation was simplified by knowing which compounds contained sulphur and/or nitrogen, and that as a result there was greater confidence in the designation of such compounds as "positively identified". An example of the important information that can be obtained with the multi-detector system described is demonstrated in Figure 4—32. Withycombe et al. (385), from the same laboratories, consider the CECD to be a more reliable quantitative nitrogen detector than the AFID. However, they say that from the standpoint of simplicity they would prefer the improved versions of the AFID for qualitative purposes. For sulphur compounds they consider no detector to be comparable to the FPD.
Volatile constituents of the crust of white bread were studied by Folkes and Gramshaw (386), using GC and FID, AFID, FPD and MS. One hundred and ninety compounds were definitively or tentatively identified, including S-compounds, N-compounds and S + N-compounds. Here, most identification data were again obtained with GC-MS, but valuable information was said to have resulted from the use of element-specific detection.

4.6.3.4. Other compounds

The preceding sections on nitrogen and sulphur compounds discussed the most important areas for selective detection in the aroma field. We will now briefly discuss some examples of the selective detection of some other groups of compounds.

Diketones and hydroxyketones

Owing to their strong electron-capturing properties (285), compounds such as diacetyl and pentane-2,3-dione can be detected with high sensitivity by electron-capture detection (ECD). According to SHILMAN and SCHAPER (387), the ECD response to diacetyl is 1200 times that of the FID. Vicinal diketones have been determined with ECD in beer by HARRISON et al. (388), STEFFEN (389), HAUKELI and LIE (390) and KOLB (313); in wine, by POSTEL and GÜVENC (391), and in milk by SCANLAN and LINDSAY (392). RONKAINEN and BRUMMER (393) isolated α-hydroxyketones (acetoin and 3-hydroxy-2-pentanone) as the 2,4-dinitrophenylhydrazones; these were then converted to the vicinal diketones by steam distillation from 2-oxo-glutaric acid solution, and determined with ECD. The method was considered suitable for fermentation solutions and beverages. POSTEL and GÜVENC (391) determined acetoin with ECD after oxidation to diacetyl; they separately determined the original diacetyl content, and calculated the acetoin content by difference.

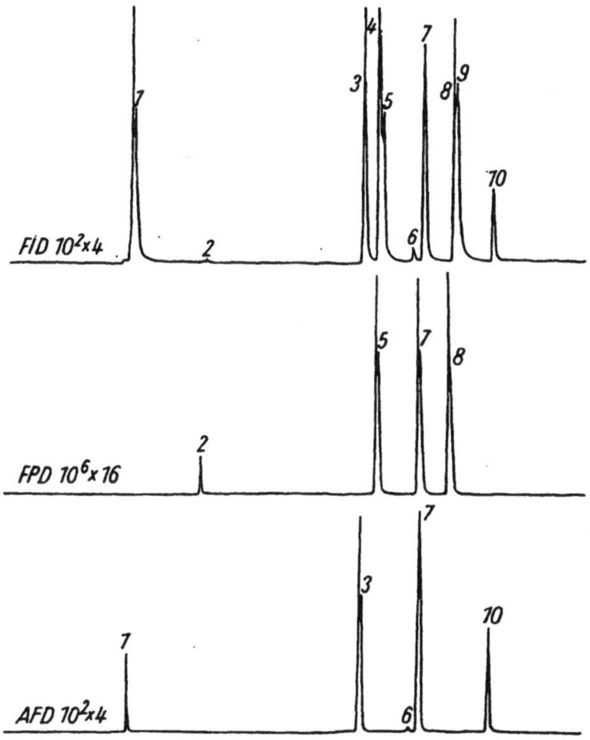

Fig. 4–32
Chromatograms of a synthetic mixture, using a three-detector gas chromatographic system (222). 1) ethanol; 2) methyl ethyl disulphide; 3) 1-pentylpyrrole; 4) furfural; 5) 2-pentylthiophene; 6) 2-methyl-5-vinylpyridine; 7) 4-methyl-5-vinylthiazole; 8) tetrahydrothiophen-3-one; 9) 5-methylfurfural; 10) pentylpyrazine.

Other carbonyl compounds

Acetone in butter can be determined with GC/ECD as the 2,4-dinitrophenylhydrazone (394). The method is thought to have potentials for wider application to carbonyl compounds in food oils and fats. Acetaldehyde in apple tissue has been determined in the same way (395). Confirmation was obtained with MS. HOSHIKA (396) sensitively and selectively determined benzaldehyde with FPD after conversion of the compound into an S-containing Schiff's base with 3-methylthio-aniline. The method is based on analyses by FPD and FID, both before and after derivatization; it might prove useful for other aldehydes as well.

Alcohols, meracptans, carboxylic acids and phenols

Several derivatization procedures have been proposed to enable the sensitive and relatively selective detection of alcohols by ECD. BURKINSHAW et al. (397) were able to detect femtogram amounts of alcohols as the pentafluorophenyldimethylsilyl ethers. WALLE and EHRSSON (319) used the heptafluorobutyryl derivatives. No applications in aroma research have been reported. KAWAHARA (398) determined organic acids, mercaptans and phenols with ECD as the pentafluorobenzyl derivatives. According to DAVIS (399), both acids and phenols, or acids alone, can be made detectable with this method, depending on the derivatization conditions. HEENAN and McCALLUM (400) have developed a method for determination of phenols with FPD as the phenyl diethyl phosphate esters. Such compounds as thymol, *iso*-eugenol and 2-methoxy-4-methylphenol could be determined sensitively and highly selectively; 10 pg of thymol phosphate ester was detectable. Analysis of the pentafluorobenzoate derivatives with ECD was about five times more sensitive, but the FPD was recommended in view of its exceptionally stable baseline and high selectivity. Apparently unaware of this publication of HEENAN and McCALLUM, DEO and HOWARD (401) have likewise proposed the formation of phenyl diethyl phosphates, and detection with the FPD. They report a (much higher) detection limit of 10—25 ng. Alcohols could be determined in the same way.

Halogen compounds

Inspection of the compilation "Volatile compounds in Food" (250) shows that about thirty volatile halogen compounds are reported as having been detected in food products. Some of these may have been contaminants from the solvents used in the analyses, or may reflect contamination of the food products from environmental sources, and their relevance as constituents of food aromas is questionable. As already discussed, halogen compounds can be detected highly sensitively with ECD, a property which is made use of extensively in the analysis of pesticides. Applications of the ECD in aroma research — except for the determination of halogen-containing derivatives — are not known to the present author, but an application of the ECD in off-flavour research is the study of musty chloroanisoles in poultry products (287, 402, 403). The precursors of the chloroanisoles, i.e. the chlorophenols, have also been determined, *viz.* after ethylation as the chlorophenetoles. KRIJGSMAN and VAN DE KAMP (291) have analyzed chlorophenols by ECD on glass capillary columns after acetylation. Workers in the field of chloroanisoles have also used mass spectrometry (50, 402, 404, 405). Obviously, availability or otherwise of such facilities often settles an investigator's decision whether or not to use selective detection.
As indicated in Table 4—12 and discussed in paragraph 4.6.2.3., halogen compounds may also be determined selectively with the electrolytic conductivity detector (174), the microcoulometric detector (238, 269, 294), the alkali flame ionization detector (195, 197, 295) ,the microwave-excited emission detector (242), and even the FID (296, 297). None of these detectors has, however, been applied to halogen compounds in aroma research.

Unsaturated compounds

HOSHIKA (406) has determined styrene in urban air by ECD after reaction with bromine. The lower detection limit was 10 pg, 500 times less than the detection limit of the FID for styrene. The method might therefore be valuable for the determination of unsaturated compounds in aroma research.

Miscellaneous groups of compounds

All the examples of the application of group-specific detection described so far in this paragraph on non-N, non-S compounds, except one, use derivatization to achieve the selective detection aimed at. The exception is the direct determination of diketones as such by ECD. Apart from the selective detection of some classes of sulphur compounds by ECD, already described under S-compounds (283), this has; until recently, been the only practicable example of direct molecular-structure-specific or chemical-class-specific detection in aroma research. (These are not really good examples of specificity, though, since many other compounds are also detected very sensitively by the ECD). One interesting, new approach to group-specific detection is based on infra-red detection. It has been shown by HAUSDORF (407) that infra-red detection can be made sufficiently sensitive to serve as a GC detection system. A non-dispersive technique is used, with interchangeable filters for the selective detection of one of the following groups: aliphatic compounds, alkenes, alkynes, aromatic compounds, esters, ketones, aldehydes, alcohols, phenols, halogenated compounds, sulphur compounds, or nitrogen compounds. The flow-through cell has a dead volume of only 7 μl, and can be coupled to support-coated open tubular GC columns. The sensitivity of this IR detector is but one thousandth of that of the FID for some compounds, but it is one tenth of the sensitivity of the FID for e.g. carbonyl, hydroxyl, and halogen compounds. An analysis of peppermint oil illustrates the potentials of the method (Fig. 4—33).
For the rest, higher specificity than just one-element specificity has only been achieved by the use of dual, triple or multiple detector systems. A remarkable group-specific detector exists outside the field of aroma research, i.e. the thermal energy analyser, which is virtually specific for N-nitroso compounds (168—170). Other direct methods of group-specific detection (i.e. methods not based on derivative formation) have not gained much practical importance.

Radio-active compounds

The use of radio-active precursor compounds has proved to be a powerful tool for elucidation of chemical or bio-chemical pathways in the formation of aroma compounds. Most of the work in this area has come from the laboratories of DRAWERT and TRESSL and their co-workers; it has been recorded in an impressive series of publications. Those authors used their radio-GC techniques e.g. in studies on the metabolic production of aroma compounds by yeasts; they were able to make important progress in the disentanglement of major and minor pathways of fermentation processes (408). Other topics studied by them include the incorporation of ^{14}C-amino acids and

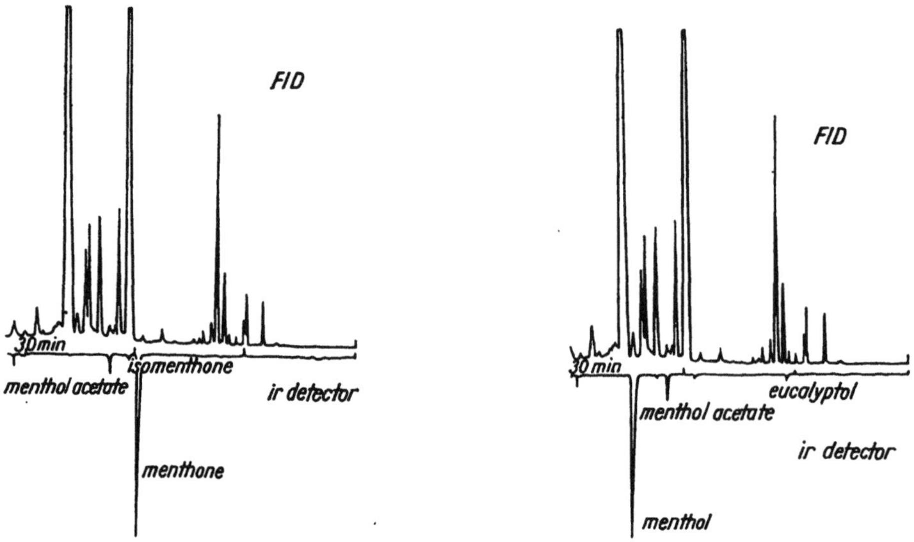

Fig. 4—33
(Left)
Selective detection of carbonyl compounds in peppermint oil with an infra-red detector (407).
Column: 50 ft × 0.50 mm i. d. SCOT, prepared with Carbowax 20 M. Column temperature programmed from 75 to 150 °C at about 4°/min. Sample volume: 0.04 µl. No inlet split. Column effluent split 1:1 to the IR-detector and a flame ionization detector. Carbonyl filter (5.6 µm).
(Right)
Selective detection of compounds containing a hydroxyl group in peppermint oil. Column and conditions as in chromatogram on left. Hydroxyl filter (9.8 µm).

other precursors into the aroma compounds of bananas and other fruits and vegetables (351, 409—415). The same authors also studied the pathways of the biosynthesis of the bitter substances of hop (416, 417).

The potentials of these radio-assay procedures can be appreciated from Figure 4—34, in which an FID chromatogram and a radiogram are compared (411). The sample is from a biogenesis test with 1-^{14}C-butyrate in banana slices, and is shown to contain 3-methylbutyl butyrate as the major metabolite. The reader should note that the radiogram and the FID chromatogram were obtained with a capillary GC column (100 m × 0.5 mm i.d.). Of the carrier gas (16 ml N_2/min), 3 ml was led to the FID and the remaining 13 ml to the radio-activity detector. The latter is shown to be fully compatible with the capillary column.

Radio-activity detection, either on-line or off-line, has also been used in studies on the formation of acetate artifacts from acetic acid during extraction and concentration procedures (418), the biosynthesis of trans-2-hexenal from linolenic acid in tea leaves (419), the formation of pyrazines in sugar-amino acid model systems (420), and the development of congeners from ethanol during whisky maturation (421).

The selective detection of radio-active reaction products in such model systems is a powerful tool for the investigator who is trying to unravel complicated reaction mechanisms; it enables him to unequivocally establish the reaction sequences leading from a precursor to its metabolites or its reaction products, and to reject other proposed mechanisms.

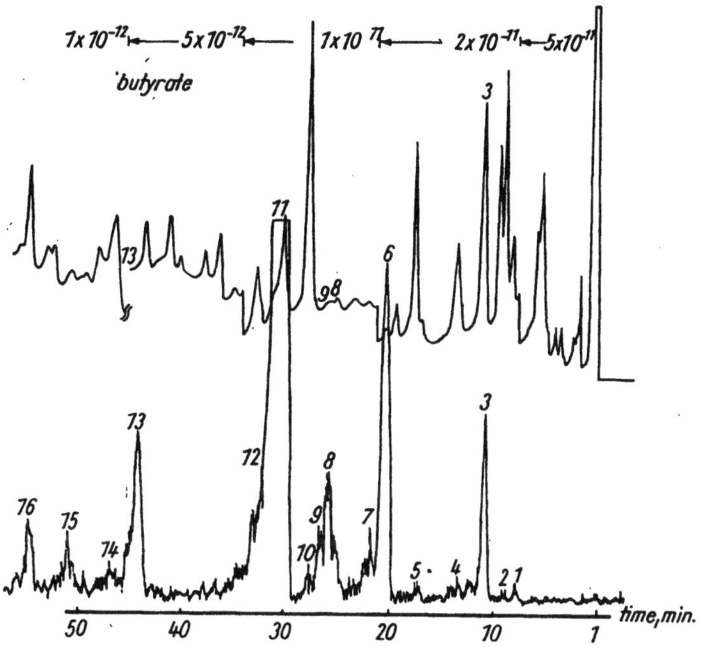

Fig. 4—34
FID-chromatogram (upper drawing) and radiogram of a biogenesis test with 1-^{14}C-butyrate on banana cuts (411).
Column: 100 m × 0.5 mm i. d., polyethylene glycol 1500. Carrier gas: 16 ml N_2/min, 3 ml to FID and 13 ml to radioactivity detector. Temperature program: 60—150 °C, at 2—5 °/min
Butyrates: 3) ethyl; 6) 2-methylpropyl; 8) 2-pentyl; 9) butyl; 11) 3-methylbutyl; 13) hexyl; 15) trans-2-hexenyl.

4.6.4. Evaluation of the importance of selective detection in aroma research

Nitrogen compounds

The most prominent nitrogen-specific detectors currently available are the alkali flame ionization or thermionic detector and the electrolytic conductivity detector. Most versions of the thermionic detector are relatively simple and inexpensive, but they often display irregular and erratic performance and have a rather critical set of interdependent operational parameters. A more sophisticated, novel construction (212, 213) is promising and may prove to be a break-through among these thermionic nitrogen detectors. Several applications of thermionic detectors in aroma research and related areas have been described in this section on selective detection. The electrolytic conductivity detector of COULSON (225, 226), or the improved microversion of HALL (234), have not been used in aroma research so far, and it is not obvious, why not (see also 236).
They are convenient and reliable in use, and — unlike most thermionic detectors, which also detect phosphorus — are fully nitrogen-specific. The improved electrolytic conductivity detector is commercially available. It is more expensive than the alkali flame ionization detectors. Other nitrogen-specific detectors are of minor importance.

The microwave-excited emission spectrometric detector may be useful for nitrogen detection, if nitrogen from ambient air, or as impurity in the gas used for the plasma discharge, can be rigidly excluded. This detector is rather expensive, but its range of applicability is much wider than that of other detectors.
Under present conditions, it can be predicted that either the new thermionic nitrogen-phosphorus detectors (212, 213, 220) or the Hall detector (234, 236) will be the preferred nitrogen detector of the years to come. An alternative way for the determination of nitrogen compounds is the introduction of sulphur or halogen into the molecule, so that other selective detectors can then be called upon.

Sulphur compounds

The early availability of a convenient and reliable sulphur-specific detector, the flame photometric detector, has made any development of other S-detectors less urgent. The vast majority of sulphur compound determinations in aroma research, and in the pesticide field as well, have been carried out with this detector. Still, several objections can be raised against it. Its response to S-compounds is non-linear; it is dependent on the type of compound, and can be quenched by co-eluting non-S compounds. Its dynamic range is limited. Unless precautions are taken, the flame is extinguished by every injection of more than 1 μl of solvent. In spite of these drawbacks, the FPD has met with great enthusiasm right from its appearance, and, accordingly, its applications in aroma research are numerous. Many examples have been given in the preceding paragraphs.
Other S-detectors that are available have also been described, but as yet their importance is low. Alternatives for the flame photometric detector as an S-detector would be the Hall electrolytic conductivity detector, which is said to be twenty times as sensitive and to have a lower noise level (234), or the alkali flame ionization detector (210, 278).

Other compounds

The formation of halogenated or nitro-group-containing derivatives may afford the possibility of sensitive and relatively selective detection of certain aroma compounds with the electron-capture detector. Halogenated compounds can also be determined with the electrolytic conductivity detector. It has even been predicted (174) that, for these compounds, the latter detector, in the modification of HALL (234), will eventually replace the electron-capture detector. Diketones can be determined as such with the electron-capture detector. Introduction of sulphur or phosphorus into a molecule will pave the way for selective detection with the flame photometric detector.
A quite different type of selective detection has been achieved by the use of radioactivity detectors. Impressive results concerning reaction mechanisms in the biogenesis of aroma compounds have been obtained by German workers in their studies with labelled compounds (see e.g. 411, 415, and the many other publications of those authors, as cited in paragraph 4.6.3.4.).

Table 4–13
Examples of the use of selective detectors in combination with capillary GC columns[a)]

reference	subject	reference No.
Thermionic detector		
KINLIN et al., 1972	roasted filberts	(339)
MILLER et al., 1973	amines in fish	(307)
MUSSINAN et al., 1973	pyrazines in beef	(310)
WALRADT, 1973	aroma mixtures	(222)
HARTIGAN et al., 1974	tobacco; air pollution	(214)
GROENEN and TEN NOEVER DE BRAUW 1975	tobacco smoke	(41)
HŘIVNÁČ et al., 1976	meat flavour	(383)
WITHYCOMBE et al., 1976	cooked pork liver	(385)
FOLKES and GRAMSHAW, 1977	bread	(386)
ISHIGURO and SUGAWARA, 1977	tobacco smoke	(311)
Flame photometric detector		
KINLIN et al., 1972	roasted filberts	(339)
WALRADT, 1973	aroma mixtures	(222)
BERTSCH et al. 1974	tobacco smoke	(357)
BERTSCH et al., 1976	tobacco smoke	(358)
BLOMBERG, 1976	tobacco smoke	(260)
HŘIVNÁČ et al., 1976	meat flavour	(383)
WITHYCOMBE et al., 1976	cooked pork liver	(385)
FOLKES and GRAMSHAW, 1977	bread	(386)
HOSHIKA, 1977	S-derivatives of amines	(322)
Electrolytic conductivity detector		
No applications with capillary columns are known [b)]		
Microcoulometric detector		
No applications with capillary columns are known		
Microwave plasma detector		
No applications with capillary columns are known[c)]		
Radioactivity detector		
DRAWERT et al., 1972	bananas	(411)
Electron capture detector		
LOVELOCK, 1961	electron capturing compounds	(285)
GEE et al., 1974	chloroanisoles	(287)
BÄCHMANN et al., 1977	chlorine compounds	(288)
BRECHBÜHLER et al., 1977	chlorinated water, pesticides, body fluids	(289)
EKLUND et al., 1977	TFA-derivatives of mono-saccharides	(290)
KRIJGSMAN en VAN DE KAMP, 1977	derivatives of chloro-phenols	(291)
REJTHAR and TESAŘÍK, 1977	pesticides	(292)
DE BOER et al., 1978	drugs	(293)
Infra-red detector		
HAUSDORF, 1977	peppermint oil, gasoline	(407)

a) For the alkali flame ionization detector and the flame photometric detector, examples have been selected from the aroma field and related areas only, but examples from other research areas (e.g. pesticides, drugs) are known; for the other detectors examples are rather scare, so that such a restriction was not made for them.

b) But potentially suited for use with capillary columns (176).

c) But potentially suited for use with capillary columns (241).

General evaluation

Prospective users of selective detection systems should realize that unambiguous identification of unknown compounds with such detectors only is no easy and straightforward task. With simple molecules and ditto samples, little doubt will generally remain about an identification, based on selective detection plus confirmed GC retention values on two columns of different polarity. The more complicated the molecule or the mixture, however, the less reliable will be such an identification. The use of high-resolution capillary columns can add considerably to the reliability of identifications. Indeed, it cannot be emphasized too much that increasing the resolution of the GC columns used is an essential contribution to the reliability of the resulting identification. For more complicated molecules, confirmed GC retention values together with an agreeing mass and/or IR spectrum are recommendable or required to warrant unambiguous identification. Especially with the mass spectrometer tuned as a selected-ion monitor (molecular-mass chromatography; mass fragmentography), a highly selective and decisive analytical technique is available. The acme of selectivity can be obtained with mass chromatography at increased resolution of the mass spectrometer, in combination with capillary column gas chromatography (see sections 4.1. and 4.2.). In all these cases, the availability of information, obtained separately by selective detection, will, of course, considerably facilitate the pertinent identification process. If MS facilities are not available, the use of two capillary column of different polarity, together with selective detection, is indicated. A large inventory of reference compounds is then required. Examples of the use of capillary columns with selective detectors are given in Table 4—13. The use of several selective detectors simultaneously (for N, S, halogen), together with flame ionization detection, is highly rewarding (see e.g. 383 to 386). Most of these authors, however, did resort to mass spectrometry for the eventual identification.

The potentials and advantages of selective detectors can be summarized as follows:

— they provide (additional) information on the identity of unknown compounds;
— they can yield important savings on mass spectrometer time;
— it is possible to prove the absence of specified compounds, or — in case of their seeming presence — to determine their maximum concentration, with a minimum of sample clean-up;
— in most cases there is no (visible) solvent peak;
— base-line drift during temperature programming generally does not occur.

The over-all shortcoming of selective detectors lies in the fact that they do not enable definitive identifications to be made, except for simple molecules.

References

1. Bassler, G. C. and R. M. Silverstein, Spectrometry in Organic Analysis, Application to small samples, J. Food Sci. **28**, 446—452 (1963)
2. Russell, G. F. and W. G. Jennings, Constituents of black pepper, Some oxygenated compounds, J. Agric. Food Chem. **17**, 1107—1112 (1969)

3. CREVELING, R. K. and W. G. JENNINGS, Volatile Components of Bartlett Pear, Higher boiling esters, J. Agric. Food Chem. **18**, 19—24 (1970)
4. VARO, P. T. and D. E. HEINZ, Volatile Components of Cumin Seed Oil, Identification and Characterization of 1,4-*p*-menthadien-7-al isolated from cumin seeds, J. Agric. Food Chem. **18**, 234—242 (1970)
5. RODIN, J. O., C. M. HIMEL, R. M. SILVERSTEIN, R. W. LEEPER and W. A. GORTNER, Volatile flavor and aroma components of pineapple, I. Isolation and tentative identification of 2,5-dimethyl-4-hydroxy-3(2H)-furanone, J. Food Sci. **30**, 280—285 (1965)
6. BONDAROVITCH, H. A., A. S. GIAMMARINO, J. A. RENNER, F. W. SHEPHARD, A. J. SHINGLER and M. A. GIANTURCO, Some aspects of the Chemistry of Tea, A contribution to the knowledge of the volatile constituents, J. Agric. Food Chem. **15**, 36—47 (1967)
7. ANDERSON, J. and E. VON SYDOW, The Aroma of Black Currants, II. Lower boiling compounds, Acta Chem. Scand. **20**, 522—528 (1966)
8. ASTON, F. W., A positive-ray spectrograph, Phil. Mag. **38**, 707—715 (1919)
9. DEMPSTER, A. J., A new method of positive ray analysis, Phys. Rev. **11**, 316—324 (1918)
10. MCFADDEN, W. H., Techniques of Combined Gas Chromatography/Mass Spectrometry: Applications in Organic Analysis, Wiley-Interscience, New York 1973, pp. 463
11. —, Mass Spectrometry of Organic Ions, ed. MCLAFFERTY, F. W., Academic Press, New York 1963, pp. 730
12. BEYNON, J. H., R. A. SAUNDERS and A. E. WILLIAMS, The Mass Spectra of Organic Molecules, Elsevier Publishing Co., Amsterdam 1968, pp. 510
13. BIEMANN, K., Mass Spectrometry: Organic Chemical Applications, McGraw-Hill, New York 1962, pp. 370
14. —, Topics in Organic Mass Spectrometry, ed. BURLINGAME, A. L., Wiley-Interscience, New York 1970, pp. 471
15. BRUNNÉE, C. und H. VOSHAGE, Massenspektrometrie, Verlag Karl Thiemig KG, München 1964, pp. 316
16. BUDZIKIEWICZ, H., C. DJERASSI and D. H. WILLIAMS, Mass Spectrometry of Organic Compounds, Holden-Day, Inc., San Francisco, Calif. 1967, pp. 690
17. MCFADDEN, W. H. and R. G. BUTTERY, in "Topics in Organic Mass Spectrometry", ed. BURLINGAME, A. L., Wiley-Interscience, New York 1970, p. 327—367
18. ISSENBERG, PH., Mass spectrometry for flavor research, Food Technol. **23**, 1435—1442 (1969)
19. TERANISHI R., P. ISSENBERG, I. HORNSTEIN and E. L. WICK, Flavor research, Marcel Dekker Inc., New York, USA 1971, pp. 315
20. ISSENBERG, PH., A. KOBAYASHI and T. J. MYSLIWY, Combined gas chromatography — mass spectrometry in flavor research. Methods and applications, J. Agric. Food Chem. **17**, 1377 to 1386 (1969)
21. KOLOR, M. G., Flavor Components, in "Biochemical Applications of Mass Spectrometry", ed. WALLER, G. R., Wiley-Interscience, New York 1972, p. 701—722
22. GOHLKE, R. S., Time-of-flight mass spectrometry and gas-liquid partition chromatography, Anal. Chem. **31**, 535—541 (1959)
23. VALOIS, P. J. DE, A simple interface in combination gas chromatography mass spectrometry, in "Application de la spectrométrie de masse (sm) et de la résonance magnétique nucléaire (rmn) dans les industries alimentaires", XV Symp. Int., Bologna, Italy 1975, 71—76 (1977)
24. MAARSE, H. and M. C. TEN NOEVER DE BRAUW, The analysis of volatile components of Jamaica rum, J. Food Sci. **31**, 951—955 (1966)
25. HEINS, J. TH., H. MAARSE, M. C. TEN NOEVER DE BRAUW and C. WEURMAN, Direct food vapour analysis and component identification by a coupled capillary GLC-MS arrangement, J. Gas Chromatogr. **4**, 395—397 (1966)
26. MCFADDEN, W. H., Techniques of Combined Gas Chromatography/Mass Spectrometry: Applications in Organic Analysis, Wiley-Interscience, New York 1973, p. 157—221
27. FREEDMAN, A. N., The gas chromatograph — mass spectrometer interface, Anal. Chim. Acta **59**, 19—31 (1972)
28. SAALFELD, F. E., The marriage of gas chromatography and mass spectrometry, Ind. Res. Augustus 1969, 58—60
29. NOEVER DE BRAUW, M. C. TEN und C. BRUNNÉE, Ergebnisse mit einer optimierten Kopplung von Gas-Chromatograph und Massenspektrometer, Z. Anal. Chem. **229**, 321—335 (1967)

30. MUNSON, M. S. B. and F. H. FIELD, Chemical ionization mass spectrometry. 1. General introduction, J. Am. Chem. Soc. **88**, 2621—2630 (1966)
31. BECKEY, H. D., Strukturbestimmung organischer Moleküle und quantitative Analysen mit dem Feldionisations-Massenspektrometer, Angew. Chem. **81**, 662—679 (1969)
32. BECKEY, H. D., Field desorption mass spectrometry: A technique for the study of thermally unstable substances of low volatility, Int. J. Mass Spectrom. Ion Phys. **2**, 500—503 (1969)
33. HORNING, E. C., M. G. HORNING, D. I. CARROLL, I. DZIDIC and R. N. STILLWELL, New picogram detection system based on a mass spectrometer with an external ionization source at atmospheric pressure, Anal. Chem. **45**, 936—943 (1973)
34. McFADDEN, W. H., Techniques of Combined Gas Chromatography/Mass Spectrometry: Applications in Organic Analysis, Wiley-Interscience, New York 1973, p. 278—330
35. McLAFFERTY, F. W., Mass spectrometer, minicomputer, liquid chromatograph: an ideal combination, (in Dutch) Chem. Weekbl. Mag. m **333**—m **337** (1976)
36. BRUNNÉE, C. und H. VOSHAGE, Massenspektrometrie, Verlag Karl Thiemig KG, München 1965, p. 27—30
37. MATTAUCH, J. and R. HERZOG, Mass spectrograph, Z. Phys. **89**, 786 (1934)
38. McFADDEN, W. H., Techniques of Combined Gas Chromatography/Mass Spectrometry: Applications in Organic Analysis, Wiley-Interscience, New York 1973, p. 408—413
39. McFADDEN, W. H., Techniques of Combined Gas Chromatography/Mass Spectrometry: Applications in Organic Analysis, Wiley-Interscience, New York 1973, p. 68—70
40. MULDERS, E. J., M. C. TEN NOEVER DE BRAUW and S. VAN STRATEN, The odour of white bread. II. Identification of components in pentane-ether extracts, Z. Lebensm. Unters. Forsch. **150**, 305—310 (1973)
41. GROENEN, P. J. and M. C. TEN NOEVER DE BRAUW, Determination of volatile N-nitrosamines in the vapour phase of the smoke from various tobacco products, Beitr. Tabaksforsch. **8**, **3**, 113—123 (1975)
42. NOEVER DE BRAUW, M. C. TEN, The combination GC-MS: an important aid in identification of components in complex mixtures (in Dutch), Chem. Weekbl. **71** (14), 25—29 (1975)
43. MAARSE, H. and M. C. TEN NOEVER DE BRAUW, Catty odour as an air pollutant (in Dutch), Chem. Weekbl. **68** (6), 11—12 (1972)
44. SCHAEFER, J., J. M. H. BEMELMANS and M. C. TEN NOEVER DE BRAUW, Investigation into the components responsible for the malodour of pig houses (in Dutch), Landbouwkd. Tijdschr. **86**, 228—232 (1974)
45. MAARSE, H. and M. C. TEN NOEVER DE BRAUW, Another catty odour compound causing air pollution, Chem. Ind. (London) 36—37 (1974)
46. NOEVER DE BRAUW, M. C. TEN and J. H. KOEMAN, Identification of chlorinated styrenes in cormorant tissues by a computerized gas chromatography-mass spectrometry system, Sci. Total Environ. **1**, 427—432 (1972/1973)
47. NOEVER DE BRAUW, M. C. TEN, C. VAN INGEN and J. H. KOEMAN, Mirex in seals, Sci. Total Environ. **2**, 196—198 (1973)
48. SICKELS, TH. W. and D. T. STAFFORD, Environmental sampling method. Rapid identification by GC/MS/DS and spectral search, Int. Lab. May/June 1977, 23—28
49. BLACK, D. R., R. A. FLATH and R. TERANISHI, Membrane molecular separators for gas chromatographic-mass spectrometric interfaces, J. Chromatogr. Sci. **7**, 284—289 (1969)
50. BEMELMANS, J. M. H. and M. C. TEN NOEVER DE BRAUW, The presence of chloroanisoles in tainted tobacco, Sci. Total Environ. **3**, 126—128 (1974)
51. RAPP, U. and M. HÖHN, Capillary GC/MS and Exact Mass Determinations, Application Note No. 26, Varian MAT GmbH Bremen (1978)
52. NOEVER DE BRAUW, M. C. TEN, C. VAN INGEN and P. J. GROENEN, Identification and quantitative determination of trace amounts of nitrosamines in cigarette smoke and cured meat products, in "Application de la spectrométrie de masse (sm) et de la résonance magnétique nucléaire (rmn) dans les industries alimentaires", XV Symp. Int., Bologna, Italy 1975, 99—111
53. GROENEN, P. J., M. W. DE COCK-BETHBEDER, R. J. G. JONK and C. VAN INGEN, Further studies on the occurrence of volatile N-nitrosamines in meat products by combined gas chromatography and mass spectrometry, Proc. 2nd Int. Symp. Nitrite Meat Prod., Zeist 1976, 227—237

54. GROENEN, P. J., R. J. G. JONK, C. VAN INGEN and M. C. TEN NOEVER DE BRAUW, Determination of eight volatile nitrosamines in thirty cured meat products with capillary gas chromatography-high resolution mass spectrometry: the presence of nitrosodiethylamine and the absence of nitrosopyrrolidine, I.A.R.C. Sci. Publ. **14**, 321—331 (1976)
55. CROSS, A. D. and R. A. JONES, An introduction to practical infra-red spectroscopy, 3rd ed., Butterworths, London 1969, pp. 102
56. NAKANISHI, K., Infra-red absorption spectroscopy, practical, Holden-Day Inc., San Francisco 1962, pp. 230
57. BELLAMY, L. J., The infra-red spectra of complex molecules, 3rd ed., Chapman & Hall, London 1975, pp. 418
58. —, Sadtler Standard Spectra, perfumes & flavors series, Sadtler Res. Lab. Inc., Philadelphia 1965, 600 spectra
59. MORRIS, W. W., High resolution infra-red spectra of fragrance and flavor compounds, 99 spectra, J. Assoc. Off. Anal. Chem. **56**, 1037—1064 (1973)
60. HORÁK, M., O. MOTL, J. PLÍVA, V. HEROUT und F. ŠORM, Die Terpene; Sammlung der Spektren und physikalischen Konstanten, Akademie-Verlag, Berlin 1960—1963
61. MITZNER, B. M., E. T. THEIMER and S. K. FREEMAN, The infra-red spectra of monoterpenes and related compounds, Appl. Spectrosc. **19**, 169—185 (1965)
62. MITZNER, B. M., V. J. MANCINI, S. LEMBERG and E. T. THEIMER, Infra-red spectra of monoterpenes and related compounds, II. Terpene alcohols, Appl. Spectrosc. **22**, 34—53 (1968)
63. MITZNER, B. M. and V. J. MANCINI, Infra-red spectra of oxygenated terpenes, Am. Perfum. Cosmet. **84**, 37—54 (1969)
64. WENNINGER, J. A., R. L. YATES and M. DOLINSKY, High resolution infra-red spectra of some naturally occurring sesquiterpene hydrocarbons, I. J. Assoc. Off. Anal. Chem. **50**, 1313—1325 (1967); II. ibid. **53**, 949—963 (1970)
65. —, Documentation of molecular spectroscopy, Butterworths, London 1956—1973, more than 20,000 rim-coded spectral cards
66. —, Selected infra-red spectral data, Thermodynamics Res. Center, College Station, Texas 1942
67. —, Sadtler Standard Spectra, total series, Sadtler Res. Lab. Inc., Philadelphia 1959
68. —, Molecular formula list of compounds; names and references to published infra-red spectra, Am. Soc. Testing Mat., Philadelphia 1969
69. SHANKARANARAYANA, M. L., K. O. ABRAHAM, B. RAGHAVAN and C. P. NATARAJAN, Infra-red identification of compounds separated by gas and thin layer chromatography: applications to flavor analysis, CRC Crit. Rev. Food Sci. Nutr. **6**, 271—315 (1975)
70. JENNINGS, W. G., Isolement et caractérisation des composés responsables de la flaveur, Ind. Aliment. Agric. **87**, 537—541 (1970)
71. COPIER, H. and J. H. VAN DER MAAS, Micro infra-red spectrometry of gas chromatographic fractions, Spectrochim. Acta, Part A **23a**, 2699—2700 (1967)
72. KLEIN, W. J. DE, Infra-red spectroscopic identification of compounds separated by gas-liquid chromatography, using a potassium bromide micropellet technique, Z. Anal. Chem. **246**, 294—297 (1969)
73. BRADY, R. F., Analysis of microgram quantities of organic vapors by combined capillary-column gas chromatography and vapor phase infrared spectrometry, Anal. Chem. **47**, 1425 to 1428 (1975)
74. CROOKS, J. E., D. L. GERRARD and W. F. MADDAMS, Simple direct combination of gas chromatography and vapor phase infra-red spectrometry, Anal. Chem. **45**, 1823—1827 (1973)
75. BROWN, R. A., J. M. KELLIHER, J. J. HEIGL and C. W. WARREN, Rapid scan infra-red spectrometer for operation with support coated open tubular or packed column gas chromatography, Anal. Chem. **43**, 353—358 (1971)
76. —, Spectrotherm Corporation, The Spectrotherm Model ST-10 GC-IR Automated System, Spectrotherm Corporation 3040 Olcott Street, Santa Clara, Calif. 95051 U.S.A.
77. LOW, M. J. D. and S. K. FREEMAN, Infra-red interferometry in flavor analysis, J. Agric. Food Chem. **16**, 525—528 (1968)
78. LOW, M. J. D., H. MARK and S. H. CHAN, Fourier transform spectroscopy in flavor analysis, V. Infra-red spectra of vapors, J. Agric. Food Chem. **22**, 488—490 (1974)

79. Low, M. J. D., Infra-red Fourier Transform spectroscopy in flavor analysis, IV. Spectra of gas chromatography fractions, J. Agric. Food Chem. **19**, 1124—1127 (1971)
80. Penzias, G. J., Gas chromatograph peaks identified on-line by a new grating infra-red spectrophotometer, Anal. Chem. **45**, 890—895 (1973)
81. Lephardt, J. O. and B. J. Bulkin, On-the-fly gas chromatography-infrared spectrometry using a cholesteric liquid crystal-effluent interface, Anal. Chem. **45**, 706—710 (1973)
82. Welti, D., Infra-red vapour spectra, Heyden & Son Ltd., London (1970)
83. Freeman, S. K., Applications of laser Raman spectroscopy to natural products research, J. Agric. Food Chem. **21**, 521—525 (1973)
84. Bulkin, B. J., K. Dill and J. J. Dannenberg, Raman spectra of trapped gas chromatograph effluents, Anal. Chem. **43**, 974—975 (1971)
85. Freeman, S. K. and D. W. Mayo, Application of Laser-excited Raman spectroscopy to organic chemistry, I. Raman spectra of some acyclic monoterpenes, Appl. Spectrosc. **23**, 610—614 (1969)
86. Oertel, R. P. and D. V. Myhre, Raman spectra-structure correlations for pyrazines; new method for obtaining spectra of trapped nanoliter gas chromatograph fractions, Anal. Chem. **44**, 1589—1593 (1972)
87. Jackman, L. M. and S. Sternhell, Applications of nuclear magnetic resonance in organic chemistry, 2nd ed., Pergamon Press, Oxford 1969, pp. 456
88. Bhacca, N. S., L. F. Johnson and J. N. Shoolery, High resolution NMR spectra catalog, Varian Associates, Palo Alto. Vol. I, 1962, pp. 40 and 368 PMR spectra. Vol. II, 1963, pp. 62 and 332 PMR spectra
89. Brügel, W., Nuclear magnetic resonance spectra and chemical structure, Vol. I, The spectral NMR parameters of compounds with analyzed spectra, Steinkopff Verlag, Darmstadt 1967, pp. 235
90. Dixon, C. W., C. T. Malone and G. R. Umbreit, A scheme for the preparative chromatographic isolation and concentration of trace components in natural products using 4-inch diameter columns, J. Chromatogr. **35**, 475—488 (1968)
91. Tsuda, T., Y. Ojika, M. Izuda, I. Fujishima and D. Ishii, Combination of GLC and nuclear magnetic resonance, J. Chromatogr. **69**, 194—197 (1972)
92. —, Varian Instruments Application Report NMR-2, Proton NMR spectra of microgram quantities, Varian Instrument Division, 611 Hansen Way, Palo Alto, California U.S.A.
93. —, Interpretive Spectroscopy, ed. Freeman, S. K., Reinhold Publ. Co., New York 1965, pp. 295
94. Silverstein, R. M. and G. C. Bassler, Spectrometric identification of organic compounds, 2nd ed., Wiley & Sons, New York 1967, pp. 256
95. Drawert, F., R. Felgenhauer und G. Kupfer, Reaktions-Gaschromatographie, Angew. Chem. **72**, 555—559 (1960)
96. Kokes, R. J., H. Tobin and P. H. Emmett, New microcatalytic-chromatographic technique for studying catalytic reactions, J. Am. Chem. Soc. **77**, 5860—5862 (1955)
97. Beroza, M., Determination of the chemical structure of microgram amounts of organic compounds by gas chromatography, Anal. Chem. **34**, 1801—1811 (1962)
98. Beroza, M. and R. A. Coad, Reaction gas chromatography, J. Gas Chromatogr. **4**, 199—216 (1966)
99. McKeag, R. G. and F. W. Hougen, Simple micro-reactor for subtractive gas chromatography, Anal. Chem. **49**, 1078—1079 (1977)
100. Nigam, R. N., A dual-channel micro-reactor for reaction gas chromatography, J. Chromatogr. **119**, 620—624 (1976)
101. Haken, J. K., D. K. M. Ho and M. K. Withers, The identification of peaks in gas chromatography by use of subtractor columns. Part 1: The removal of aldehydes and ketones, J. Chromatogr. Sci. **10**, 566—569 (1972)
102. Cronin, D. A., The use of precolumn reactions for the selective removal of compounds bearing alcohol or carbonyl functional groups in the gas chromatographic analysis of mixtures, J. Chromatogr. **64**, 25—37 (1972)
103. Ashes, J. R. and J. K. Haken, Anomolous behavior of some aminoalkyl polysiloxanes as stationary phases in gas chromatography, Anal. Chem. **45**, 1131—1136 (1973)

104. Appleyard, J. and J. K. Haken, An evaluation of polyalkylimines as specific carbonyl abstractors in gas chromatography, J. Chromatogr. **99**, 319—328 (1974)
105. Moore, B. P. and W. V. Brown, 3,4,5-Trimethoxybenzylhydrazine, an efficient subtraction reagent for aldehydes and ketones in gas chromatography, J. Chromatogr. **121**, 279—284 (1976)
106. Withers, M. K., The identification of peaks in gas chromatography by the use of subtractor columns. II The action of free fatty acid phase on aldehydes, J. Chromatogr. **66**, 249—253 (1972)
107. Bierl, B. A., M. Beroza and W. T. Ashton, Reaction loops for reaction gas chromatography. Subtraction of alcohols, aldehydes, ketones, epoxides and acids and carbon-skeleton chromatography of polar compounds, Mikrochim. Acta **1969**, 637—653
108. Andreikova, L. G. and L. A. Kogan, Identification and quantitative determinations of phenols in mixtures by reaction-gas chromatography, Zh. Anal. Khim. **29**, 344—346 (1974)
109. Regnier, F. E. and J. C. Huang, Identification of some oxygen containing functional groups by reaction gas chromatography, J. Chromatogr. Sci. **8**, 267—271 (1970)
110. —, I.A.R.C. monographs on the evaluation of the carcinogenic risk of chemicals to man, International Agency for Research on Cancer, Lyon; vol. 1, 80—86 (1972); ibid vol. 2, 41—47 (1974)
111. Hefendehl, F. W., Die Abscheidung von Terpenalkoholen durch Reaktions-Gaschromatographie mit Borsäure, Naturwissenschaften **51**, 138—139 (1964)
112. Haken, J. K., The micro-analyses of solvent mixtures by reaction gas chromatography, J. Oil Col. Chem. Assoc. **56**, 28—33 (1973)
113. Walsh, J. T. and Ch. Merritt Jr., Qualitative functional group analysis of gas chromatographic effluents, Anal. Chem. **32**, 1378—1381 (1960)
114. Casu, B. and L. Cavallotti, A simple device for qualitative functional group analysis of gas chromatographic effluents, Anal. Chem. **34**, 1514—1516 (1962)
115. Hoff, J. E. and E. D. Feit, New technique for functional group analysis in gas chromatography, Anal. Chem. **36**, 1002—1008 (1964)
116. Drawert, F., W. Heimann, R. Emberger und R. Tressl, Gas-Chromatographische Untersuchung pflanzlicher Aromen. II. Anreicherung, Trennung und Identifizierung von Apfelaromastoffen, Chromatographia **2**, 57—66 (1969)
117. Tressl, R., F. Drawert und W. Heimann, Über die Biogenese von Aromastoffen bei Pflanzen und Früchten. V. Mitteilung: Anreicherung, Trennung und Identifizierung von Bananenaromastoffen, Z. Lebensm. Unters. Forsch. **142**, 249—263 (1970)
118. Kaminsky, E., E. Wasowicz and R. Przybylski, Detection of the functional groups of volatile organic compounds by gas chromatography, (in Polish), Chem. Anal. (Warsaw) **17**, 1307—1320 (1972)
119. Bassette, R. and C. H. Whitnah, Removal and identification of organic compounds by chemical reactions in chromatographic analysis, Anal. Chem. **32**, 1098—1100 (1960)
120. Weurman, C., Sampling in airborne odorant analysis, in "Human responses to environmental odors", Academic Press Inc., New York, San Francisco, London 1974, p. 263—328
121. Lindeman, L. P., Combination of mass spectrometer, gas chromatograph and catalytic hydrogenation for the analysis of olefins, Am. Chem. Soc., Div. Petrol. Chem., Preprints **7** (3), 15—22 (1962)
122. Teeter, R. M., C. F. Spencer, J. W. Green and L. H. Smithson, Analysis of alpha olefins using a gas chromatograph — mass spectrometer combination, J. Am. Oil Chem. Soc. **43**, 82—86 (1966)
123. Sevenants, M. R. and W. G. Jennings, Volatile components of peach II, J. Food Sci. **31**, 81—86 (1966)
124. Stanley, G. and K. E. Murray, A technique for the hydrogenolysis of sub-microgram amounts of material collected from a gas chromatograph, J. Chromatogr. **60**, 345—350 (1971)
125. Beroza, M. and R. Sarmiento, Instantaneous hydrogenation of unsaturated esters, alcohols, ethers, ketones and other compound types and determination of their separation factors, Anal. Chem. **38**, 1042—1047 (1966)
126. Schomburg, G., R. Rienäcker and R. G. Schaefer, Application of a pulse-type microreactor connected to a capillary chromatograph for the investigation of thermal and catalytic conversions, Chromatographia **5**, 532—539 (1972)

127. Mounts, T. C. and H. J. Dutton, Micro vapor-phase hydrogenation accessory for gas chromatographic analysis of fatty acid esters of glyceride oils, Anal. Chem. **37**, 641—644 (1965)
128. Brownlee, R. G. and R. M. Silverstein, A micro-preparative gas chromatograph and a modified carbon skeleton determination, Anal. Chem. **40**, 2077—2079 (1968)
129. Kepner, R. E. and H. Maarse, Hydrogenolysis of terpenes in the injection port of a gas chromatograph. I. Monoterpenes, J. Chromatogr. **66**, 229—237 (1972)
130. Cronin, D. A. and J. Gilbert, Hydrogenation and ozonolysis of submicrogram quantities of unsaturated organic compounds eluted from gas chromatographic columns, J. Chromatogr. **87**, 387—400 (1973)
131. Stanley, G. and B. H. Kennett, Reaction gas chromatography of microgram and submicrogram samples using sealed glass capillaries, J. Chromatogr. **75**, 304—307 (1973)
132. Beroza, M. and R. A. Coad, Reaction gas chromatography, in "The practice of gas chromatography", eds. Ettre, L. S. and A. Zlatkis, Interscience Publishers, New York, London, Sydney 1967, p. 461—510
133. Kubeczka, K. H., Vorrichtung zur Isolierung, Anreicherung und chemischen Characterisierung gaschromatographisch getrennter Komponenten im μg-Bereich, J. Chromatogr. **31**, 319—325 (1967)
134. Widén, K. G., Application of carbon-skeleton chromatography to the study of the constituents of essential oils. I. Evaluation of some catalysts and experimental procedure, Farm. Aikak **82**, 65—75 (1973)
135. Okamoto, T. and T. Onaka, Reaction gas chromatography determination of skeletal structure by means of reaction gas chromatography, Chem. Pharm. Bull. (Tokyo) **11**, 1086—1088 (1963)
136. Mizrahi, I. and I. C. Nigam, Reaction gas chromatography. II. Dehydrogenation of monoterpene compounds on platinum-alumina catalyst, J. Chromatogr. **25**, 230—241 (1966)
137. Nigam, I. C., Reaction gas chromatography. I. Catalytic aromatization of alicyclic and heterocyclic compounds, J. Chromatogr. **24**, 188—192 (1966)
138. Bungert-Hansing, I. und H. Jork, Strukturaufklärung sesquiterpenoider Pflanzeninhaltsstoffe. I. Mikrodehydrierung von Alkoholen des Selinan-, Eremophilan- und Cadinan-Typs, Z. Anal. Chem. **268**, 203—205 (1974)
139. Stahl, E. und T. K. B. Müller, Schwefel- und Selendehydrierungen im Mikrogrammbereich, Z. Anal. Chem. **271**, 257—264 (1974)
140. Stahl, E. und T. K. B. Müller, Katalytische Dehydrierung im Mikrogrammbereich, Z. Anal. Chem. **274**, 257—264 (1975)
141. Stahl, E., A thermomicro procedure for rapid extraction and direct application in thin-layer chromatography, Analyst **94**, 723—727 (1969)
142. Thompson, C. J., H. J. Coleman, C. C. Ward and H. T. Rall, Desulfurization as a method of identifying sulfur compounds, Anal. Chem. **32**, 424—430 (1960)
143. Thompson, C. J., H. J. Coleman, R. L. Hopkins, C. C. Ward and H. T. Rall, Identification of oxygen compounds in gas-liquid chromatographic fractions by catalytic deoxygenation, Anal. Chem. **32**, 1762—1765 (1960)
144. Beroza, M. and R. Sarmiento, Determination of the carbon skeleton and other structural features of organic compounds, Anal. Chem. **35**, 1353—1357 (1963)
145. Beroza, M. and R. Sarmiento, New catalyst and technique for analyzing fatty acids, their esters and long chain compounds by carbon-skeleton chromatography, Anal. Chem. **37**, 1040—1041 (1965)
146. Thompson, C. J., H. J. Coleman, R. L. Hopkins and H. T. Rall, Hydrogenolysis — An identification tool, J. Gas Chromatogr. **5**, 1—10 (1967)
147. Cronin, D. A. and J. Gilbert, A technique for the detection of nitrogen-containing compounds in gas chromatographic eluates by means of hydrogenolysis and colour reactions, J. Chromatogr. **89**, 209—214 (1974)
148. Stránský, Z., J. Grúz and E. Ružička, Parent hydrocarbon determinations by means of gas chromatography, J. Chromatogr. **59**, 158—160 (1971)
149. Grúz, J., Über die Einrichtung für chromatographische Strukturbestimmung von organischen Substanzen, Acta Univ. Palacki Olomuc., Fac. Rerum Nat. **41**, 151—153 (1973)
150. Kleipool, R. J. C., A. C. Tas, H. Maarse, R. Neeter and H. T. Badings, Reaction of hydrogen sulphide with 2-alkenals, Z. Lebensm. Unters. Forsch. **161**, 231—238 (1976)

151. Kleipool, R. J. C., A. C. Tas and S. van Straten, The identification of two natural beverage components, Lebensm. Wiss. Technol. **9**, 296—298 (1976)
152. Galfré, A. et G. Guiochon, Détermination du squelette carboné de terpènes par hydrogénation catalytique et chromatographie en phase gazeuse, Recherches **18**, 13—20 (1971)
153. Maarse, H., Hydrogenolysis of terpenes in the injection port of a gas chromatograph. II. Sesquiterpenes, J. Chromatogr. **106**, 369—384 (1975)
154. Beroza, M. and B. A. Bierl, Ozone generator for microanalysis, Mikrochim. Acta **1969**, 720—723
155. Moore, B. P. and W. V. Brown, Gas liquid chromatographic identification of ozonolysis fragments as a basis for micro-scale structure determinations, J. Chromatogr. **60**, 157—166 (1971)
156. Beroza, M. and B. A. Bierl, Rapid determination of olefin position in organic compounds in microgram range by ozonolysis and gas chromatography, Anal. Chem. **39**, 1131—1135 (1967)
157. Beroza, M. and B. A. Bierl, Apparatus for ozonolysis of microgram to milligram amounts of a compound, Anal. Chem. **38**, 1976—1977 (1966)
158. White, E. R., B. M. Sutton, J. E. Blank, E. Moeckel and J. E. Zarembo, Application of ozonolysis and gas chromatography for structure elucidation of substituted vulpinic acids, Anal. Chem. **44**, 1582—1585 (1972)
159. Nickell, E. Ch. and O. S. Privett, A simple, rapid micromethod for the determination of the structure of unsaturated fatty acids via ozonolysis, Lipids **1**, 166—170 (1966)
160. Davison, V. L. and H. J. Dutton, Quantitative determination of double bond positions by ozonization pyrolysis, Anal. Chem. **38**, 1302—1305 (1966)
161. Dhont, J. H., Pyrolysis and gas chromatography for the detection of the benzene ring in organic compounds, Nature **192**, 747—748 (1961)
162. Dhont, J. H., Identification of alcohols by pyrolysis, Analyst (London) **89**, 71—74 (1964)
163. Merritt Jr., Ch. and C. DiPietro, Qualitative analysis of gas chromatographic eluates by means of vapor phase pyrolysis. I. Small molecule pyrograms, Anal. Chem. **44**, 57—59 (1972)
164. Levy, R. L., Trends and advances in design of pyrolysis units for gas chromatography, J. Gas Chromatogr. **5**, 107—113 (1967)
165. Fanter, D. L., J. Q. Walker and C. J. Wolf, Pyrolysis — gas chromatography of hydrocarbons, Anal. Chem. **40**, 2168—2175 (1968)
166. Harris, W. E., Reaction gas chromatography and fatty acids, J. Chromatogr. Sci. **13**, 514—515 (1975)
167. Clark, R. G. and H. E. Nursten, The sensory analysis and identification of volatiles from walnut (*Juglans regia* L.) headspace, J. Sci. Food Agric. **28**, 69—77 (1977)
168. Fine, D. H., F. Rufeh, D. Lieb and D. P. Rounbehler, Description of the thermal energy analyzer (TEA) for trace determination of volatile and non-volatile N-nitroso compounds, Anal. Chem. **47**, 1188—1191 (1975)
169. Fine, D. H., D. P. Rounbehler, A. Rounbehler, A. Silvergleid, E. Sawicki, K. Krost and G. A. DeMarrais, Determination of dimethylnitrosamine in air, water, and soil by thermal energy analysis: measurements in Baltimore, Md., Environ. Sci. Technol. **11**, 581—584 (1977)
170. Fine, D. H. and D. P. Rounbehler, Trace analysis of volatile N-nitroso compounds by combined gas chromatography and thermal energy analysis, J. Chromatogr. **109**, 271—279 (1975)
171. James, A. T. and A. J. P. Martin, Gas-liquid partition chromatography: the separation and micro-estimation of volatile fatty acids from formic acid to dodecanoic acid, Biochem. J. **50**, 679—690 (1952)
172. Kamer, J. H. van de, K. W. Gerritsma and E. J. Wansink, Gas-liquid partition chromatography: the separation and micro-estimation of volatile fatty acids from formic acid to dodecanoic acid, Biochem. J. **61**, 174—176 (1955)
173. —, Symposium on substance-selective detectors in chromatography, Div. Anal. Chem., Amer. Chem. Soc., San Francisco, Calif. 2—3 September, 1976, J. Chromatogr. **134**, 1—165 (1977)

174. Cochrane, W. P. and R. Greenhalgh, Evaluation and comparison of selective gas chromatographic detectors for the analysis of pesticide residues, Chromatographia **9**, 255—265 (1976)
175. Mellor, N., Thermionic detectors in gas chromatography, Selective detection of phosphorus, nitrogen and sulphur compounds, J. Chromatogr. **123**, 396—399 (1976)
176. Adlard, E. R., A review of detectors for gas chromatography, Crit. Rev. Anal. Chem. **5**, 13—36 (1975)
177. Aue, W. A., Detectors for use in GC analysis of pesticides, J. Chromatogr. Sci. **13**, 329—333 (1975)
178. Leathard, D. A., Qualitative analysis by gas chromatography, Adv. Chromatogr. **13**, 266—303 (1975)
179. Pigliucci, R., W. Averill, J. E. Purcell and L. S. Ettre, The routine use of selective gas chromatographic detectors, Chromatographia **8**, 165—175 (1975)
180. Pellizzari, E. D., Electron capture detection in gas chromatography, J. Chromatogr. **98**, 323—362 (1974)
181. Murphy, M., The development and application of specific detectors, J. Chromatogr. Sci. **11**, 557—558 (1973)
182. Natusch, D. F. S and T. M. Thorpe, Element selective detectors in gas chromatography, Anal. Chem. **45**, 1185A—1194A (1973)
183. Selucky, M. L., Specific gas chromatography detectors, Part II: Electrolytic conductivity detector, Chromatographia **5**, 359—366 (1972)
184. Brazhnikov, V. V., M. V. Gur'ev and K. I. Sakodynskii, Thermionic detectors in gas chromatography, Chromatogr. Rev. **12**, 1—41 (1970)
185. Krejčí, M. and M. Dressler, Selective detectors in gas chromatography, Chromatogr. Rev. **13**, 1—60 (1970)
186. O'Donnell, J. F., Element-selective detectors for gas chromatography, Amer. Lab. February 1969, no page numbering
187. Cram, S. P. and R. S. Juvet Jr., Gas Chromatography, Anal. Chem. **48**, 411R—442R (1976); for Detectors, see 418R—420R
188. Cram, S. P. and R. S. Juvet Jr., Gas Chromatography, Anal. Chem. **44**, 213R—241R (1972); for Detectors see 225R—227R
189. Juvet Jr., R. S. and S. P. Cram, Gas Chromatography, Anal. Chem. **46**, 101R—124R (1974); for Detectors see 107R—111R
190. Dickes, G. J. and P. V. Nicholas, Gas chromatography in food analysis, 1st ed., Butterworths, London etc. 1976, p. 43—60
191. Ševčík, J., Detectors in gas chromatography, 1st ed., Elsevier Scientific Publishing Company, Amsterdam 1975, pp. 192
192. David, D. J., Gas chromatographic detectors, 1st ed., Wiley-Interscience, New York 1974, pp. 295
193. Hartmann, C. H., Gas chromatography detectors, Anal. Chem. **43**, 113A—125A (1971)
194. Jentzsch, D. und E. Otte, Detektoren in der Gas-Chromatographie, 1st ed., Akademische Verlagsgesellschaft, Frankfurt am Main 1970, pp. 454
195. Karmen, A., Specific detection of halogens and phosphorus by flame ionization, Anal. Chem. **36**, 1416—1421 (1964)
196. Giuffrida, L., A flame ionization detector highly selective and sensitive to phosphorus — A sodium thermionic detector, J. Assoc. Off. Anal. Chem. **47**, 293—300 (1964)
197. Karmen, A. and L. Giuffrida, Enhancement of the response of the hydrogen flame ionization detector to compounds containing halogens and phosphorus, Nature **201**, 1204—1205 (1964)
198. Giuffrida, L., N. F. Ives and D. C. Bostwick, Gas chromatography of pesticides — Improvements in the use of special ionization detection systems, J. Assoc. Off. Anal. Chem. **49**, 8—21 (1966)
199. Aue, W. A., C. W. Gehrke, R. C. Tindle, D. L. Stalling and C. D. Ruyle, Application of the alkali-flame detector to nitrogen containing compounds, J. Gas Chromatogr. **5**, 381—382 (1967)
200. Ives, N. F. and L. Giuffrida, Investigation of thermionic detector response for the gas chromatography of P, N, As, and Cl organic compounds, J. Assoc. Off. Anal. Chem. **50**, 1—4 (1967)

201. HARTMANN, C. H., Alkali-flame detector for organic nitrogen compounds, J. Chromatogr. Sci. **7**, 163—167 (1969)
202. BREIMER, D. D. and J. M. VAN ROSSUM, Rapid and sensitive gas chromatographic determination of hexobarbital in plasma of man using a nitrogen detector, J. Chromatogr. **88**, 235—243 (1974)
203. DONIKE, M., J. JAENICKE, D. STRATMANN und W. HOLLMANN, Chromatografischer Nachweis von Stickstoff-haltigen Pharmaka in wäßrigen Lösungen mit dem Stickstoffdetektor, J. Chromatogr. **52**, 237—250 (1970)
204. DONIKE, M. und D. STRATMANN, Temperatur-programmierte GC Analyse stickstoffhaltiger Pharmaka: die Reproduzierbarkeiten der Retentionszeiten und der Mengen bei automatischer Injektion. II. Die Screeningprozedur für flüchtige Dopingmittel bei den Spielen der XX. Olympiade München 1972, Chromatographia **7**, 182—189 (1974)
205. RIEDMANN, M., Gaschromatographisches Screening von Nitrosaminen in Lebensmitteln mit dem Stickstoff-Flammenionisationsdetektor, J. Chromatogr. **88**, 376—380 (1974)
206. RIEDMANN, M., Specific gas chromatographic determination of phenothiazines and barbiturate tranquillizers with the nitrogen flame ionization detector, J. Chromatogr. **92**, 55—59 (1974)
207. MARTENS, F. K., M. A. MARTENS, T. SÖYLEMOZOGLU and A. HEYNDRICKX, Significance of the rubidium bromide thermionic detector equipped with a gate electrode in the analysis of halogenated dithiocarbamate derivatives, J. Chromatogr. **140**, 86—92 (1977)
208. VERGA, G. R. and F. POY, Gas chromatography of nitrogen- and phosphorus-containing compounds. A new, high-sensitivity, variable selectivity detector, J. Chromatogr. **116**, 17—27 (1976)
209. GOUGH, T. A. and K. SUGDEN, A study of the stability of a nitrogen-selective thermionic detector, J. Chromatogr. **86**, 65—71 (1973)
210. HOODLESS, R. A., M. SARGENT and R. D. TREBLE, Optimization of the performance of a three-electrode thermionic detector for the detection of phosphorus, sulphur and nitrogen compounds by gas chromatography, J. Chromatogr. **136**, 199—212 (1977)
211. CHAMBERLAIN, A. T., A simple method for rapid alkali-flame ionization detector optimization, J. Chromatogr. **116**, 180—181 (1976)
212. KOLB, B. and J. BISCHOFF, A new design of a thermionic nitrogen and phosphorus detector for GC, J. Chromatogr. Sci. **12**, 625—629 (1974)
213. KOLB, B., M. AUER and P. POSPISIL, Reaction mechanism in an ionization detector with tunable selectivity for carbon, nitrogen and phosphorus, J. Chromatogr. Sci. **15**, 53—63 (1977)
214. HARTIGAN, M. J., J. E. PURCELL, M. NOVOTNÝ, M. L. MCCONNELL and M. L. LEE, Analytical performance of a novel nitrogen sensitive detector and its application with glass open tubular columns, J. Chromatogr. **99**, 339—348 (1974)
215. VASILIADES, J. and K. C. BUSH, Gas liquid chromatographic determination of therapeutic and toxic levels of amitryptiline in human serum with a nitrogen-sensitive detector, Anal. Chem. **48**, 1708—1711 (1976)
216. LAWRY, J. D., L. J. WILLIAMSON and V. A. RAISYS, Micro method for the gas chromatographic determination of serum theophylline utilizing an organic nitrogen sensitive detector, J. Chromatogr. **143**, 83—88 (1977)
217. BAILEY, E., M. FENOUGHTY and L. RICHARDSON, Automated high-resolution gas chromatographic analysis of psychotropic drugs in biological fluids using open-tubular glass capillary columns. I. Determination of Nomifensine in human plasma, J. Chromatogr. **131**, 347—355 (1977)
218. BAKER, J. K., Identification and chemical classification of drugs based on the relative response of a nitrogen selective detector and a flame ionization detector in gas chromatographic analysis, Anal. Chem. **49**, 906—908 (1977)
219. LUBKOWITZ, J. A., J. L. GLAJCH, B. P. SEMONIAN and L. B. ROGERS, Study of the nitrogen response mode of the thermionic rubidium silicate detector, J. Chromatogr. **133**, 37—47 (1977)
220. BURGETT, C. A., D. H. SMITH and H. B. BENTE, The nitrogen-phosphorus detector and its applications in gas chromatography, J. Chromatogr. **134**, 57—64 (1977)

221. CRAVEN, A., Simplified version of the AFID for nitrogen mode operation, Anal. Chem. **42**, 1679—1680 (1970)
222. WALRADT, J. P., Simultaneous use of nitrogen, sulphur, and flame ionization detectors as an ancillary aid in GCMS, in "Techniques of combined gas chromatography/mass spectrometry: applications in organic analysis", ed. MCFADDEN, W. H., 1st ed., John Wiley and Sons, New York 1973, p. 414—416
223. COWARD, R. F. and P. SMITH, The detection of nitrogenous compounds using a thermionic detector, J. Chromatogr. **61**, 329—333 (1971)
224. STEWART, W. H., An easily constructed alkali flame ionization detector, Anal. Chem. **44**, 1547—1548 (1972)
225. COULSON, D. M., Electrolytic conductivity detector for gas chromatography, J. Gas Chromatogr. **3**, 134—137 (1965)
226. COULSON, D. M., Selective detection of nitrogen compounds in electrolytic conductivity gas chromatography, J. Gas Chromatogr. **4**, 285—287 (1966)
227. PATCHETT, G. G., Evaluation of the electrolytic conductivity detector for residue analysis of nitrogen-containing pesticides, J. Chromatogr. Sci. **8**, 155—158 (1970)
228. RHOADES, J. W. and D. E. JOHNSON, Gas chromatography and selective detection of N-nitrosamines, J. Chromatogr. Sci. **8**, 616—617 (1970)
229. RHOADES, J. W. and D. E. JOHNSON, Method for the determination of N-nitrosamines in tobacco smoke condensate, J. Natl. Cancer Inst. **48**, 1841—1843 (1972)
230. SEN, N. P., L. A. SCHWINGHAMER, B. A. DONALDSON and W. F. MILES, N-Nitrosodimethylamine in fish meal, J. Agric. Food Chem. **20**, 1280—1281 (1972)
231. ESSIGMAN, J. M. and P. ISSENBERG, Gas chromatographic determination of volatile nitrosamines in food, J. Food Sci. **37**, 684—688 (1972)
232. GREENHALGH, R. and W. P. COCHRANE, Comparative gas chromatographic response of organophosphorus compounds containing nitrogen and nitrogen compounds with alkali-flame and electrolytic conductivity detectors, J. Chromatogr. **70**, 37—44 (1972)
233. PALFRAMAN, J. F., J. MACNAB and N. T. CROSBY, An evaluation of the alkali-flame ionization detector and the Coulson electrolytic conductivity detector in the analysis of N-nitrosamines in foods, J. Chromatogr. **76**, 307—319 (1973)
234. HALL, R. C., A highly sensitive and selective micro-electrolytic conductivity detector for gas chromatography, J. Chromatogr. Sci. **12**, 152—160 (1974)
235. VON RAPPARD, E., G. EISENBRAND and R. PREUSSMANN, Selective detection of N-nitrosamines by gas chromatography using a modified microelectrolytic conductivity detector in the pyrolytic mode, J. Chromatogr. **124**, 247—255 (1976)
236. PAPE, B. E., D. H. RODGERS and T. C. FLYNN, The electrolytic conductivity detector as an element-selective gas chromatographic detector, J. Chromatogr. **134**, 1—24 (1977)
237. COULSON, D. M. and L. A. CAVANAGH, Automatic chloride analyzer, Anal. Chem. **32**, 1245 to 1247 (1960)
238. COULSON, D. M., L. A. CAVANAGH, J. K. DE VRIES and B. WALTHER, Microcoulometric gas chromatography of pesticides, J. Agric. Food Chem. **8**, 399—402 (1960)
239. MARTIN, R. L., Fast and sensitive method for determination of nitrogen — Selective nitrogen detector for gas chromatography, Anal. Chem. **38**, 1209—1213 (1966)
240. CASSILL, C. C., R. P. STANOVICK and R. F. COOK, A specific gas chromatographic method for residues of organic nitrogen pesticides, Residue Reviews **26**, 63—87 (1968)
241. MCCORMACK, A. J., S. C. TONG and W. D. COOKE, Sensitive selective gas chromatography detector based on emission spectrometry of organic compounds, Anal. Chem. **37**, 1470—1476 (1965)
242. MCLEAN, W. R., D. L. STANTON and G. E. PENKETH, A quantitative tunable element-selective detector for gas chromatography, Analyst (London) **98**, 432—442 (1973)
243. DAGNALL, R. M., T. S. WEST and P. WHITEHEAD, Use of the microwave-excited emissive detector for gas chromatography for quantitative measurement of inter-element ratios, Anal. Chem. **44**, 2074—2078 (1972)
244. SIMPSON, C. F., A multielement selective detector for gas chromatography, Process Biochem. **10**, 13—16 + 23 (1975)

245. JONG, H. J. DE, Microwave discharge and afterflow excited emission spectra of organic molecules and their fragments in the 2000—7000 Å region — Application to gas chromatographic analysis, Thesis, University of Leiden 1973
246. DALEN, J. P. J. VAN, P. A. DE LEZENNE COULANDER and L. DE GALAN, Optimization of the microwave-induced plasma as an element-selective detector for non-metals, Anal. Chim. Acta **94**, 1—19 (1977)
247. DRAEGER, H. and B. DRAEGER, Detection or determination of compounds containing phosphorus and (or) sulfur in air or other gases, Ger. 1.133.918 (1962), Chem. Abstr. **57**, 14444A (1962)
248. CRIDER, W. L., Hydrogen flame emission spectrophotometry in monitoring air for sulfur dioxide and sulfuric acid aerosol, Anal. Chem. **37**, 1770—1773 (1965)
249. BRODY, S. S. and J. E. CHANEY, Flame photometric detector — The application of a specific detector for phosphorus and for sulfur compounds sensitive to nanogram quantities, J. Gas Chromatogr. **4**, 42—46 (1966)
250. VAN STRATEN, S., FL. DE VRIJER and J. C. DE BEAUVESER, Volatile compounds in food, 4th ed., Central Institute for Nutrition and Food Research TNO, Zeist, The Netherlands 1977, pp. 325
251. GRICE, H. W., M. L. YATES and D. J. DAVID, Response characteristics of the Melpar flame photometric detector, J. Chromatogr. Sci. **8**, 90—94 (1970)
252. MIZANY, A. I., Some characteristics of the Melpar flame photometric GC detector in the sulfur mode, J. Chromatogr. Sci. **8**, 151—154 (1970)
253. SELUCKY, M. L., Specific gas chromatography detectors, Chromatographia **4**, 425—434 (1971)
254. STUBBS, R. C., Response of the flame photometric detector to sulfur compounds, Amer. Gas Assoc. Operation Section Proc. D53—D55 (1971)
255. DEVONALD, B. H., R. S. SERENIUS and A. D. MCINTYRE, Evaluation of the flame photometric detector for analysis of sulphur compounds, Pulp and Paper Magazine of Canada **73**, 50—53 (1972)
256. SUGIYAMA, T., Y. SUZUKI and T. TAKEUCHI, Intensity characteristics of S_2 emission for sulfur compounds with flame photometric detector, J. Chromatogr. Sci. **11**, 639—641 (1973)
257. GREENHALGH, R. and M. A. WILSON, Optimization and response of the Pye flame photometric detector to some insecticides in the phosphorus and sulphur modes, J. Chromatogr. **128**, 157—162 (1976)
258. CRIDER, W. L. and R. W. SLATER, Flame-luminescence intensification and quenching detector (FLIQD) in gas chromatography, Anal. Chem. **41**, 531—533 (1969)
259. ZEHNER, J. M. and R. A. SIMONAITIS, Improved technique for gas chromatographic analysis of S-containing organic compounds using the flame photometric detector, J. Chromatogr. Sci. **14**, 348—350 (1976)
260. BLOMBERG, L., Gas chromatographic separation of some sulphur compounds on glass capillary columns using flame photometric detection, J. Chromatogr. **125**, 389—397 (1976)
261. RUPPRECHT, W. E. and T. R. PHILLIPS, The utilization of fuel-rich flames as sulphur detectors, Anal. Chim. Acta **47**, 439—449 (1969)
262. PATTERSON, P. L., R. L. HOWE and A. ABU-SHUMAYS, Dual-flame photometric detector for sulphur and phosphorus compounds in gas chromatograph effluents, Anal. Chem. **50**, 339 to 345 (1978)
263. PATTERSON, P. L., Comparison of quenching effects in single- and dual-flame photometric detectors, Anal. Chem. **50**, 345—349 (1978)
264. BURGETT, C. A. and L. E. GREEN, Improved flame photometric detection without solvent flameout, J. Chromatogr. Sci. **12**, 356—359 (1974)
265. FARWELL, S. O. and R. A. RASMUSSEN, Limitations of the FPD and ECD in atmospheric analysis — A review, J. Chromatogr. Sci. **14**, 224—234 (1976)
266. DÖHLER, W., L. HUBER und P. SEIFRIED, Die Anwendung Flammenphotometrischer Detektoren in der Analytik schwefelhaltiger Substanzen, GIT Fachz. Lab. **21**, 937—943 (1977)
267. MARUYAMA, M. and M. KAKEMOTO, Behavior of organic sulfur compounds in flame photometric detectors, J. Chromatogr. Sci. **16**, 1—7 (1978)
268. BURNETT, C. H., D. F. ADAMS and S. O. FARWELL, Relative FPD responses for a systematic group of sulfur-containing compounds, J. Chromatogr. Sci. **16**, 68—73 (1978)

269. BURCHFIELD, H. P., D. E. JOHNSON, J. W. RHOADES and R. J. WHEELER, Selective detection of phosphorus, sulfur, and halogen compounds in the gas chromatography of drugs and pesticides, J. Gas Chromatogr. **3**, 28—34 (1965)

270. ADAMS, D. F., G. A. JENSEN, J. P. STEADMAN, R. K. KOPPE and T. J. ROBERTSON, Improved sulfur-reacting microcoulometric cell for gas chromatography, Anal. Chem. **38**, 1094—1096 (1966)

271. GIUFFRIDA, L. and N. F. IVES, Microcoulometric gas chromatography: improvements in instrumentation, J. Assoc. Off. Anal. Chem. **52**, 541—545 (1969)

272. ŠEVČÍK, J., Coulometric titration of compounds containing sulphur and chlorine, after their gas chromatographic separation and detection by the flame ionization detector, Chromatographia **4**, 102—105 (1971)

273. COCHRANE, W. P., B. P. WILSON and R. GREENHALGH, Determination of sulphur- and chlorine-containing pesticides with an electrolytic conductivity detector, J. Chromatogr. **75**, 207—218 (1973)

274. DRESSLER, M. and J. JANÁK, Detection of sulphur compounds with an alkali flame ionization detector, J. Chromatogr. Sci. **7**, 451—452 (1969)

275. DRESSLER, M., V. MARTINŮ and J. JANÁK, Response of the alkali flame ionization detector to silicon, tin, and lead compounds, J. Chromatogr. **59**, 429—433 (1971)

276. AUE, W. A., K. O. GERHARDT and S. LAKOTA, Determination of heteroelement content in organics by alkali flame gas-liquid chromatography, J. Chromatogr. **63**, 237—247 (1971)

277. ŠEVČÍK, J., The detection mechanism and the response character of the thermionic detector using an alkali salt, Chromatographia **6**, 139—148 (1973)

278. HOODLESS, R. A., M. SARGENT and R. D. TREBLE, Sulphur response of the alkali flame-ionization detector, Analyst (London) **101**, 757—758 (1976)

279. BACHE, C. A. and D. J. LISK, Selective emission spectrometric determination of nanogram quantities of organic bromine, chlorine, iodine, phosphorus, and sulfur compounds in a helium plasma, Anal. Chem. **39**, 786—789 (1967)

280. HOUPT, P. M. et H. COMPAAN, Une nouvelle méthode spectrographique (émission) pour l'identification de traces de matières organiques contenant des halogènes et du mercure dans les fractions obtenues par chromatographie en phase gazeuse, Analusis **1**, 27—33 (1972)

281. DAGNALL, R. M., T. S. WEST and P. WHITEHEAD, A simple atmospheric microwave-excited emissive detector for gas chromatography. Anal. Chim. Acta **60**, 25—35 (1972)

282. KOJIMA, T., Y. SEO and J. SATO, Selective gas-chromatographic detector with ion-selective membrane electrodes. IV. Determination of thiophenic compound peaks in the selective gas-chromatographic detection of sulfur compounds using ion-selective electrodes as detectors, Bunseki Kagaku **23**, 1389—1393 (1974) (Chem. Abstr. **82**, 149028r (1975))

283. OAKS, D. M., H. HARTMANN and K. P. DIMICK, Analysis of sulfur compounds with electron capture/hydrogen flame dual channel gas chromatography, Anal. Chem. **36**, 1560—1565 (1964)

284. LOVELOCK, J. E. and S. R. LIPSKY, Electron affinity spectroscopy — A new method for the identification of functional groups in chemical compounds separated by gas chromatography, J. Amer. Chem. Soc. **82**, 431—433 (1960)

285. LOVELOCK, J. E., Affinity of organic compounds for free electrons with thermal energy — Its possible significance in biology, Nature **189**, 729—732 (1961)

286. AUE, W. A. and S. KAPILA, The electron capture detector — Controversies, Comments and Chromatograms, J. Chromatogr. Sci. **11**, 255—263 (1973)

287. GEE, M. G., D. G. LAND and D. ROBINSON, Simultaneous analysis of 2,3,4,6-tetrachloroanisole, pentachloroanisole and the corresponding chlorophenols in biological tissue, J. Sci. Food Agric. **25**, 829—834 (1974)

288. BÄCHMANN, K., W. EMIG, J. RUDOLPH und D. TSOTSOS, Die gleichzeitige Verwendung von FID und ECD im Anschluß an eine Kapillarsäule, Chromatographia **10**, 684—686 (1977)

289. BRECHBÜHLER, B., L. GAY and H. JAEGER, A micro electron capture detector for temperature programmed analysis with capillary columns in a wide range of applications, in "Glass capillary chromatography", Second Intern. Symp., Hindelang 1977, ed. KAISER, R. E., Inst. of Chromatography, Bad Durkheim 1977, p. 53—72

290. Eklund, G., B. Josefsson and C. Roos, Gas-liquid chromatography of monosaccharides at the picogram level using glass capillary columns; trifluoroacetyl derivatization and electron-capture detection, J. Chromatogr. **142**, 575—585 (1977)
291. Krijgsman, W. and C. G. van de Kamp, Determination of chlorophenols by capillary gas chromatography, J. Chromatogr. **131**, 412—416 (1977)
292. Rejthar, L. and K. Tesařík, Performance characteristics of a system containing a capillary column and an electron-capture detector, J. Chromatogr. **131**, 404—407 (1977)
293. De Boer, A. G., J. Röst-Kaiser, H. Bracht and D. D. Breimer, Assay of underivatized nitrazepam and clonazepam in plasma by capillary gas chromatography applied to pharmacokinetic and bioavailability studies in humans, J. Chromatogr. **145**, 105—114 (1978)
294. Dressmann, R. C. and E. F. McFarren, A sample-bottle purging method for the determination of vinyl chloride in water at submicrogram per liter levels, J. Chromatogr. Sci. **15**, 69—72 (1977)
295. Karmen, A., Specific detection of subnanogram quantities of organophosphorus by flame ionization, J. Gas Chromatogr. **3**, 336—339 (1965)
296. Karagözler, A. E. and C. F. Simpson, Direct quantitative analysis using the flame-ionization detector, J. Chromatogr. **150**, 329—338 (1978)
297. Gough, T. A., M. A. Pringuer and C. J. Woollam, Response of a hydrogen-rich flame ionization detector to some chlorinated hydrocarbons, J. Chromatogr. **150**, 533—536 (1978)
298. Popjak, G., A. E. Lowe, D. Moore, L. Brown and F. A. Smith, Scintillation counter for the measurement of radioactivity of vapors in conjunction with gas-liquid chromatography, J. Lipid Res. **1**, 29—39 (1959)
299. Cacace, F. and I. UlHaq, Radiometric analysis of tritiated organic compounds by means of vapor phase chromatography, Science **131**, 732—733 (1960)
300. Cacace, F., A. Guerino and I. UlHaq, Gas-chromatographic separation and radiometric analysis of high-boiling substances marked with C^{14}, Ann. Chimica (Rome) **50**, 915—919 (1960)
301. Winkelman, J. W. and A. Karmen, Use of an ionization chamber for measuring radioactivity in gas chromatography effluents, Anal. Chem. **34**, 1067—1071 (1962)
302. Drawert, F. und O. Bachmann, Neuere Methoden zur Trennung und kontinuierlichen Messung von ^{14}C-Verbindungen in der Gasphase, Angew. Chem. **75**, 717—722 (1963)
303. Aliprandi, B. and F. Cacace, Labeling of organic compounds by substitution with tritium oxide, Ann. Chimica (Rome) **50**, 931—934 (1960)
304. Drawert, F., A. Rapp und H. Ullemeyer, Radio-gaschromatographische Analyse von Hydriervorgängen, Restaktivitäten von Trennsäulen, Veresterungen, Extraktionen und der Stoffwechsel-Leistungen von Hefen unter Verwendung einer Apparatur nach dem Baustein-Prinzip, Chem.-Ztg./Chem. Apparatur **88**, 379—384 (1964)
305. Miller III, A., R. A. Scanlan, J. S. Lee and L. M. Libbey, Quantitative and selective gas chromatographic analysis of dimethyl- and trimethylamine in fish, J. Agric. Food Chem. **20**, 709—711 (1972)
306. Miller III, A., R. A. Scanlan, L. M. Libbey, H. Petropakis and A. F. Anglemier, Quantitative determination of dimethyl- and trimethylamine in fish protein concentrate, J. Agric. Food Chem. **21**, 451—453 (1973)
307. Miller III, A., R. A. Scanlan, J. S. Lee and L. M. Libbey, Volatile compounds produced in sterile fish muscle (*Sebastes melanops*) by *Pseudomonas putrefaciens*, *Pseudomonas fluorescens* and an *Achromobacter* species, Appl. Microbiol. **26**, 18—21 (1973)
308. Dumont, J. P., S. Roger et J. Adda, Mise en évidence d'un composé à hétérocycle azoté responsable d'un défaut d'arôme dans le Gruyère de Comte, Le Lait **55**, 479—487 (1975)
309. Cronin, D. A. and P. Stanton, 2-Methoxy-3-sec-butylpyrazine — An important contributor to carrot aroma, J. Sci. Food Agric. **27**, 145—151 (1976)
310. Mussinan, C. J., R. A. Wilson and I. Katz, Isolation and identification of pyrazines in pressure-cooked beef, J. Agric. Food Chem. **21**, 871—872 (1973)
311. Ishiguro, S. and S. Sugawara, Comparison of volatile N-containing compounds in the smoke of lamina and midrib of flue-cured tobacco leaves, Agric. Biol. Chem. **41**, 377—382 (1977)
312. Vickroy, D. G., The characterization of cigarette smoke from Cytrel® smoking products and its comparison to smoke from flue-cured tobacco, Beitr. Tabakforsch. **8**, 415—421 (1976)

313. KOLB, B., Application of an automated head-space procedure for trace analysis by gas chromatography, J. Chromatogr. **122**, 553—568 (1976)
314. GUERIN, M. R., G. OLERICH and A. D. HORTON, Routine gas chromatographic component profiling of cigarette smoke for the identification of biologically significant constituents, J. Chromatogr. Sci. **12**, 385—391 (1974)
315. HOSHIKA, Y., Gas chromatographic determination of lower aliphatic primary amines as their fluorine-containing Schiff bases, Anal. Chem. **49**, 541—543 (1977)
316. DAY Jr., E. W., T. GOLAB and J. R. KOONS, Determination of micro quantities of the C_1—C_4 primary and secondary amines by electron affinity detection, Anal. Chem. **38**, 1053—1057 (1966)
317. WALLE, T., Quantitative gas chromatography of picogram quantities of amines as 2,4-dinitrophenyl derivatives, Acta Pharm. Suecica **5**, 367—384 (1968)
318. MOSIER, A. R., C. E. ANDRE and F. G. VIETS Jr., Identification of aliphatic amines volatilized from cattle feedyard, Environ. Sci. Technol. **7**, 642—644 (1973)
319. WALLE, T. and H. EHRSSON, Quantitative gas chromatographic determination of picogram quantities of amino and alcoholic compounds by electron capture detection. I. Preparation and properties of the heptafluorobutyryl derivatives, Acta Pharm. Suecica **7**, 389—406 (1970) (Chem. Abstr. **73**, 123556a (1970))
320. WALLE, T. and H. EHRSSON, Quantitative gas chromatographic determination of picogram quantities of amino and alcoholic compounds by electron capture detection. II. Application to plasma and urine samples, Acta Pharm. Suecica **8**, 27—38 (1971) (Chem. Abstr. **75**, 18134e (1971))
321. BROOKS, J. B., W. B. CHERRY, L. THACKER and C. C. ALLEY, Analysis by gas chromatography of amines and nitrosamines produced in vivo and in vitro by *Proteus mirabilis*, J. Infect. Diseases **126**, 143—153 (1972)
322. HOSHIKA, Y., Gas chromatographic separation of lower aliphatic primary amines as their sulphur-containing Schiff bases using a glass capillary column, J. Chromatogr. **136**, 253—258 (1977)
323. DREWS, B., G. BÄRWALD und H. J. NIEFIND, Gaschromatographische Bestimmung flüchtiger Schwefelverbindungen, Monatsschr. Brau. **22**, 140—144 (1969)
324. HARRISON, G. A. F. and C. M. COYNE, Flame photometric determination at the parts per thousand million level of some volatile sulphur compounds, J. Chromatogr. **41**, 453—455 (1969)
325. BÄRWALD, G. und G. MIGLIO, Flüchtige Schwefelverbindungen in Naturhopfen, Hopfenextrakt und Isoextrakt, Monatsschr. Brau. **23**, 288—289 (1970)
326. RICHARDSON, P. and M. MOCEK, Determination of sulfur compounds in beer by the flame photometric method, Proc. Am. Soc. Brew. Chem. 128—135 (1971)
327. RICHARDSON, P. J. and M. MOCEK, Artifacts in GLC determination of volatile sulphur compounds in beer, J. Inst. Brew. **79**, 26—28 (1973)
328. McCOWEN, N. M., S. R. PALAMAND and W. A. HARDWICK, Use of flame photometric detector for the determination of sulfur compounds in beer and brewing raw materials, Proc. Am. Soc. Brew. Chem. 136—148 (1971)
329. JANSEN, H. E., J. STRATING and W. M. WESTRA, The gas chromatographic determination of traces of hydrogen sulphide and alkane thiols in beer, J. Inst. Brew. **77**, 154—159 (1971)
330. SCHREIER, P., F. DRAWERT und A. JUNKER, Flüchtige Schwefelverbindungen des Bieraromas, Brauwissenschaft **27**, 205—207 (1974)
331. SCHREIER, P., F. DRAWERT und A. JUNKER, Gaschromatographisch-massenspektrometrische Untersuchung flüchtiger Inhaltsstoffe des Weines. II. Thioäther-Verbindungen des Weinaromas, Z. Lebensm. Unters. Forsch. **154**, 279—284 (1974)
332. SCHREIER, P., F. DRAWERT und A. JUNKER, Gaschromatographische Bestimmung der Inhaltsstoffe von Gärungsgetränken. IV. Schwefelhaltige Aromakomponente des Bieres, Brauwissenschaft **28**, 73—79 (1975)
333. SCHREIER, P., F. DRAWERT und A. JUNKER, Gaschromatographisch-massenspektrometrische Untersuchung flüchtiger Inhaltsstoffe des Weines. IV. Nachweis sekundärer Amide in Wein, Z. Lebensm. Unters. Forsch. **157**, 34—37 (1975)
334. SCHREIER, P. und F. DRAWERT, Schwefelhaltige Aromastoffe, in "Geruch- und Geschmackstoffe", ed. DRAWERT, F., 1st ed., Verlag Hans Carl, Nürnberg 1975, p. 113—131

335. Pickett, J. A., Studies on flavour-active sulphur-components of hops and beer, Proc. Anal. Div. Chem. Soc. **13**, 215—217 (1976)
336. Ronkainen, P., J. Denslow and O. Leppänen, The gas chromatographic analysis of some volatile sulfur compounds, J. Chromatogr. Sci. **11**, 384—390 (1973)
337. Miglio, G., Über die Entstehung flüchtiger Schwefelkomponenten beim Erhitzen aus schwefelhaltigen Aminosäuren und anderen Verbindungen, Branntweinwirtschaft **111**, 381—385 (1971)
338. Wilson, R. A., C. J. Mussinan, I. Katz and A. Sanderson, Isolation and identification of some sulfur chemicals present in pressure-cooked beef, J. Agric. Food Chem. **21**, 873—876 (1973)
339. Kinlin, T. E., R. Muralidhara, A. O. Pittet, A. Sanderson and J. P. Walradt, Volatile components of roasted filberts, J. Agric. Food Chem. **20**, 1021—1028 (1972)
340. Heidemann, F. H. and J. Wismer-Pedersen, Flavor components of cured pork products, Proc. 22nd Europ. Meeting of Meat Research Workers, Malmö 1976, Swedish Meat Research Centre, Kävlinge, Sweden 1976, p. A6: 1—6
341. Persson, T. and E. C. F. von Sydow, Aroma of canned beef — Gas chromatographic and mass spectrometric analysis of the volatiles, J. Food Sci. **38**, 377—385 (1973)
342. Persson, T. and E. C. F. von Sydow, The aroma of canned beef — Processing and formulation aspects, J. Food Sci. **39**, 406—413 (1974)
343. Qvist, I. H. and E. C. F. von Sydow, Unconventional proteins as aroma precursors — Chemical analysis of the volatile compounds in heated soy, casein, and fish protein model systems, J. Agric. Food Chem. **22**, 1077—1084 (1974)
344. Qvist, I. H. and E. C. F. von Sydow, Unconventional proteins as aroma precursors — Chemical analysis of the volatile compounds in unheated and heated rapeseed protein model systems, J. Agric. Food Chem. **24**, 437—442 (1976)
345. Qvist, I. H., E. C. F. von Sydow and C. Å. Åkesson, Unconventional proteins as aroma precursors — Instrumental and sensory analysis of the volatile compounds in a canned meat product containing soy or rapeseed protein, Lebensm. Wiss. Technol. **9**, 311—320 (1976)
346. Manning, D. J., Sulphur compounds in relation to Cheddar cheese flavour, J. Dairy Res. **41**, 81—87 (1974)
347. Freeman, G. G., Distribution of flavour components in onion (*Allium cepa* L.), leek (*Allium porrum*) and garlic (*Alium sativum*), J. Sci. Food Agric. **26**, 471—481 (1975)
348. Freeman, G. G., R. J. Whenham, R. Self and J. Eagles, Volatile flavour components of parsley leaves (*Petroselinum crispum* (Mill.) Nyman), J. Sci. Food Agric. **26**, 465—470 (1975)
349. Badings, H. T., H. Maarse, R. J. C. Kleipool, A. C. Tas, R. Neeter and M. C. ten Noever de Brauw, Formation of odorous compounds from hydrogen sulphide and 2-butenal, Z. Lebensm. Unters. Forsch. **161**, 53—59 (1976)
350. Tressl, R., D. Bahri, M. Holzer and T. Kossa, Formation of flavor components in asparagus — 2. Formation of flavor components in cooked asparagus, J. Agric. Food Chem. **25**, 459—463 (1977)
351. Tressl, R., M. Holzer and M. Apetz, Formation of flavor components in asparagus — 1. Biosynthesis of sulfur-containing acids in asparagus, J. Agric. Food Chem. **25**, 455—459 (1977)
352. Kawabata, M., K. Ohtsuki, H. Kokura and Y. Wakahara, Determination of dimethylsulfide in the headspace vapor of green tea by gas chromatography, Agric. Biol. Chem. **41**, 2285—2287 (1977)
353. Groenen, P. J. and L. J. van Gemert, Flame photometric detection of volatile sulphur compounds in smoke from various types of cigarettes, J. Chromatogr. **57**, 239—246 (1971)
354. Guerin, M. R., Detection of sulfur-containing compounds in the gas phase of cigarette smokes, Anal. Lett. **4**, 751—759 (1971)
355. Horton, A. D. and M. R. Guerin, Quantitative determination of sulfur compounds in the gas phase of cigarette smoke, J. Chromatogr. **90**, 63—70 (1974)
356. Guerin, M. R., G. Olerich and A. D. Horton, Routine gas chromatographic component profiling of cigarette smoke for the identification of biologically significant constituents, J. Chromatogr. Sci. **12**, 385—391 (1974)
357. Bertsch, W., F. Shumbo, R. C. Chang and A. Zlatkis, Preparation of high resolution nickel open tubular columns, Chromatographia **7**, 128—134 (1974)

358. Bertsch, W., F. Hsu and A. Zlatkis, Heart cutting technique in high resolution gas chromatography applied to sulfur compounds in cigarette smoke, Anal. Chem. **48**, 928—931 (1976)
359. Stevens, R. K., A. E. O'Keeffe and G. C. Ortmann, Absolute calibration of a flame photometric detector to volatile sulfur compounds at subpart-per-million levels, Environ. Sci. Technol. **3**, 652—655 (1969)
360. Stevens, R. K., J. D. Mulik, A. E. O'Keeffe and K. J. Krost, Gas chromatography of reactive sulfur gases in air at the parts-per-billion level, Anal. Chem. **43**, 827—831 (1971)
361. Stevens, R. K. and A. E. O'Keeffe, Modern aspects of air pollution monitoring, Anal. Chem. **42**, 143A—148A (1970)
362. Tourres, D. A., Trace analysis of hydrogen sulphide and sulphur dioxide by rapid, on line, gas chromatography, Chromatographia **5**, 441—448 (1972)
363. Rasmussen, R. A., Analysis of trace organic sulfur compounds in air, Amer. Lab. December (1972), no page numbering
364. Barynin, J. A. M. and M. J. G. Wilson, Outdoor experiments on smell, Atmosph. Environ. **6**, 1—11 (1972)
365. Pecsar, R. E. and C. H. Hartmann, Automated gas chromatographic analysis of sulfur pollutants, J. Chromatogr. Sci. **11**, 492—502 (1973)
366. Bruner, F., A. Liberti, M. Possanzini and I. Allegrini, Improved gas chromatographic method for the determination of sulfur compounds at the ppb level in air, Anal. Chem. **44**, 2070—2074 (1972)
367. Bruner, F., P. Ciccioli and F. Di Nardo, Further developments in the determination of sulfur compounds in air by gas chromatography, Anal. Chem. **47**, 141—144 (1975)
368. Bruner, F., P. Ciccioli and G. Bertoni, Analysis of sulphur compounds in environmental samples with specific detection and selective columns, J. Chromatogr. **120**, 200—202 (1976)
369. Souza, T. L. C. De, and S. P. Bhatia, Development of calibration systems for measuring total reduced sulfur and sulfur dioxide in ambient concentrations in the parts-per-billion range, Anal. Chem. **48**, 2234—2240 (1976)
370. Ciccioli, P., G. Bertoni, E. Brancaleoni, R. Fratarcangeli and F. Bruner, Evaluation of organic pollutants in the open air and atmospheres in industrial sites using graphitized carbon black traps and gas chromatographic-mass spectrometric analysis with specific detectors, J. Chromatogr. **126**, 757—770 (1976)
371. Goode, K. A., Gas chromatographic determination of sulphur compounds in North Sea natural gas by a flame photometric detector, J. Inst. Petrol. **56**, 33—41 (1970)
372. Hoshika, Y. and Y. Ida, Gas chromatographic determination of sulphur compounds in town gas, J. Chromatogr. **134**, 423—432 (1977)
373. Banwart, W. L. and J. M. Bremner, Identification of sulfur gases evolved from animal manures, J. Environ. Qual. **4**, 363—366 (1975)
374. Schaefer, J., Sampling, characterization and analysis of malodours, Agric. Environ. **3**, 122—127 (1977)
375. Blanchette, A. R. and A. D. Cooper, Determination of hydrogen sulfide and methyl mercaptan in mouth air at the parts-per-billion level by gas chromatography, Anal. Chem. **48**, 729—731 (1976)
376. Schait, A. and J. Cuzněr, The use of the microcoulometric detector in the brewing laboratory, Proc. Am. Soc. Brew. Chem. 29—39 (1970)
377. Martin, R. L. and J. A. Grant, Determination of sulfur-compound distributions in petroleum samples by gas chromatography with a coulometric detector, Anal. Chem. **37**, 644—649 (1965)
378. Martin, R. L. and J. A. Grant, Determination of thiophenic compounds by types in petroleum samples, Anal. Chem. **37**, 649—657 (1965)
379. Applebury, T. E. and M. J. Schaer, Analysis of kraft pulp mill gases by process gas chromatography, J. Air Pollut. Control Assoc. **20**, 83—86 (1970)
380. Vitenberg, A. G., L. M. Kuznetsova, I. L. Butaeva and M. D. Inshakov, Gas chromatographic determination of trace amounts of sulfur compounds in industrial effluents, Anal. Chem. **49**, 128—133 (1977)

381. SCHILLER, R. G. and R. B. BRONSKY, Gas chromatographic analysis for hydrogen sulfide, organic sulfides, mercaptans, and carbon dioxide in hydrocarbon mixtures using an electrolytic conductivity detector, J. Chromatogr. Sci. **15**, 541—545 (1977)
382. NOVOTNÝ, J. and A. MÜLLER, Gas chromatographic determination of sulphur compounds in gases using a single-flame thermionic detector, J. Chromatogr. **148**, 211—218 (1978)
383. HŘIVNÁČ, M., W. FRISCHKNECHT and L. ČECHOVÁ, Gas chromatographic multidetector coupled to a glass capillary column, Anal. Chem. **48**, 937—940 (1976)
384. TRESSL, R., M. HOLZER and M. APETZ, Biogenesis of volatiles in fruit and vegetables, in "Aroma Research", Proc. int. Symp. Aroma Research, Zeist (Neth.) 1975, eds. MAARSE, H. and P. J. GROENEN, Pudoc, Wageningen (Neth.) 1975, p. 41—62
385. WITHYCOMBE, D. A., J. P. WALRADT and A. HRUZA, Simultaneous detection of nitrogen and sulfur containing flavor volatiles, in "Phenolic, sulfur, and nitrogen compounds in food flavors", eds. CHARALAMBOUS, G. and I. KATZ, ACS Symposium Series **26**, 85—95 (1976), Amer. Chem. Soc., Washington 1976
386. FOLKES, D. J. and J. W. GRAMSHAW, Volatile constituents of white bread crust, J. Food Technol. **12**, 1—8 (1977)
387. SHILMAN, A. and L. W. SCHAPER, The analysis of small concentrations of biacetyl, J. Gas Chromatogr. **4**, 486 (1966)
388. HARRISON, G. A. F., W. J. BYRNE and E. COLLINS, The determination of diacetyl and 2,3-pentanedione in beer head space by gas chromatography, J. Inst. Brew. **71**, 336—341 (1965)
389. STEFFEN, P., Die Bestimmung der vicinalen Diketone im Bier, Nahrung **13**, 697—702 (1969)
390. HAUKELI, A. D. and S. LIE, The influence of 2-acetohydroxy acids on the determination of vicinal diketones in beer and during fermentation, J. Inst. Brew. **77**, 538—543 (1971)
391. POSTEL, W. und U. GÜVENC, Gaschromatographische Bestimmung von Diacetyl, Acetoin und 2,3-Pentandion in Wein, Z. Lebensm. Unters. Forsch. **161**, 35—44 (1976)
392. SCANLAN, R. A. and R. C. LINDSAY, Quantitative determination of diacetyl by electron capture, J. Food Sci. **33**, 440—441 (1968)
393. RONKAINEN, P. and S. BRUMMER, Analysis of α-hydroxy ketones by gas chromatography, J. Chromatogr. **45**, 341—345 (1969)
394. LOVE, J. L. and R. G. C. DONNELLY, Electron capture gas-liquid chromatographic determination of acetone in butter, J. Assoc. Off. Anal. Chem. **59**, 1140—1141 (1976)
395. SMAGULA, J. M. and W. J. BRAMLAGE, Determination of acetaldehyde in apple tissue by 2,4-dinitrophenylhydrazone formation, J. Food Sci. **42**, 48—51 (1977)
396. HOSHIKA, Y., Selective and sensitive gas chromatographic determination of benzaldehyde, J. Chromatogr. **129**, 436—439 (1976)
397. BURKINSHAW, P. M., E. D. MORGAN and C. F. POOLE, Detection of alcohols at the femtogram level as pentafluorophenyldimethylsilyl ethers, J. Chromatogr. **132**, 548—551 (1977)
398. KAWAHARA, F. K., Gas chromatographic analysis of mercaptans, phenols, and organic acids in surface waters with use of pentafluorobenzyl derivatives, Environ. Sci. Technol. **5**, 235 to 239 (1971)
399. DAVIS, B., Crown ether catalyzed derivatization of carboxylic acids and phenols with pentafluorobenzylbromide for electron capture gas chromatography, Anal. Chem. **49**, 832—834 (1977)
400. HEENAN, M. P. and N. K. MCCALLUM, A new method for the gas chromatographic detection of phenols, J. Chromatogr. Sci. **12**, 89—90 (1974)
401. DEO, P. G. and P. H. HOWARD, Phosphorylation of alcohols/phenols for gas-liquid chromatographic separation and flame photometric detection, J. Assoc. Off. Anal. Chem. **61**, 210—213 (1978)
402. CURTIS, R. F., C. DENNIS, J. M. GEE, M. G. GEE, N. M. GRIFFITHS, D. G. LAND, J. L. PEEL and D. ROBINSON, Chloroanisoles as a cause of musty taint in chickens and their microbiological formation from chlorophenols in broiler house litters, J. Sci. Food Agric. **25**, 811—828 (1974)
403. PARR, L. J., M. G. GEE, D. G. LAND, D. ROBINSON and R. F. CURTIS, Chlorophenols from wood preservatives in broiler house litter, J. Sci. Food Agric. **25**, 835—841 (1974)
404. ENGEL, C., A. P. DE GROOT and C. WEURMAN, Tetrachloroanisol — A source of musty taste in eggs and broilers, Science **154**, 270—271 (1966)

405. BEMELMANS, J. M. H. and M. C. TEN NOEVER TE BRAUW, Chloroanisoles as off-flavor components in eggs and broilers, J. Agric. Food Chem. **22**, 1137—1138 (1974)
406. HOSHIKA, Y., Gas chromatographic determination of styrene as its dibromide, J. Chromatogr. **136**, 95—103 (1977)
407. HAUSDORF, H. H., Selective infrared detectors for chromatography, J. Chromatogr. **134**, 131—146 (1977)
408. DRAWERT, F., A. RAPP und H. ULLEMEYER, Radio-gaschromatographische Untersuchung der Stoffwechselleistungen von Hefen (*Saccharomyces* und *Schizosaccharomyces*) in der Bildung von Aromastoffen, Vitis **6**, 177—197 (1967)
409. TRESSL, R. und F. DRAWERT, II. Anwendung physikalisch-chemischer Methoden auf Probleme der Biogenese von Aromastoffen, Ernähr. Umsch. **17**, 396—400 (1970)
410. TRESSL, R., R. EMBERGER, F. DRAWERT und W. HEIMANN, Über die Biogenese von Aromastoffen bei Pflanzen und Früchten. XI. Mitt.: Einbau von ^{14}C-Leucin und -Valin in Bananenaromastoffe, Z. Naturforsch. Teil B **25**, 704—707 (1970)
411. DRAWERT, F., R. EMBERGER, R. TRESSL and U. PRENZEL, Application of reaction radio-gas-chromatography to problems of flavour analysis, Chromatographia **5**, 319—323 (1972)
412. TRESSL, R., F. DRAWERT und U. PRENZEL, Anwendung der Reaktions-Radio-Gas-Chromatographie zur Untersuchung der Biogenese von Aromastoffen, Chromatographia **6**, 7—13 (1973)
413. TRESSL, R. and F. DRAWERT, Biogenesis of banana volatiles, J. Agric. Food Chem. **21**, 560—565 (1973)
414. HEIMANN, W., K.-H. FRANZEN, A. RAPP und H. ULLEMEYER, Untersuchungen über Entstehungsweise flüchtiger Substanzen bei der Lipoxygenase-Linolsäure-Reaktion, Z. Lebensm. Unters. Forsch. **162**, 109—114 (1976)
415. TRESSL, R., D. BAHRI et M. HOLZER, Sur la formation et l'analyse par chromatographie en phase gazeuse et spectrométrie de masse de constituants aromatiques dans quelques végétaux, Ann. Falsif. Expert. Chim. **69**, 715—723 (1976)
416. DRAWERT, F. und J. BEIER, Einbau von ^{14}C-Essigsäure in Hopfenbitterstoffe, Phytochemistry **13**, 2749—2753 (1974)
417. DRAWERT, F. und J. BEIER, Über die Biosynthese der Hopfenbitterstoffe II. Mitt.: Gas-Chromatographie und Reaktions-Radio-Gas-Chromatographie von Bitterstoffen, Vorstufen und Metaboliten, Chromatographia **7**, 273—282 (1974)
418. MIGLIO, G., Die Solventextraktion und ihre Problematik bei der Anwendung auf Destillate in der gaschromatographischen Analyse, Branntweinwirtschaft **111**, 494—496 (1971)
419. HATANAKA, A., T. KAJIWARA and J. SEKIYA, Biosynthesis of *trans*-2-hexenal in chloroplasts from *Thea sinensis*, Phytochemistry **15**, 1125—1126 (1976)
420. KOEHLER, P. E., M. E. MASON and J. A. NEWELL, Formation of pyrazine compounds in sugar-amino acid model systems, J. Agric. Food Chem. **17**, 393—396 (1969)
421. REAZIN, G. H., S. BALDWIN, H. S. SCALES, H. W. WASHINGTON and A. ANDREASEN, Determination of the congeners produced from ethanol during whisky maturation, J. Assoc. Off. Anal. Chem. **59**, 770—776 (1976)

5. Coordination of sensory and instrumental analysis

5.1. Introduction

L. J. VAN GEMERT

In a journal's editorial note, STEWART (1) called establishing the importance of the volatile components in foods by means of sensory analysis the challenge in flavour research. BURR (2) stated that flavour research is a special field within the broad area of research on biologically active compounds. According to MOULTON (3), the final approach for the determination of the compounds responsible for the odour of a food is a bioassay based on stimulation of the human nose, and NURSTEN (4) wrote that there can be no scientific evaluation of flavour without sensory assessment. All these statements clearly indicate the vital importance of sensory analysis in flavour research, if that should be more than just the identification and quantification of chemical compounds.

STAHL (5) was one of the first to discuss the use of sensory panels in relation to instrumental analysis of odorous compounds of foods. Since then, numerous articles have been published in this connection. However, there are only a few review articles concerning the relation or coordination between the two types of analysis. In a review, WICK (6) pointed out the differences in sensitivity between the GC detector (FID) and the human nose. Sensory techniques in relation to instrumental measurements were discussed by PANGBORN (7). Discussions on different aspects of the relation between sensory and instrumental analysis were presented at a few symposia (8–10). In 1971, a book on the principles and techniques of flavour research was published, that contains a chapter on sensory analytical methods (11).

On the problem of reproducing a natural food aroma by means of identified volatile compounds, literature shows a diversity of approaches. For instance, according to POLAK (12), the reconstitution of aromas can be achieved by a flavourist using his experience and knowledge in combination with the results of physico-chemical studies. This method could be called pragmatic and intuitive. Another view on this subject is given by MOSKOWITZ (13); he describes a stepwise procedure for what he calls an optimum strategy towards the duplication of a flavour. Here, the use of psychophysical and psychometric analysis is essential, and the psychologist rather than the flavourist plays the key role. This approach is of a much more basic nature. In this respect the method described by VON SYDOW (14) may be called intermediate. Here the statistical testing of psychophysical models for correlation plays a central part.

Of course, the sensory method to be used depends on the purpose of the investigation and on other factors: for example, the amount of material available (15). On the other hand, one should remain aware that sensory analysis is an interdisciplinary science with its own pitfalls and difficulties. This fact has been ignored in many flavour studies. Just smelling and tasting in an informal way, without any thoughts to reproducibility and similar factors, cannot justifiably be called sensory analysis.

It is, however, not intended to describe the methodology and background of sensory analysis. The reader may be referred to the literature (e.g. 16–21). An extensively annotated bibliography on physiology, psychology, methods and applications in connection with sensory evaluation of food has been prepared by DRAKE and JOHANSSON (22, 23).

In the following sections of this chapter, first of all the use of sensory analysis as a guidance for instrumental analysis will be described. Secondly, a number of odour properties and sensory methods relevant to their measurement will be discussed.

5.2. Guidance for instrumental analysis

5.2.1. Samples

Selection of samples is of paramount importance. CHANG (24) stated that if the aim of an investigation is the determination of the chemical compounds responsible for the

aroma of a food product, it is absolutely necessary to start with samples having the desired aroma without any off-flavour or taint. And when, for instance, an off-flavour has to be examined, one has to be certain that the samples do indeed have the off-flavour, preferably to a high degree.
When studying natural products such as fruits and vegetables, which may occur in a large number of varieties, one should be aware of differences between these. An example is provided by the study of McCarthy et al. (25) concerning two banana varieties; distinct differences in flavour profiles and in vapour chromatograms were observed.
From the above it may be clear that adequate sensory measurements are needed to select the right samples and that these samples have to be described as precisely as possible.

5.2.2. Isolates and concentrates

As was pointed out in 2.1.6., it is necessary to perform sensory analysis in order to ascertain whether or not the isolate or concentrate contains the pertinent odorous compounds. Weurman (15), discussing vapour sampling for GC analysis, stated that an odour will always be found to have altered after its manipulation. This dictum undoubtedly relates to such manipulations as isolation and concentration by extraction or distillation. It is very unlikely that the odour of a food product will remain the same during preparation for chromatographic analysis. The changed substrate will result in a changed composition of the vapour phase, even if the volatile compounds have been isolated quantitatively and concentrated without any loss or artifact formation. From this point of view, the requirement formulated by Chang (24) is questionable. He says that fractionation and identification should only be allowed to proceed if an isolate is obtained which is judged by a sensory panel to have an odour closely reminiscent of the genuine flavour of the original food.
In the literature, mention of sensory evaluation of isolates and concentrates by appropriate methods is rare. One example is the study by Soedarmo Moeljohardjo (26) on the aroma of trassi (fermented fish product). Another was reported by Schultz et al. (27), who compared the organoleptic properties of extracts from an apple essence obtained with use of five different extractants. Triangle tests in combination with preference tests were conducted by Pyne and Wick (28) in comparing freshly pulped tomatoes and distillates from them. Recently, Dijkstra and Wikén (29) evaluated the distillation of a mushroom extract by measuring the number of flavour units (see 5.3.2.1.) of extract, distillate and residue. Very often it is only stated, without performing proper sensory analysis, that the isolation and concentration procedures led to a sample with a good enough odour resemblance to the original product (e.g. 30—33). A possible reason for such meagre procedures can be the already mentioned, very often found, alteration of an odour during isolation and concentration; it makes simple means such as triangle tests pointless. This test is used to determine whether samples are different or not. In fact, results of sophisticated descriptive methods are needed much more. Furthermore, it is often difficult to collect sufficient material for performing proper sensory tests, or to taste or smell the isolate or concentrate because of the poisonous or odorous nature of the solvent used. However, the restriction last mentioned

can be overcome with the aid of paper strips. But this technique requires the availability of persons with a thorough knowledge of the odour and flavour of the product plus experience in this type of assessment.
In some cases, depending on the type of product and the isolation procedure, it may be worth while to recombine the isolated volatile compounds with the residue and to compare the product thus obtained with the original one. The same principle can be applied after concentration. In this way it can be checked if the composition of the volatile compounds has changed, assuming the substrate has not.
Fractionation of a concentrate into groups on the basis of functionality is described by PALMER (34); it is accomplished by elution from silica gel at 0 °C with Freon or ether in Freon. After evaporation of most of the solvent, fractions with a characteristic odour are located by sniffing. Group separation has been used to determine whether or not compound groups such as carbonyls, sulphides, thiols, acids or bases are important — as a group — for the original product odour (e.g. 35—41). For a description and evaluation of the experimental procedures the reader is referred to the publications mentioned above and to 2.1.5. and 4.5.
Finally, VON SYDOW (14, 42) recommends the use of head space sampling techniques to avoid isolates and concentrates of volatile compounds being not representative for the odour of the product. But, even with a technique as mild as this one, alterations can occur (15).

5.2.3. Chromatographic effluent

Ever since the use of GC in aroma research, people will have smelled the components emerging from the column. DAY et al. (43) used the observed odour of eluting compounds for tentative identification. The use of professional perfumers as a human sensor at the exit of a GC column for identification purposes has been studied by FULLER et al. (44). But nowadays, due to the further development of identification techniques, smelling of GC effluents is used more specifically for odour characterization.
Since the general use of destructive detectors such as the FID, it has become rather common in aroma studies to equip a gas chromatograph with a sniffing port in parallel with the detector. Sniffing at the exit of a GC column is a difficult and often rather uncomfortable task. It can be facilitated for a seated observer by a convenient location of the sniffing port. And by protecting his nose, for instance with teflon tubings, from contact with hot surfaces. Moreover, to decrease the irritating effects of hot dry carrier gas, the effluent can be mixed with humid air (45—48). Use of a tape recorder, or the aid of an assistant, can further promote the observer's concentration on odour assessments (38, 44, 48—52). Preferably, the assessor should not see the chromatogram.
VON SYDOW and co-workers (50—52) describe the following standard procedure. Odour assessments are made at the maximum of each discernible gas chromatographic peak or shoulder and, besides, every 15 seconds. The maximum continuous sniffing period allowed for each assessor is 20 minutes. In principle, the same method is applied in our laboratory (53) with, however, one modification. Since the frequency of compounds emerging from the GC column decreases during a run, the time between two assessments increases during a run from 10 to 30 seconds. Naturally, this time interval should be adapted to experimental conditions. According to ACREE et al. (47), linger-

ing of strong odours can be prevented by mixing the effluent with a large volume of rapidly moving air. A 500-fold dilution did not appear to result in a severe reduction of the odour intensity as compared with the outcome of directly sniffing the effluent.

Stanley and Guadagni (54), Teranishi et al. (11) and McGugan and Howsam (55) correctly pointed out that it should be ascertained whether indeed all relevant odorous compounds emerge from the column during the GC analysis. The suggestion was made to trap the total effluent and to compare its odour with that of the unchromatographed sample by a few trials with two or three experienced panelists (55). Clark and Nursten (41) in this connection applied triangle tests.

A rather disquieting phenomenon has been observed by Williams and Tucknott (56). They state to have encountered situations in which the characteristic odour of the sample being examined was associated with relatively major components which, upon identification, proved to be virtually odourless. The hypothesis was put forward that significant odourous compounds, many of which are of a very polar nature, are absorbed by active surfaces near the exit of the gas chromatograph and are washed off when a major peak elutes.

Sniffing at the end of a column is widely practised in aroma studies. Nevertheless, several authors have raised objections. One of the first was Burr (2), who considered it an extremely tedious procedure with the possibility of errors due to olfactory fatigue. Objections were also expressed by others (24, 57—60). These objections can be formulated as follows: (a) rapidly eluting early peaks, which allow little time for odour assessment; (b) differences in concentration between the compound in the eluting gas stream and in the head space of the original food; (c) only one observer can assess the effluent at a time (unless multiple port devices are employed), and (d) the observer may be affected by olfactory fatigue and may become weary, and therefore inaccurate, in the course of a long run.

For these and similar reasons, it may be preferable to sniff GC fractions after trapping rather than "on-the-fly".

5.2.4. Chromatographic fractions

In principle all kinds of trapping devices — which are discussed in 3.2.6.3. — may be used in order to obtain fractions for sensory assessment. The following devices and methods have actually been used:

(a) cooled glass capillaries (61);
(b) cooled glass capillaries containing activated alumina or Celite 545 (57, 58) or lactose (62);
(c) glass medicine dropper tubes containing a 6 mm layer of granular silica gel (63);
(d) bubbling the effluent in water (24, 49, 60), other deodorized media, e.g. liquid paraffin (64), or even coffee (65), and
(e) collecting effluent in a syringe (59).

All these different devices and techniques have apparently been found satisfactory. Nevertheless, one should be cautious with the application of any trapping technique, as efficiencies can vary much (see 3.2.6.3.). Parliment and Scarpellino (66) present

in detail a scheme which, according to them, permits the flavour chemist to assess the compounds of major organoleptic importance in a systematic fashion.
Fractionation of odour components in connection with gel permeation chromatography is discussed by NOBLE (67).

5.2.5. Identified compounds and synthetic mixtures

In contrast to the situation with isolates, concentrates and (trapped) GC fractions, sufficient material of identified compounds is normally available for adequate sensory analysis. A few remarks must be made: (a) it is very important to work with pure compounds — a detailed account on purification of odorous compounds, especially by zone refining, is given by TURK and TURK (68); (b) the compounds should have storage stability or be used directly after preparation, and (c) the choice of the diluent — air, water, liquid paraffin, etc. — and the concentration can be critical, as these factors considerably influence the organoleptic properties of the compounds.

5.3. Odour measurements

5.3.1. Introduction

The most important properties of an odour are its intensity and quality (e.g. 69, 70). However, the following more refined subdivision (71) is feasible: *detectability* — the lowest concentration which creates an olfactory sensation —; *intensity* — the strength of the perceived sensation —; *quality* — the characteristic properties or kind of odour —, and *acceptability* — hedonic value —.
Although also other properties have been put forward, for instance adaptation or olfactory fatigue (72—74), the four properties just mentioned are suitable for discussions on sensory analytical methods, with the emphasis on odour measurements, in relation to instrumental analysis.
A general discussion on the human senses, including a functional analysis of perceptual phenomena, can be found elsewhere (20).

5.3.2. Detectability

5.3.2.1. General

The threshold concept plays a key role in detectability measuring. Sensory thresholds have been studied extensively in psychophysics, and appropriate methods had been developed (e.g. 75) before thresholds entered the field of aroma research. Threshold concentrations can be simply described as the lowest concentrations at which an odorous compound is just perceptible by the sense of smell.
PATTON and JOSEPHSON (76) were the first to practise threshold measurements in aroma research for determining the significance of the individual odorous compounds in foods. Their premise was that compounds with a concentration in a food exceeding

the threshold concentration are usually significant to its aroma, whereas those occurring below that threshold are not. GUADAGNI and co-workers (49, 54, 77, 78) developed the concept of the "odor unit" — the number of odour units is equal to the ratio of component concentration to threshold concentration — as a quantitative measure for determination of the relative contribution of components to total odour intensity. Working on Delicious apple essence, they showed that the components with a characteristic apple aroma also had the highest numbers of odour units. Research along this line on potato chips (79) revealed that the compound with the highest number of odour units, i.e. methional, also received the highest odour similarity score in relation to the odour of fresh potato chips.
ROTHE and THOMAS (80) and ROTHE (81) independently developed the concept of the "aroma value"; it is equal to GUADAGNI's number of odour units. Other names have also been used, for instance "odour value" (82), "unit flavour base" (83) and "flavour unit" (e.g. 29).
KÖSTER and WOUTERS (84) criticized the use of aroma values, etc., by stating that equal aroma values of different compounds do not necessarily represent an equal odour intensity. They partly based their dictum on the work of CAIN (85). Very recently, FRIJTERS (86) repeated this criticism. In some reviews (87, 88), this objection and others are discussed. It is there concluded that the use of threshold values still raises many problems; it only seems suitable for limited application. Nevertheless, aroma values can give a lot of information (89).
The discussion so far has been tacitly limited to threshold values determined in odourless media such as water and paraffin oil. HARRISON and co-workers (90—92) determined threshold values in beer. Here, the term "flavour index" is used as a measure of relative flavour importance (91). The flavour index is equal to the aroma value. However, it is now concluded that compounds showing a low flavour index are the most important and require closest production control. The most abundant source so far of flavour threshold values, about 240, is the work of MEILGAARD (93, 94) All these values, though, have solely been determined in beer.

5.3.2.2. Threshold determinations

Many different methods have been used for determining sensory thresholds for purposes of aroma research. The differences concern:

(a) sensory technique (sniffing, tasting, spraying in mouth, etc.);
(b) type of threshold (detection or recognition);
(c) solvent or diluent (water, paraffin oil, vegetable oil, beer, etc.);
(d) samples (temperature, amount available, etc.);
(e) experimental design (single stimulus, paired comparison, triangle test, time between samples, etc.);
(f) number and selection criteria of subjects (sensitivity, experience, etc.);
(g) interpretation of results (statistical testing, 50%-threshold value, etc.).

For a general discussion on threshold determinations, the reader is referred to the literature (e.g. 16, 18, 20, 70, 75). Some recent publications concerning threshold values of odorous compounds will be mentioned below.

Teranishi et al. (95) discuss the threshold values obtained over the years by the method described by Guadagni et al. (96). In this discussion, odour thresholds in air and water are compared with the molecular structure of aldehydes, ketones, pyrazines and thiazoles. Salo (97) observed synergistic, additive and suppressive effects based on odour threshold values determined for individual flavour compounds of whisky and mixtures of these. Griffiths (73) observed some very low threshold values in water for several trichloroanisoles, which could not be confirmed (98). Especially for 2,3,6-trichloroanisole, a much higher value was obtained. However, this value has recently been revised (99). An interesting aspect of the paper by Griffiths (73) are the differences found in the fatigue effect caused by some chloroanisoles.
As we already mentioned, Meilgaard (93, 94) has made an extensive study of the volatile compounds of beer. Besides threshold data, he also provided descriptions

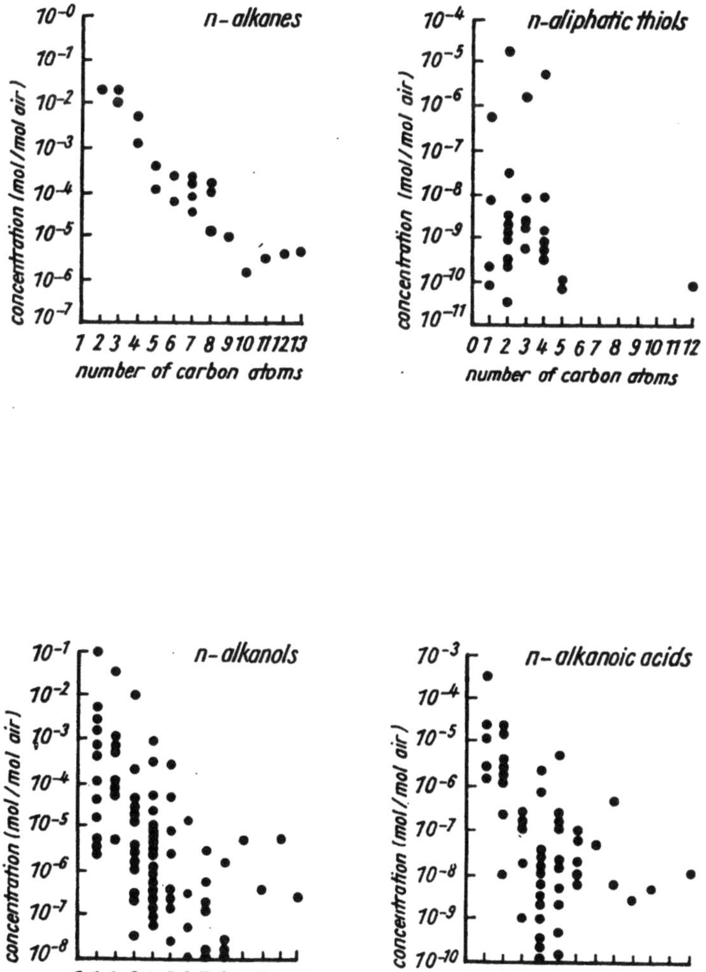

Fig. 5—1
Survey of compiled detection odour thresholds in air (107).

concerning the flavour quality of these compounds in beer. Significant interactions, including repression, enhancement and synergism were observed. ROTHE et al. (87) presented flavour or taste threshold values in water for a number of homologue series of compounds. Threshold concentrations were found in the range of 500 to 0.00002 ppm.
Recently, three different calculation methods for analysis of threshold data were compared (100). It was concluded that none was sufficiently superior to the other two. These three methods were extreme-value analysis (101, 102), probit analysis (e.g. 103), and a special computer program developed by DRAKE (104). This program has been used by ERIKSSON et al. (105) for fitting individual odour threshold curves, resulting in remarkable differences between subjects.
STAHL (106) published a compilation of odour and taste threshold values from the literature, using about 1200 data. Though VAN GEMERT and NETTENBREIJER (107) exclusively compiled odour threshold values in air and water, their report covers about 2100 data. The said workers made every effort to deal with original data only, thus avoiding any misleading duplication within the collection. A review of this compilation, as far as the odour threshold values in air are concerned, was presented by VAN GEMERT and MAARSE (108).
In order to show the wide ranges of reported threshold values for the same compounds, the compiled detection threshold values (107) for a number of homologous series are visualized in Figure 5—1.

5.3.3. Intensity

Direct measurement of intensities has often been practised on samples, prior to instrumental analysis, with the purpose of establishing significant statistical correlations between sensory and instrumental data. These measurements can concern total odour or flavour intensities, or intensities of certain aspects of odour and flavour. However, this type of correlation techniques will not be discussed here.
The GC effluent is usually described in qualitative terms only, but occasionally the odour intensities of the eluting compounds are also assessed (33, 48, 50—52, 109). The method applied is a 3- to 6-point category scale, ranging, for instance, from very weak to very strong. Trapped GC fractions have also been assessed on odour intensity with category scaling (34, 66). Information pertaining to the use of different types of scales for assessing intensities can be found elsewhere (e.g. 70, 110, 111). In this context it should be mentioned that Committee E-18 of the American Society for Testing and Materials (ASTM) has recently published recommended procedures for assessing odour intensities based on the so-called butanol scale (112—114). The odour intensity of an experimental sample is then matched against the odour intensities of a series of concentrations of 1-butanol, the reference odorant. The assigned matching points are averaged and the results are reported in parts per million 1-butanol.

5.3.4. Quality

5.3.4.1. General

JENNINGS and SEVENANTS (115), in connection with their study on Bartlett pears, launched the idea of "character impact compounds", which they defined as: natural

components of the substance in question that individually possesses aroma suggestive of and differentiating that substance. Subsequent research on peach (116, 117) and apricot (118) did not, however, reveal any character impact compounds.
DAY and LIBBEY (119) concluded from their study of Cheddar cheese flavour that the aroma is due to a balanced combination of many compounds and cannot be related to a few discrete compounds. SELF et al. (120) proposed that differences in aroma among certain foods, are not rooted in the presence of one or more unique compounds, but rather in the relative quantitative pattern of low-boiling substances. Just the same, two recent publications review some character impact compounds in fruits and vegetables (121), and a few character impact compounds containing sulphur (122).

5.3.4.2. Descriptive analysis

Odour quality can be measured by various methods (e.g. 70), the descriptive systems of which have been practised most frequently. A well-known descriptive system is the so-called flavour profile method (123). For each different series of samples, however, a separate list of descriptors has to be developed. The same applies to the method recently developed by STONE et al. (124), called the Quantitative Descriptive Analysis. A more general purpose list of descriptors has been made by HARPER and co-workers (125, 126) after an extensive review of the relevant literature (127). This list of 44 terms has been extended by VON SYDOW and co-workers (50, 51) and DRAVNIEKS (128); the latter extended it to comprise 136 terms (see Table 5—1). Furthermore, a subcommittee of the ASTM is compiling a thesaurus of odour descriptors; it presents about 200 critical words as major entries (113).
Another list of descriptors, with references to chemical compounds, has been made by ZAUSCH (129). SCHUTZ (130) used a matching-standards technique for the description of irradiation odour. LAND et al. (131) applied HARPER's list to some stimuli proposed as odour standards. CLAPPERTON et al. (132) developed a list of descriptors specifically adapted to beer. WILLIAMS et al. did the same for cider and perry (133) and Cox's Orange Pippin apples (134). Land (135) discusses "directed odour description" and mentions possible improvements of this method; MOSKOWITZ (111) also discusses the descriptive systems.
Usually, in order to get an impression of the relative importance, odour and flavour qualities of GC effluent and fractions, and those of identified compounds, are described in terms freely chosen. Several recent reviews on food aromas (136—139), feature long lists of identified compounds with their descriptions. But according to, for example, PALAMAND, the odour and flavour quality of compounds may depend on the concentration as well as on the environment or medium; he gives detailed examples (140). Such phenomena may be one of the reasons for the sometimes contradictory reports on odour and flavour qualities of given compounds. A case in point is 3,4-dimethylthiophene. According to BOELENS et al. (141), the flavour character of this compound is distinctly that of fried onion; it was also identified in this product. GALETTO and HOFFMAN (142) stated that this compound did not have any fried onion character, either smelled or tasted. Furthermore, they found a threshold value that was about 5,000 times higher.

Table 5—1
List of 136 descriptors, as developed by DRAUNIEKS (128), for characterization of various types of odours

Fragrant	Oily, fatty	Aromatic	Fruity (citrus)
Sweaty	Like mothballs	Meaty (cooked)	Fruity (others)
Almond-like	Like gasoline, solvent	Sickening	Putrid, foul, decayed
Burnt, smoky	Cooked vegetables	Musty, earthy, moldy	Woody, resinous
Herbal, green, cut grass	Sweet	Sharp, pungent, acid	Musk-like
Etherish, anesthetic	Fishy	Camphor-like	Soapy
Sour, acid, vinegar	Spicy	Light	Garlic, onion
Like blood, raw meat	Paint-like	Heavy	Animal
Dry, powdery	Rancid	Cool, cooling	Vanilla-like
Like ammonia	Minty, peppermint	Warm	Fecal (like manure)
Disinfectant, carbolic	Sulfidic	Metallic	Floral
Perfumery	Yeasty	Eucalyptus	Strawberry-like
Malty	Cheesy	Buttery	Stale
Cinnamon-like	Honey-like	Like burnt paper	Cork-like
Popcorn	Anise (licorice)	Cologne	Lavender
Incense	Turpentine (pine oil)	Caraway	Cat-urine-like
Melona (cantaloupe, honey-dew)	Fresh green vegetables	Orange (fruit)	Bark-line, birch bark
Tar-like	Medicinal	Household gas	Rose-like
Peanut butter	Celery	Leather-like	Nutty (walnut etc.)
Violets	Burnt candle	Pear (fruit)	Fried fat
Tea-leaves	Mushroom-like	Stale tobacco smoke	Wet paper-like
Wet wool, wet dog	Pineapple (fruit)	Raw cucumber	Coffee-like
Chalky	Fresh cigarette smoke	Raw potato	Peach (fruit)
Mouse-like	Laurel leaves	Beery (beer-like)	Oak wood, cognac
Pepper-like	Scorched milk	Cedarwood-like	Grapefruit
Bean-like	Sewer odor	Coconut-like	Grape-juice-like
Banana-like	Sooty	Rope-like	Eggy (fresh eggs)
Burnt ruber	Crushed weeds	Seminal (sperm-like)	Bitter
Geranium leaves	Rubbery (new rubber)	Like cleaning fluid (carbona)	Cadaverous (like dead animal)
Urine-like	Bakery (fresh bread)	Cardboard-like	Raisin-like
Lemon (fruit)	Seasoning (meat)	Crushed grass	Maple (as in syrup)
Dirty linen-like	Apple (fruit)	Chocolate	Hay
Kippery (smoked fish)	Soupy	Molasses	Kerosene
Caramel	Grainy (as grain)	Sauerkraut-like	Clove-like

The very quality of an odour is undoubtedly one of its most important properties. Thus, measurement of it can be essential for getting an indication of the importance of individual compounds, or groups of compounds, for the aroma of a food. However, judicious application of descriptive methods is certainly not easy and takes a lot of time. Illustrative for the difficulties in assigning descriptive words to odours, at least for untrained subjects, is the following. LAWLESS and CAIN (143) concluded from their recognition tests with odours that "odour memory" has unique features. They found that the odour memory system greatly differs from the verbal memory system, and more specifically that descriptive terms are more easily forgotten than odours. The complexity of the measurement of odour quality can be realized from CLARK and NURSTEN's study of the sensory term "nutty" (144). The upshot was a subdivision into five types (coconut, almond, walnut, peanut and roasted peanut), each of which was anchored to specific compounds.
Finally, it should be remarked that, according to WILLIAMS (133), the use of descriptive methods, along with the description of odour qualities, can be very useful for the development of meaningful terminology which should warrant unambiguous communication of information on the aroma of food products.

5.3.5. Acceptability

Sensory measurements concerning preferences and acceptabilty are extensively performed to serve consumer research. In aroma research, though, assessments are rarely based on hedonic value. One of the exceptions is DRAVNIEKS' work on body odours (128). Using odour quality descriptors, the GC effluent was also assessed by a 7-point hedonic category scale (145). Information regarding the theoretical background for odour preferences can be found in the literature (146, 147).

5.4. Concluding remarks

It has been stated that "even at its best, sensory testing is limited by the taster's subjectivity" (148). But this "subjectivity" is the very thing that should be measured. And that can only be done by sensory analysis. HARRIES (149) said that perhaps high variance is synonymous with a high level of sensory information. Of course, instrumental methods are objective, accurate and economical (e.g. 148), but they do not a priori yield valid information. Only sensory methods, if performed correctly, result in valid measurements of the sensations perceived. Odour and flavour are sensations (see Chapter 1.), not mixtures of chemical substances. Too often chemical studies of volatile compounds are erroneously identified with flavour or aroma research (150).
Nevertheless, if one wants to create or modify odours and flavours, it is essential to know the chemical composition plus the relations between components and perceived sensations. For this purpose sensory analysis, as a guidance to instrumental analysis, started about twenty years ago. In the beginning it was thought, or hoped, that rather simple concepts such as the number of odour units (49, 54, 77, 78) or character impact compounds (115, 121, 122) could be used to study the role of indi-

vidual compounds in connection with the flavour of foods. However, it has meanwhile been shown from research in the field of psychophysics that knowledgeable prediction of odour and flavour properties of mixtures from those of the individual compounds is very complex (151—156). Research on flavour compounds in beer (93, 94) and in whisky (97, 157) likewise revealed a complex situation.

A different approach to the role of individual compounds in mixtures is that with the aid of sophisticated statistical techniques, such as multivariate analysis. Here, the sensory data derived from food samples are related to the instrumental data (e.g. 10). Although application of these techniques will certainly give more insight in the flavour of foods in relation to chemical composition, a pertinent discussion would be beyond the framework of this monograph. However, proper use of less advanced or sophisticated techniques like the ones discussed in this chapter, can hopefully result in very useful information.

References

1. STEWART, G. F., The challenge in flavor research, Food Technol. (Chicago) **17**, 5 (1963)
2. BURR, H. K., Gas chromatography in flavor research: its uses and limitations, Food Technol. (Chicago) 18, 60—62 (1964)
3. MOULTON, D. G., Physiological aspects of olfaction, J. Food Sci. **30**, 908—916 (1965)
4. NURSTEN, H. E., The scientific evaluation of flavours, Flavour Ind. **1**, 521—522 (1970)
5. STAHL, W. H., Gas chromatography and mass spectrometry in the study of flavor, in "Chemistry of natural food flavors — a symposium", eds. MITCHELL Jr., J. H., N. J. LEINEN, E. M. MRAK and S. D. BAILEY, National Academy of Science, Washington 1957, p. 58—75
6. WICK, E. L., Chemical and sensory aspects of the identification of odor constituents in foods: a review, Food Technol. (Chicago) **19**, 827—833 (1965)
7. PANGBORN, R. M., Flavor perception. Relation of sensory to instrumental measurements, in "Progress in Food Science and Technology. Key to man's future", eds. TILGNER, D. J. and A. MORIJS, Proc. second Int. Congress Food Sci. Technol. 1966, p. 303—317
8. —, Correlation of subjective-objective methods in the study of odors and taste, American Society for Testing and Materials STP 440, American Society for Testing and Materials, Philadelphia 1968, pp. 107
9. —, "Proceedings of the International Symposium on Aroma Research", eds. MAARSE, H. and P. J. GROENEN, Pudoc, Wageningen 1975, pp. 245
10. —, "Correlating sensory objective measurements — new methods for answering old problems", eds. POWERS, J. J. and H. R. MOSKOWITZ, American Society for Testing and Materials STP 594, American Society for Testing and Materials, Philadelphia 1976, pp. 134
11. TERANISHI, R., I. HORNSTEIN, PH. ISSENBERG and E. WICK, Flavor Research, Principles and Techniques, Marcel Dekker, Inc., New York 1971, pp. 315
12. POLAK, E. H., The reconstitution of flavors, J. Am. Oil Chem. Soc. **45**, 680A—684A (1968)
13. MOSKOWITZ, H. R., Towards a resolution of dual approaches in flavor analysis, Lebensm. Wiss. Technol. **8**, 237—242 (1975)
14. SYDOW, E. VON, Contribution of individual chemical compounds to the aroma of foods, in "Geruchs- und Geschmacksstoffe", ed. DRAWERT, F., Verlag Hans Carl, Nürnberg 1975, p. 187—200
15. WEURMAN, C., Sampling in airborne odorant analysis, in "Human responses to environmental odors", eds. TURK, A., J. W. JOHNSTON and D. G. MOULTON, Academic Press, New York 1974, p. 263—328
16. AMERINE, M. A., R. M. PANGBORN and E. B. ROESSLER, Principles of sensory evaluation of food, Academic Press, New York 1965, pp. 602

17. —, Basic principles of sensory evaluation, American Society for Testing and Materials STP 433, American Society for Testing and Materials, Philadelphia 1968, pp. 105
18. —, Manual on sensory testing methods, American Society for Testing and Materials STP 434, American Society for Testing and Materials, Philadelphia 1968, pp. 77
19. Kiermeier, F. und U. Haevecker, Sensorische Beurteilung von Lebensmitteln, J. F. Bergmann, München 1972, pp. 101
20. Harper, R., Human senses in action, Churchill Livingstone, Edinburgh 1972, pp. 358
21. —, Sensorische Prüfung von Nahrungsmitteln, Arbeiten der DLG, Band 140, DLG-Verlag, Frankfurt (Main) 1974, pp. 112
22. Drake, B. and B. Johansson, Sensory evaluation of food. Annotated bibliography, SIK-Rapport No. 255, Vols. 1 and 2, Svenska Institutet för Konserveringsforskning, Göteborg 1969, pp. 148
23. Drake, B. and B. Johansson, Sensory evaluation of food. Annotated bibliography. Supplement 1968—1973, SIK-Rapport No. 350, Vols. 1 and 2, Svenska Institutet för Konserveringsforskning, Göteborg 1974, pp. 235
24. Chang, S. S., Overcoming problems in flavor component identification, Food Technol. (Chicago) **27**, no. 4, 27—39 (1973)
25. McCarthy, A. I., J. K. Palmer, C. P. Shaw and E. E. Anderson, Correlation of gas chromatographic data with flavor profiles of fresh banana fruit, J. Food Sci. **28**, 379—384 (1963)
26. Soedarmo Moeljohardjo, D., On the volatile flavour compounds of cooked trassi, a cured shrimp paste condiment of the Far East, Thesis, Agricultural Univ. of Wageningen 1972, 15—17
27. Schultz, T. H., R. A. Flath, D. R. Black, D. G. Guadagni, W. G. Schultz and R. Teranishi, Volatiles from Delicious apple essence — extraction methods, J. Food Sci. **32**, 279—283 (1967)
28. Pyne, A. W. and E. L. Wick, Volatile components of tomatoes, J. Food Sci. **30**, 192—200 (1965)
29. Dijkstra, F. Y. and T. O. Wikén, Studies on mushroom flavours 1. Organoleptic significance of constituents of the cultivated mushroom, *Agaricus bisporus*, Z. Lebensm. Unters. Forsch. **160**, 255—262 (1976)
30. Wick, E. L., J. E. Hoff, S. A. Goldblith and B. E. Proctor, The application of radiation-distillation to the production and isolation of components of beef irradiated flavor, J. Food Sci. **26**, 258—260 (1961)
31. Heinz, D. E., R. M. Pangborn and W. G. Jennings, Pear aroma: relation of instrumental and sensory techniques, J. Food Sci. **29**, 756—761 (1964)
32. Murray, K. E., J. Shipton, F. B. Whitfield, B. H. Kennett and G. Stanley, Volatile flavor components from green peas (*Pisum Sativum*) 1. Alcohols in unblanched frozen peas, J. Food Sci. **33**, 290—294 (1968)
33. Dirinck, P., L. Schreyen, F. van Wassenhove and N. Schamp, Flavour quality of tomatoes, J. Sci. Food Agric. **27**, 499—508 (1976)
34. Palmer, J. K., Separation of components of aroma concentrates on the basis of functional group and aroma quality, J. Agric. Food Chem. **21**, 923—925 (1973)
35. Minor, L. J., A. M. Pearson, L. E. Dawson and B. S. Schweigert, Chicken flavor: the identification of some chemical components and the importance of sulfur compounds in the cooked volatile fraction, J. Food Sci. **30**, 686—696 (1965)
36. Klose, A. A., H. H. Palmer, H. Lineweaver and A. A. Campbell, Direct olfactory demonstration of fractions of chicken aroma, J. Food Sci. **31**, 638—642 (1966)
37. Ziemba, Z. and Y. Mälkki, Changes in odour components of canned beef due to processing, Lebensm. Wiss. Technol. **4**, 118—122 (1971)
38. Barylko-Pikielna, N., M. Daniewski and Z. Mielniczuk, Gas chromatographic-sensory parallel characterization of methionine-glucose reaction products and their relation to meat aroma, Nahrung **18**, 125—132 (1974)
39. Pareles, S. R. and S. S. Chang, Identification of compounds responsible for baked potato flavor, J. Agric. Food Chem. **22**, 339—340 (1974)
40. Manning, D. J. and J. C. Price, Cheddar cheese aroma — the effect of selectively removing specific classes of compounds from cheese headspace, J. Dairy Res. **44**, 357—361 (1977)
41. Clark, R. G. and H. E. Nursten, The sensory analysis and identification of volatiles from walnut (*Juglans regia* L.) headspace, J. Sci. Food Agric. **28**, 69—77 (1977)

42. SYDOW, E. VON, Flavor — a chemical or psychophysical concept ?, Food Technol. (Chicago) **25**, no. 1, 40—44 (1971)
43. DAY, E. A., D. A. FORSS and S. PATTON, Flavor and odor defects of gamma-irradiated skim-milk II. Identification of volatile components by gas chromatography and mass spectrometry, J. Dairy Sci. **40**, 932—941 (1957)
44. FULLER, G. H. ,R. STELTENKAMP and G. A. TISSERAND, The gas chromatograph with human sensor: perfumer model, Ann. N.Y. Acad. Sci. **116**, 711—724 (1964)
45. DRAVNIEKS, A. and A. O'DONNELL, Principles and some techniques of high-resolution headspace analysis, J. Agric. Food Chem. **19**, 1049—1056 (1971)
46. TASSAN, C. G. and G. F. RUSSELL, Sensory and gas chromatographic profiles of coffee beverage headspace volatiles entrained on porous polymers, J. Food Sci. **39**, 64—68 (1974)
47. ACREE, T. E., R. M. BUTTS, R. R. NELSON and C. Y. LEE, Sniffer to determine the odor of gas chromatographic effluents, Anal. Chem. **48**, 1821—1822 (1976)
48. MURRAY, K. E., J. SHIPTON, F. B. WHITFIELD and J. H. LAST, The volatiles of off-flavoured unblanched green peas (*Pisum sativum*), J. Sci. Food Agric. **27**, 1093—1107 (1976)
49. GUADAGNI, D. G., S. OKANO, R. G. BUTTERY and H. K. BURR, Correlation of sensory and gas-liquid chromatographic measurements of apple volatiles, Food Technol. (Chicago) **20**, 518—521 (1966)
50. SYDOW, E. VON, J. ANDERSSON, K. ANJOU, G. KARLSSON, D. LAND and N. M. GRIFFITHS, The aroma of bilberries (*Vaccinium myrtillus* L.) II. Evaluation of the press juice by sensory methods and by gas chromatography and mass spectrometry, Lebensm. Wiss. Technol. **3**, 11—17 (1970)
51. SYDOW, E. VON and G. KARLSSON, The aroma of black currants V. The influence of heat measured by odour quality assessment techniques, Lebensm. Wiss. Technol. **4**, 152—157 (1971)
52. QVIST, I. H., E. VON SYDOW and C. Å. ÅKESSON, Unconventional proteins as aroma precursors: instrumental and sensory analysis of the volatile compounds in a canned meat product containing soy or rape-seed protein, Lebensm. Wiss. Technol. **9**, 311—320 (1976)
53. SCHAEFER, J. and L. J. VAN GEMERT, unpublished results, 1977
54. STANLEY, W. L. and D. G. GUADAGNI, Correlation of subjective and objective analysis in research on processed fruit and vegetable flavor, Int. Fruchtsaft Union. Ber. Wiss. Technol. Komm. **8**, 123—140 (1967)
55. MCGUGAN, W. A. and S. G. HOWSAM, Loss of flavor components in glc columns and a test for its occurrence, J. Agric. Food Chem. **20**, 1089 (1972)
56. WILLIAMS, A. A. and O. G. TUCKNOTT, Misleading information from odour evaluation of gas chromatographic effluents in aroma analysis, Chem. Ind. (London), 125 (1977)
57. CLARK, R. G. and D. A. CRONIN, A technique for the trapping of flavour compounds eluted from gas chromatographic columns for organoleptic evaluation, J. Sci. Food Agric. **25**, 1053 (1974)
58. CLARK, R. G. and D. A. CRONIN, A new technique for trapping and sensory evaluation of flavour volatiles, J. Sci. Food Agric. **26**, 1009—1019 (1975)
59. TUCKNOTT, O. G. and A. A. WILLIAMS, A simple method for obtaining more reliable information on the odours of components eluting from gas chromatographic columns, Chem. Ind. (London), 784 (1974)
60. PARLIMENT, T. H., A technique for the organoleptic evaluation of gas chromatographic effluents, Chem. Ind. (London), **418** (1976)
61. BEMELMANS, J. M. H. and N. L. A. TE LOO, Analysis of an earthy taint in a malt distillate, in "Proc. 4th Nordic Symposium Sensory Properties of Foods", Skövde, March 18—20 (1976), p. 129—141
62. GRAMSHAW, J. W., Use of lactose in flavour assessment of gas chromatographic effluents, Chem. Ind. (London), 1072 (1976)
63. ROSEN, A. A., C. I. MASHNI and R. S. SAFFERMAN, Recent developments in the chemistry of odour in water; the cause of earthy/musty odour, Water Treat. Exam. **19**, 106—119 (1970)
64. BADINGS, H. T., personal communication, 1977
65. PARLIMENT, T. H., W. CLINTON and R. SCARPELLINO, Trans-2-Nonenal: coffee compound with novel organoleptic properties, J. Agric. Food Chem. **21**, 485—487 (1973)
66. PARLIMENT, T. H. and R. SCARPELLINO, Organoleptic techniques in chromatographic food flavor analysis, J. Agric. Food Chem. **25**, 97—99 (1977)

67. NOBLE, A. C., Gel chromatographic fractionation of flavor components for their individual sensory evaluation, J. Agric. Food Chem. **24**, 321–323 (1976)
68. TURK, A. and J. TURK, The purity of odorant substances, in "Methods in olfactory research", eds. MOULTON, D. G., A. TURK and J. W. JOHNSTON, Academic Press, London 1975, p. 63–73
69. ENGEN, T., The sense of smell, Ann. Rev. Psychol. **24**, 187–206 (1973)
70. KÖSTER, E. P., Human psychofysics in olfaction, in "Methods in olfactory research", eds. MOULTON, D. G., A. TURK and J. W. JOHNSTON, Academic Press, London 1975, p. 345–374
71. —, Methods for measuring and evaluating odorous air pollutants at the source and in the ambient air, The Third Karolinska Institute Symposium on Environmental Health, Nord. Hyg. Tidsk. **51**, no. 2, 1–77 (1970)
72. KÖSTER, E. P., Adaptation and cross-adaptation in olfaction, Thesis, University of Utrecht 1971, pp. 212
73. GRIFFITHS, N. M., Sensory properties of the chloroanisoles, Chem. Senses Flavor **1**, 187–196 (1974)
74. LAING, D. G., On the measurement of smell, CSIRO Food Res. Q. **35**, 88–95 (1975)
75. GUILFORD, J. P., Psychometric methods, McGraw-Hill, New York 1954, pp. 597
76. PATTON, S. and D. V. JOSEPHSON, A method for determining significance of volatile flavor compounds in foods, Food. Res. **22**, 316–318 (1957)
77. FLATH, R. A., D. R. BLACK, D. G. GUADAGNI, W. H. MCFADDEN and T. H. SCHULTZ, Identification and organoleptic evaluation of compounds in Delicious apple essence, J. Agric. Food Chem. **15**, 29–35 (1967)
78. GUADAGNI, D. G., Requirements for coordination of instrumental and sensory techniques, in "Correlation of subjective-objective methods in the study of odors and taste", American Society for Testing and Materials STP 440, American Society for Testing and Materials, Philadelphia 1968, p. 36–48
79. GUADAGNI, D. G., R. G. BUTTERY and J. G. TURNBAUGH, Odour thresholds and similarity ratings of some potato chip components, J. Sci. Food Agric. **23**, 1435–1444 (1972)
80. ROTHE, M. and B. THOMAS, Aromastoffe des Brotes — Versuch einer Auswertung chemischer Geschmacksanalysen mit Hilfe des Schwellenwertes, Z. Lebensm. Unters. Forsch. **119**, 302 to 310 (1963)
81. ROTHE, M., Papierchromatographische Untersuchungen über einige Geschmacksgebende Carbonylverbindungen im Brot, Wiss. Veröff. Dtsch. Ges. Ernähr. **10**, 128–135 (1963)
82. MULDERS, E. J., The odour of white bread. IV. Quantitative determination of constituents in the vapour and their odour values, Z. Lebensm. Unters. Forsch. **151**, 310–317 (1973)
83. KEITH, E. S. and J. J. POWERS, Determination of flavor threshold levels and sub-threshold, additive, and concentration effects, J. Food Sci. **33**, 213–218 (1968)
84. KÖSTER, E. P. und O. WOUTERS, Entwicklung der sensorischen Methodik, Ernähr. Umsch. **17**, 349–354 (1970)
85. CAIN, W. S., Odor intensity differences in the exponent of the psychophysical function, Percept. Psychophys. **6**, 349–354 (1969)
86. FRIJTERS, J. E. R., Some psychophysical notes on the use of the odour unit number, Proc. 2nd Weurman Flavour Research Symposium, Norwich, United Kingdom 1978, in print
87. ROTHE, M., G. WÖLM, L. TUNGER und H. J. SIEBERT, Schwellenkonzentrationen von Aromastoffen und ihre Nutzung zur Auswertung von Aromaanalysen, Nahrung **16**, 483–495 (1972)
88. ROTHE, M., Aroma values — a useful concept ?, in "Proceedings of the International Symposium on Aroma Research", eds. MAARSE, H. and P. J. GROENEN, Pudoc, Wageningen 1975, p. 111–119
89. SYDOW, E. VON, Can a sensory panel be replaced by an instrument ?, IFST Proc. **7**, 190–192 (1974)
90. HARRISON, G. A. F., Investigations on beer flavour and aroma by gas chromatography, Eur. Brew. Conv. Proc. Congr. 247–256 (1963)
91. HARRISON, G. A. F., The flavour of beer — A review, J. Inst. Brew. (London) **76**, 486–495 (1970)
92. HARRISON, G. A. F. and E. COLLINS, Determination of taste thresholds for a wide range of volatile and nonvolatile compounds in beer, Proc. Am. Soc. Brew. Chem., 83–87 (1968)
93. MEILGAARD, M. C., Flavor chemistry of beer. Part I: flavor interaction between principal volatiles, Tech. Q. Master Brew. Assoc. Am. **12**, 107–117 (1975)

94. Meilgaard, M. C., Flavor chemistry of beer. Part II: flavor and threshold of 239 aroma volatiles, Tech. Q. Master Brew. Assoc. Am. **12**, 151—168 (1975)
95. Teranishi, R., R. G. Buttery and D. G. Guadagni, Odor quality and chemical structure in fruit and vegetable flavors, Ann. N.Y. Acad. Sci. **237**, 209—216 (1974)
96. Guadagni, D. G., R. G. Buttery and S. Okano, Odor thresholds of some organic compounds associated with food flavors, J. Sci. Food Agric. **14**, 761—765 (1963)
97. Salo, P., Use of odour thresholds in sensorial testing and comparisons with instrumental analysis, in "Proceedings of the International Symposium on Aroma Research", eds. Maarse, H. and P. J. Groenen, Pudoc, Wageningen 1975, p. 121—130
98. Gemert, L. J. van, unpublished results, 1976
99. Griffiths, N. M. and G. R. Fenwick, Odour properties of chloroanisoles — effects of replacing chloro- by methyl groups, Chem. Senses Flavor **2**, 487—491 (1977)
100. Powers, J. J. and G. O. Ware, Comparison of sigmplot, probit and extreme-value methods for the analysis of threshold data, Chem. Senses Flavor **2**, 241—253 (1976)
101. Powers, J. J. and M. C. Quinlan, Flavor detection threshold values for ethyl caprylate and phenyl ethyl alcohol and estimates of the percent population having greater sensitivity, J. Food Sci. **39**, 724—725 (1973)
102. Lillard, D. A. and J. J. Powers, Aqueous odor thresholds of organic pollutants in industrial effluents, Bul. EPA-660/4-75-002, National Environmental Research Center, Environmental Protection Agency, Corvallis 1975, pp. 26
103. Gee, M. G., N. M. Griffiths and R. G. Fenwick, Potential metabolites of phenolic disinfectants: preparation and odour properties of some anisoles, Pestic. Sci. **5**, 703—708 (1974)
104. Drake, B., A fortran program, 'SIGMPLOT', for fitting sigmoid curves to threshold data and for plotting the results, Chem. Senses Flavor **1**, 519—531 (1975)
105. Eriksson, C. E., B. Lundgren and K. Vallentin, Odor detectability of aldehydes and alcohols originating from lipid oxidation, Chem. Senses Flavor **2**, 3—15 (1976)
106. —, "Compilation of odor and taste threshold values data", ed. Stahl, W. H., American Society for Testing and Materials Data Series DS 48, American Society for Testing and Materials, Philadelphia 1973, pp. 250
107. Gemert, L. J. van and A. H. Nettenbreijer, Compilation of odour threshold values in air and water, National Institute for Water Supply, Voorburg, The Netherlands, and Central Institute for Nutrition and Food Research TNO, Zeist, The Netherlands 1977, pp. 79
108. Gemert, L. J. van and H. Maarse, A compilation of human odour threshold values in air, in "Proc. 4th Nordic Symposium Sensory Properties of Foods", Skövde, March 18—20 (1976), p. 117—127
109. Persson, T. and E. von Sydow, Aroma of canned beef: gas chromatographic and mass spectrometric analysis of volatiles, J. Food Sci. **38**, 377—385 (1973)
110. Cain, W. S. and H. R. Moskowitz, Psychophysical scaling of odor, in "Human responses to environmental odors", eds. Turk, A., J. W. Johnston and D. G. Moulton, Academic Press, New York 1974, p. 1—32
111. Moskowitz, H. R., Psychophysical and psychometric approaches to sensory evaluation, CRC Crit. Rev. Food Sci. Nutr. **9**, 41—79 (1977)
112. Moskowitz, H. R., A. Dravnieks, W. S. Cain and A. Turk, Standardized procedure for expressing odor intensity, Chem. Senses Flavor **1**, 235—237 (1974)
113. Sisca, V. K., Pugh!, Lebensm. Wiss. Technol. **7**, 190 (1974)
114. —, Standardized recommended practices for referencing suprathreshold odor intensity, American Society for Testing and Materials, E 544—75 (1975)
115. Jennings, W. G. and M. R. Sevenants, Volatile esters of Bartlett pear. III, J. Food Sci. **29**, 158—163 (1964)
116. Jennings, W. G. and M. R. Sevenants, Volatile components of peach, J. Food Sci. **29**, 796—801 (1964)
117. Sevenants, M. R. and W. G. Jennings, Volatile components of peach. II, J. Food Sci. **31**, 81—86 (1966)
118. Tang, C. S. and W. G. Jennings, Volatile components of apricot, J. Agric. Food Chem. **15**, 24—28 (1967)
119. Day, E. A. and L. M. Libbey, Cheddar cheese flavor: gas chromatographic and mass spectral analysis of the neutral components of the aroma fraction, J. Food Sci. **29**, 583—589 (1964)

120. SELF, R., J. C. CASEY and T. SWAIN, The low-boiling volatiles of cooked foods, Chem. Ind. (London), 863—864 (1963)
121. SALUNKHE, D. K. and J. Y. DO, Biogenesis of aroma constituents of fruits and vegetables, CRC Crit. Rev. Food Sci. Nutr. 8, 161—190 (1977)
122. SHANKARANARAYANA, M. L., B. RAGHAVAN, K. O. ABRAHAM and C. P. NATARAJAN, Volatile sulfur compounds in food flavors, CRC Crit. Rev. Food Technol. 4, 395—435 (1974)
123. CAUL, J. F., The profile method of flavor analysis, Adv. Food Res. 7, 1—40 (1957)
124. STONE, H., J. SIDEL, S. OLIVER, A. WOOLSEY and R. C. SINGLETON, Sensory evaluation by Quantitative Descriptive Analysis, Food Technol. (Chicago) 28, no. 11, pp. 24, 26, 28—29, 32, 34 (1974)
125. HARPER, R., D. G. LAND, N. M. GRIFFITHS and E. C. BATE-SMITH, Odour qualities: a glossary of usage, Br. J. Psychol. 59, 231—252 (1968)
126. HARPER, R., E. C. BATE-SMITH, D. G. LAND and N. M. GRIFFITHS, A glossary of odour stimuli and their qualities, Perf. Essent. Oil Rec. 59, 22—37 (1968)
127. HARPER, R., E. C. BATE-SMITH and D. G. LAND, Odour description and odour classification, J. & A. Churchill Ltd., London 1968, pp. 191
128. DRAVNIEKS, A., Evaluation of human body odors: methods and interpretations, J. Soc. Cosmet. Chem. 26, 551—571 (1975)
129. ZAUSCH, T. Z., Repräsentanz von sensorischen Eindrücken durch einzelne einheitliche Riechstoffe als Standard-Bezugssubstanzen, Riechst. Aromen Körperpflegem. 20, 479—480, 483—484 (1970) und 21, 3—4, 6, 8, 10 (1971)
130. SCHUTZ, H. G., A matching-standards method for characterizing odor qualities, Ann. N.Y. Acad. Sci. 116, 517—526 (1964)
131. LAND, D. G., R. HARPER and N. M. GRIFFITHS, An evaluation of the odour qualities of some stimuli proposed as standards for odour research, Flavour Ind. 1, 842—846 (1970)
132. CLAPPERTON, J. F., C. E. DALGLIESH and M. C. MEILGAARD, Progress towards an international system of beer flavour terminology, J. Inst. Brew. (London) 82, 7—13 (1976)
133. WILLIAMS, A. A., The development of a vocabulary and profile assessment method for evaluating the flavour contribution of cider and perry aroma constituents, J. Sci. Food Agric. 26, 567—582 (1975)
134. WILLIAMS, A. A. and C. S. CARTER, A language and procedure for the sensory assessment of Cox's Orange Pippin apples, J. Sci. Food Agric. 28, 1090—1104 (1977)
135. LAND, D. G., Techniques for assessing odour: uses and limitations, in "Proceedings of the International Symposium on Aroma Research", eds. MAARSE, H. and P. J. GROENEN, Pudoc, Wageningen 1975, p. 131—138
136. MAGA, J. A. and C. E. SIZER, Pyrazines in foods, CRC Crit. Rev. Food Technol. 4, 39—115 (1973)
137. MAGA, J. A., The role of sulfur compounds in food flavor Part I: Thiazoles, CRC Crit. Rev. Food Sci. Nutr. 6, 153—176 (1975)
138. MAGA, J. A., The role of sulfur compounds in food flavor Part II: Thiophenes, CRC Crit. Rev. Food Sci. Nutr. 6, 241—270 (1975)
139. MAGA, J. A., The role of sulfur compounds in food flavor Part III: Thiols, CRC Crit. Rev. Food Sci. Nutr. 7, 147—192 (1976)
140. PALAMAND, S. R., Descriptive flavor terminology, Tech. Q. Master Brew. Assoc. Am. 11, 90—93 (1974)
141. BOELENS, M., P. J. DE VALOIS, H. J. WOBBEN and A. VAN DER GEN, Volatile flavor compounds from onion, J. Agric. Food Chem. 19, 984—991 (1971)
142. GALETTO, W. G. and P. G. HOFFMAN, Synthesis and flavor evaluation of some alkylthiophenes. Volatile components of onion, J. Agric. Food Chem. 24, 852—854 (1976)
143. LAWLESS, H. T. and W. S. CAIN, Recognition memory for odors, Chem. Senses Flavor 1, 331—338 (1975)
144. CLARK, R. G. and H. E. NURSTEN, Analysis of the sensory term "nutty" and a list of compounds claimed to be nutty, Int. Flavours Food Addit. 8, 197—201 (1977)
145. DRAVNIEKS, A., L. KEITH, B. K. KROTOSZYNSKI and J. SHAH, Vaginal odors: GLC assay method for evaluating odor changes, J. Pharm. Sci. 63, 36—40 (1974)
146. KÖSTER, E. P., Psycho-physique des préférences olfactives, Parfums Cosmet. Savons Fr. 3, 139—143 (1973)

147. ENGEN, T., Method and theory in the study of odor preferences, in "Human responses to environmental odors", eds. TURK, A., J. W. JOHNSTON and D. G. MOULTON, Academic Press, New York 1974, p. 121—141
148. DUPUY, H. P., E. T. RAYNER and J.I. WADSWORTH, Correlations of flavor score with volatiles of vegetable oil, J. Am. Oil Chem. Soc. 53, 628—631 (1976)
149. HARRIS, J. M., Complex sensory assessment, J. Sci. Food Agric. 24, 1571—1581 (1973)
150. PANGBORN, R. M., Recent developments in sensory analysis, Lecture presented at the Second Congress of the European Chemoreception Research Organization, Reading (1976)
151. KÖSTER, E. P., Relative intensity of odour mixtures at supra-threshold level, Olfactologia 1, 29—41 (1968)
152. KÖSTER, E. P. and P. MACLEOD, Psychophysical and electrophysical experiments with binary mixtures of acetophenone and eugenol, in "Methods in olfactory research", eds. MOULTON, D. G., A. TURK and J. W. JOHNSTON, Academic Press, London 1975, p. 431—444
153. BERGLUND, B., U. BERGLUND and T. LINDVALL, Psychological processing of odor mixtures, Psychol. Rev. 83, 432—441 (1976)
154. MOSKOWITZ, H. R., Multidimensional scaling of odorants and mixtures, Lebensm. Wiss. Technol. 9, 232—238 (1976)
155. LAWLESS, H. T., The pleasantness of mixtures in taste and olfaction, Sensory Processes 1, 227—237 (1977)
156. MOSKOWITZ, H. R. and C. D. BARBE, Profiling of odor components and their mixtures, Sensory Processes 1, 212—226 (1977)
157. SALO, P., Interactions among volatile aroma compounds in a whisky imitation, Lebensm. Wiss. Technol. 6, 52—58 (1973)

6. Applications in other fields of research

H. MAARSE

6.1. Introduction

Aroma research workers have much contributed to the development of methods for the isolation, concentration, separation and identification of volatile compounds. They were forced to get as much information as possible from small amounts of material, since the most important aroma compounds practically always occur in very low concentrations. Especially in the late fifties and the early sixties it became clear that, generally speaking, aroma research workers were more interested in improvement of methods and development of new apparatus and techniques than in application of research results. Some are now using their experience in other research areas; the authors of this monograph are no exception. Workers in other research areas actually avail themselves of experience and knowledge obtained in aroma research.

This chapter will present a number of examples of studies of volatile nonaroma compounds. The purpose of this chapter is, on the one hand, to show how methods originating from aroma work have found application in nonaroma studies and, on the other hand, to find out if methods developed in such studies might be of advantage to aroma researchers. In this respect some brief suggestions will be made at the end of each paragraph.

6.2. Research on odour pollution

Air pollution is known to be a problem in highly industrialized and densely populated areas. Although odour pollution is only one aspect of the total problem, it has come to be regarded as a permanent nuisance in some regions; especially in those where intensified methods of livestock breeding are practised on a large scale.
In odour pollution investigations, the same analytical methods may be employed as in food aroma research, although sampling methods naturally tend to be different.

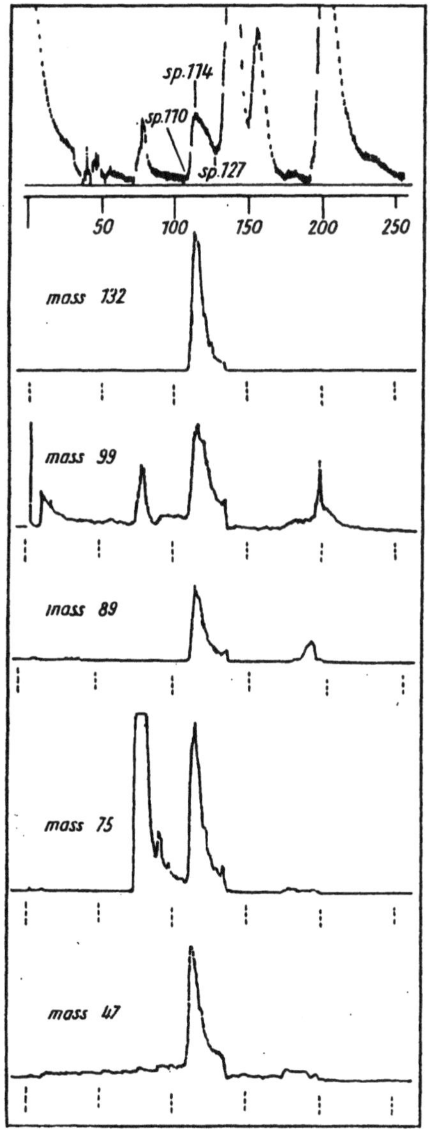

Fig. 6—1
Total ion current chromatogram and 5 mass fragmentograms of the main mass fragments of the suspected catty odour compound (1).

We should first like to present an example of an odour pollution investigation in which the sampling technique did not differ from familiar techniques in food aroma research. This was an investigation into the cause of a "catty odour" (1), which in 1971 occurred in a large part of the urban agglomeration of the western part of Holland. This odour came from the water of the Rode Vaart (Red Canal). It was already known at that time (2) that a catty odour can be caused by 4-mercapto-4-methyl-2-pentanone, a compound which is easily formed from hydrogen sulphide and mesityl oxide. Water samples from the Rode Vaart were extracted with pentane, and the

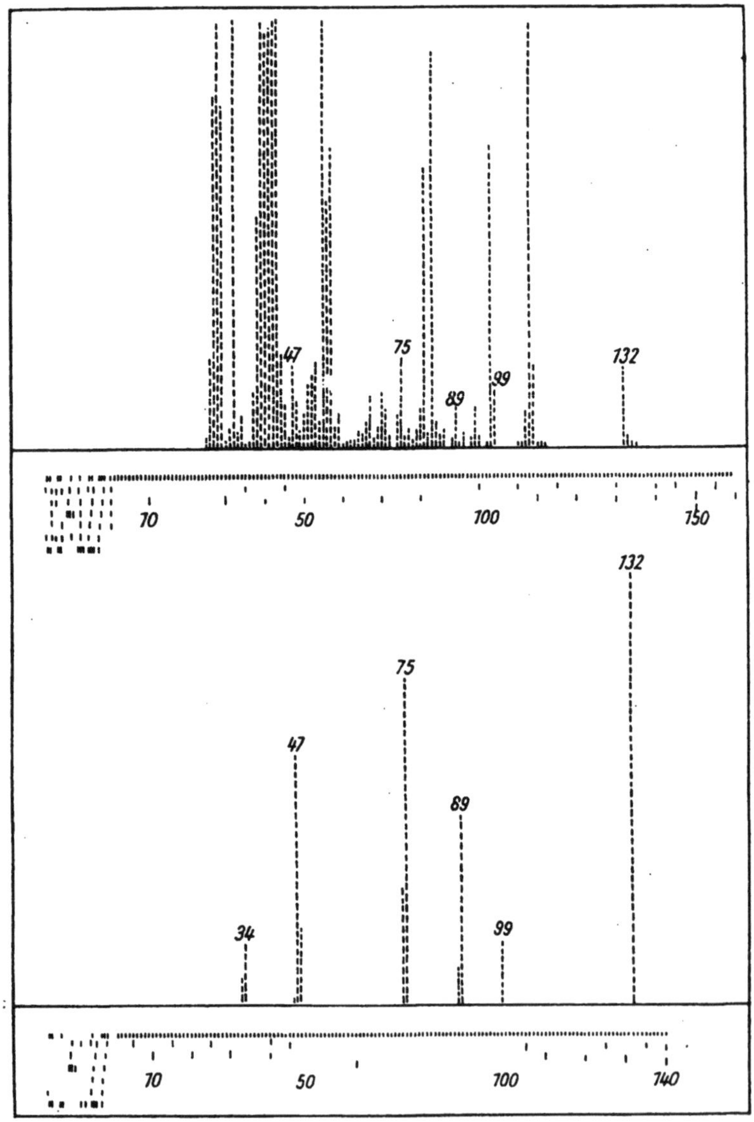

Fig. 6—2
Mass spectrum 114 (see Fig. 6—1) not corrected (upper spectrum) and corrected by subtraction of spectra 110 and 127 (lower spectrum) (1).

extract was concentrated. The concentrate was analyzed by a GC-MS-computer combination. The compounds that left the exit of the GC column entered the mass spectrometer via a so-called membrane separator (see 4.2.2.2.). The total-ion-current chromatogram and a number of mass fragmentograms, selected according to the main mass fragments of the suspected catty odour compound, were recorded (Fig. 6--1).

At the time that the catty odour was smelled at the exit bypass, all expected fragments were indeed present. The complete mass spectrum pertaining to this point (Spectrum 114) was registered (Fig. 6—2, upper spectrum). From this spectrum, and also from the gas chromatogram, it appeared that at this point in the chromatogram several compounds were present. By subtraction of spectra 110 and 127, a mass spectrum was obtained (Fig. 6—2, lower spectrum), which corresponded with the spectrum of the thiol mentioned above. Retention time and smell of the synthesized thiol and the Rode Vaart compound being also identical, it was thus proved that the catty odour of the water from the Rode Vaart was caused by 4-mercapto-4-methyl-2-pentanone.

When a malodour derives from a specific production or manufacturing process, an aroma specialist often has the advantage of knowing the aroma composition of the pertinent products, although certain process odours may derive from components not prominent in the end product. Specifically in the case of food industries, the classes of compounds responsible for an odour nuisance are often known beforehand to the aroma researcher with reasonable certainty; his problem then centres on a reliable determination of the quantitative ratios among the components.

In a study concerning the malodour of the cocoa industry, methods were developed for the quantitative measurement of two compound groups which are important for the aroma of cocoa, i.e. saturated carbonyl compounds and pyrazines. Quantitative determinations of these compounds, directly in a total vapour sample, suffer interference from the chief constituents or from compounds with the same chromatographic

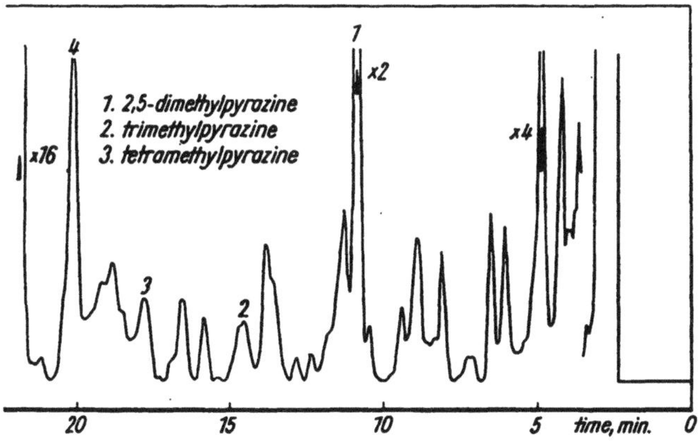

Fig. 6—3
Gas chromatogram of the compounds collected on a Tenax trap from the process gases emitted during the roasting of cocoa nibs in an industrial roast apparatus (3). Gas chromatographic conditions: 50 m × 0.75 mm i. d. stainless steel column, stationary phase Carbowax 20 M, temperature isothermal at 70° for 6 min., then programmed at 2 °C per min.

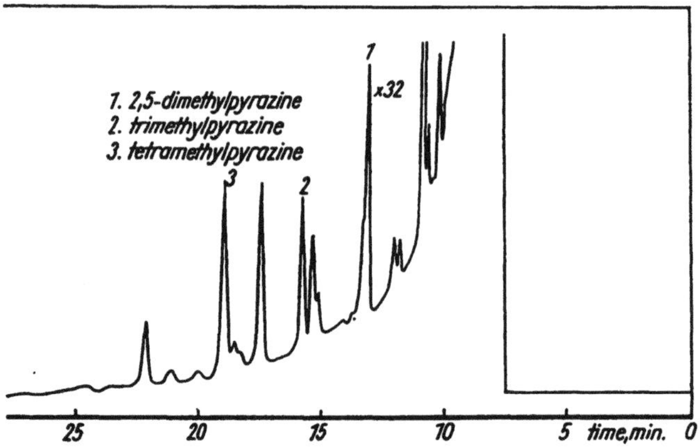

Fig. 6—4
Gas chromatogram of the pyrazines collected from process gases emitted during the roasting of cocoa nibs in an industrial roast apparatus (3). Gas chromatographic conditions: 150 m × 0.75 mm i.d. stainless steel column, stationary phase Carbowax 20 M, column temperature 120 °C.

behaviour. We, therefore, trapped these compounds as specific groups (3). Figures 6—3 and 6—4 demonstrate the advantage of this method over direct determination in a total vapour sample. In the chromatogram of the total vapour sample (Fig. 6—3), too many interfering compounds are present to allow quantitative analysis of the pyrazines.
Quantification of malodour sensations is desirable generally, in view of the fact that legal regulations are often sought for. The amount of malodour emitted at the processing site must be assessed by means of test persons. Such assessments are costly and are therefore replaced by instrumental methods whenever possible. A good correlation should, however, be guaranteed between the amount of malodour and the instrumentally measured concentrations of one or more selected compounds in the emitted gases. The amount of malodour is expressed as the odour number, i.e. the factor by which the emission gas has to be diluted so that 50% of the test persons cannot discern it from the air used for dilution.
A practical example of this approach will now be discussed. Both in the Netherlands and in other countries, increasing numbers of countryside dwellers of non-peasant origin complain about the odours from establishments for intensified breeding of livestock (pigs, calves, laying hens). It was therefore decided to start an investigation into the identities of the malodorous components in the exhaust air of swine buildings (4), in order to find some among them whose concentrations would correlate sufficiently closely with the sensorially determined overall malodour intensity (odour number) and, moreover, be amenable to a routine quantitative analysis method. The following compounds have been selected on the basis of the above two criteria: phenol; p-cresol; acetic-, propionic-, butyric-, iso-butyric-, valeric- and isovaleric acids. Other essential odorants such as indole and skatole occur in too low concentrations for quantitative determination. Since then, the exhaust air of approx. twenty swine buildings has been analyzed for the odour number and the concentrations of the above compounds. Among these, the concentration of p-cresol showed the closest correlation with the odour number. Although the correlation coefficient is still rather low (0.71), the average odour number can be predicted from a sufficient number of p-cresol deter-

minations. Similarly, the efficiency of malodour-combating installations can be estimated from p-cresol values before and after instalment.
In measurements such as these, the sampling method is of key importance. The concentrations of the malodours in swine building exhaust as well as in most emission gases, are so low that direct measurement is impossible and cumulative sampling is required. In the above-described case, it has proved possible to trap the acids and phenols quantitatively in a Pavelka absorption vessel filled with 2 N NaOH. An air volume of 100 l sufficed to obtain the chromatogram of Figure 6—5.

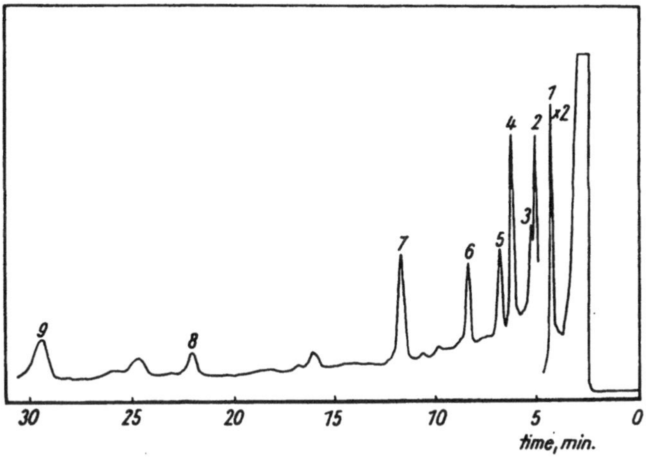

Fig. 6—5
Gas chromatogram of lower aliphatic acids and phenols collected from 100 l swine-building exhaust. Gas chromatographic conditions: 50 m × 0.75 mm i.d. stainless steel column, stationary phase F.F.A.P., column temperature 150 °C, formic acid added to carrier gas N_2. 1. acetic acid; 2. propionic acid; 3. isobutyric acid; 4. butyric acid; 5, isovaleric acid; 6. valeric acid: 7. hexanoic acid (internal standard); 8. phenol; 9. p-cresol.

Some of the techniques outlined above — collection of specific groups of compounds, cumulation of dilute vapours — should be applicable in aroma research as well.

6.3. Insect pheromone studies

The use of insecticides has long been the only large-scale method for combating insect pests. Although positive results obtained with these chemicals are evident, and their application will continue in the near future, some problems have arisen. One of these is the persistence of some insecticides and their metabolites in our food, in soil, in water and in man. Besides looking for non-persistent chemicals, some research took a closer look at the insect itself, because it seemed to be a promising source of ideas for the development of insect control chemicals (6). Insects depend on a variety of chemicals for their survival, and many of these are nontoxic to mammals but detrimental to an insect population when applied counter to their biological functions. An outstanding example is that of the pheromones. The concentrations in which these chemicals occur are very low and their identification was possible only with the aid of specially designed micro methods. A number of these methods, which were originally developed by BEROZA and co-workers, are described in Chapter 4.5. These methods might have been expected to find application in aroma research, too, but not many examples are known. Aroma research and pheromone research closely resemble each other: only small amounts of active compounds are available and assessment of their activity is not possible with instruments but only with humans and animals, respectively.

An interesting example is the study of the identity of sex pheromones of the American Cockroach (*Periplaneta americana* L.). Already in 1963, JACOBSON and BEROZA (7) assigned a structure to this sex pheromone, but had to withdraw it in 1965 (8) since a synthesized compound having the proposed structure yielded no response in American Cockroach males. PERSOONS et al. (9) obtained, from filter paper containing excretion products of many thousands of virgin females, two compounds which both evoked sexual excitation in the male cockroach. One of these compounds was identified in 1977, using UV, IR, NMR, MS and micro-hydrogenation (9, 10, 11). The structure of this compound, named periplanone-B, is given in Figure 6—6.

It is *suggested* that reaction gas chromatography (see Chapter 4.5.), especially micro-hydrogenation, used extensively in pheromone studies, be used more widely in aroma research.

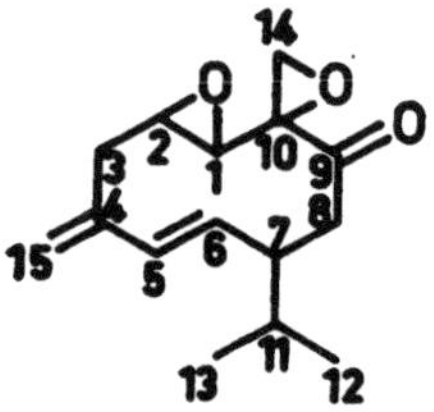

Fig. 6—6
Structure of periplanone-B (10)

6.4. Tobacco smoke

One of the most complex mixtures of volatile compounds is presented by the smoke of tobacco. Over the years, GROB has thoroughly analyzed this mixture using his outstanding glass capillary columns (see Chapter 3.2.). In one of his studies (12) a chromatogram was obtained presenting over a thousand peaks. From mass spectrometry and auxiliary techniques, GROB concluded that one peak on the average consisted of three substances; thus the actual number of compounds in tobacco smoke is at least 3,000. It is natural that for such a complex mixture specific GC detectors (see Chapter 4.6.) have been put to use in order to simplify the chromatograms. GROENEN (13) and GROENEN and VAN GEMERT (14) used an alkali flame ionization detector and a flame photometer detector (FPD) to compare the vapour phase of smoke from cigarettes made from different tobacco varieties. Cigarettes made from bright, Orient, dark and blended tobaccos, respectively, were found to produce rather similar patterns of major nitrogenous components (13). Large and small cigars also gave mutually similar patterns. In the vapour phase of cigar smoke, nitriles and N-methyl pyrrole were more abundant than in cigarette smoke; pyrrole concentration was much alike. These results have not yet been corroborated by mass spectrometric identification, but the compounds mentioned are known constituents of tobacco smoke. Of the 37 sulphur compounds detected by FPD (14), four were formed from air-cured tobacco in concentrations 4—15 times higher than from flue-cured tobacco (Fig. 6—7). Injection of only a fraction of one puff, i.e. 5 ml out of 35 ml, sufficed for

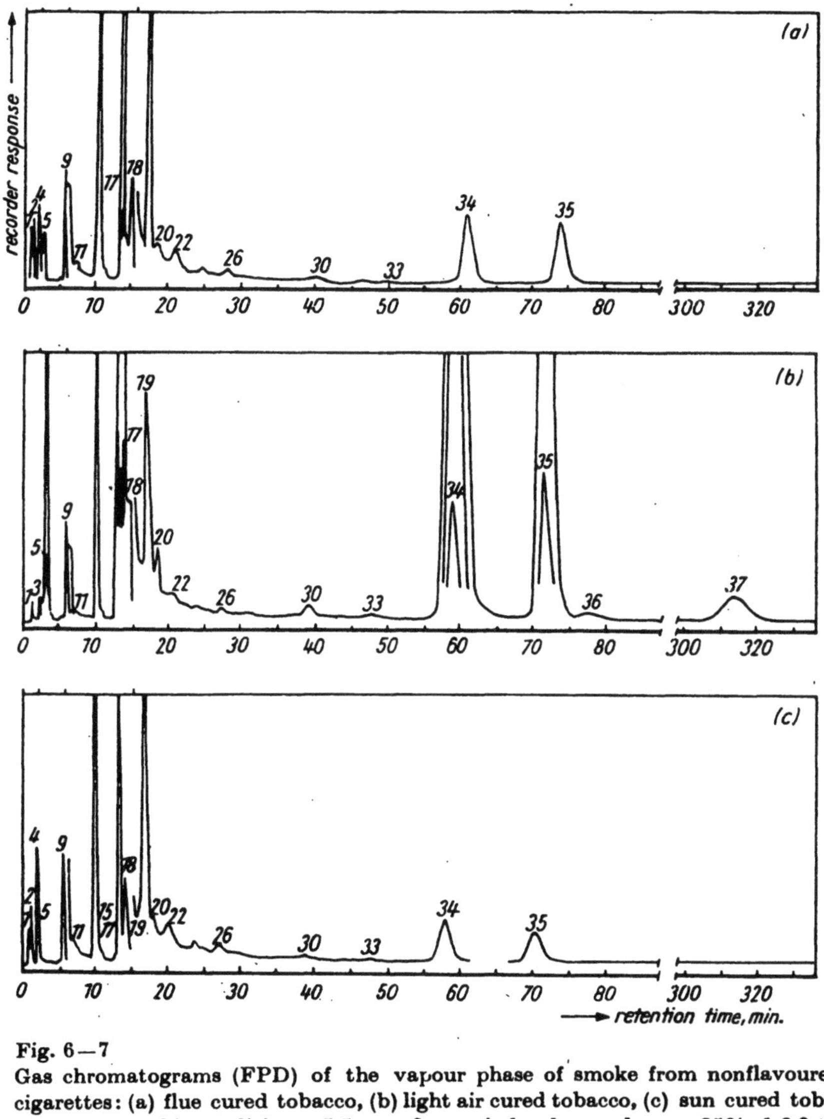

Fig. 6—7
Gas chromatograms (FPD) of the vapour phase of smoke from nonflavoured model cigarettes: (a) flue cured tobacco, (b) light air cured tobacco, (c) sun cured tobacco. Gas chromatographic conditions: 5.5 m $\times$ 3 mm i. d. glass column, 25% 1,2,3-tris-2-cyanoethoxypropane on Chromosorb W, A.W., 60—80 mesh, column temperature 75 °C (14).

the GC-FPD analysis. BERTSCH et al. (15, 16) performed cigarette smoke analysis with nickel capillary columns and an FPD, BLOMBERG (17) used glass capillary columns. A chromatogram from the work of BLOMBERG is shown in Figure 6—8.

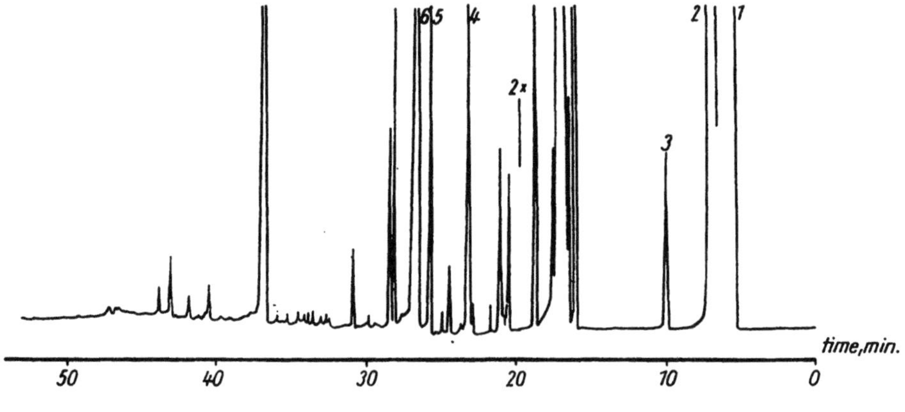

Fig. 6—8
Gas chromatoram (FPD) of the vapour of tobacco smoke (17). Gas chromatographic conditions: 90 m × 0.25 mm i.d. glass column, treated with Carbowax 20 M, coated with SF 96; temperature at time of injection, 70 °C; programmed to 195 °C at 5 °C per min. 1. hydrogen sulphide; 2. carbonyl sulphide; 3. sulphur dioxide; 4. thiophene; 5. unknown; 6, dimethyl disulphide.

With the nickel column, more peaks of higher molecular weight could be observed, but BLOMBERG ascribes this to reactions of e.g. H_2S with other smoke constituents. HORTON and GUERIN (18) have also demonstrated that ageing and contact with metal surfaces enhances the formation of higher-molecular-weight sulphur compounds. These observations indicate the need for inert chromatographic systems and sampling devices. It is also likely that artifacts are formed in the injection ports of gas chromatographs (17). Collecting the smoke on a Tenax trap might also cause artifacts and losses. The injection of fresh tobacco smoke by means of a glass syringe is assumed to minimize the formation of artifacts.

It is *suggested* to pay attention to chemical reactions of sulphur compounds in concentrates, traps, injection ports of gas chromatographs, etc., and to the means for preventing them. Selective detection (Chapter 4.6.) may be a useful ancillary technique for identification of unknown aroma compounds.

6.5. Volatile nitrosamines

Volatile nitrosamines are very potent carcinogens for animals and, most probably, for humans. Their presence has been shown in food products albeit mostly in very small amounts. Since they are volatile and occur in very low concentrations, their isolation and concentration can be carried out, except for a set of thorough safety measures, by similar methods as with aroma compounds, and final identification and quantitative analysis can be performed by GC combined with specific detection. This latter step illustrates some typical difficulties deriving from the extremely low concentrations at which these substances occur.

In the quantitative analysis of nitrosamines in tobacco smoke, the use of a nitrogen-specific detector (AFID) gave unreliable and high results, because of the presence of interfering N compounds as witnessed by a GC-MS test using low-resolution MS (19).

In a study of dimethylnitrosamine (DMNA) in cured meat products, even a mass spectrometer, low-resolution, proved a not infallible specific detector; the molecular ion of DMNA (M = 74) coincided with an isotopic fragment peak (trimethylsilyl) from a silicone anti-foam (20, 21, 22). To eliminate the possibility of such falsely positive results, silicone anti-foam was replaced by $MgSO_4$ (22).
The combination of capillary column gas chromatography and medium-resolution MS (CGC-MRMS) gives highly reliable results. The mass spectrometer is set at a resolution adequate for the elimination of interfering masses. Depending on the complexity of the sample, resolutions varying from 4,000 to 40,000 can be necessary. The mass spectrometer is used as a highly specific single-ion detector; this means that only the ion species with the characteristic accurate mass of the compound to be determined, is recorded. Interference from compounds with the same nominal mass is largely eliminated; it can be further minimized by use of a second column with a stationary phase of different polarity. Using this method, Groenen et al. (22, 23, 24) determined eight nitrosamines in meat products. An example is given in Figure 6—9; in the upper

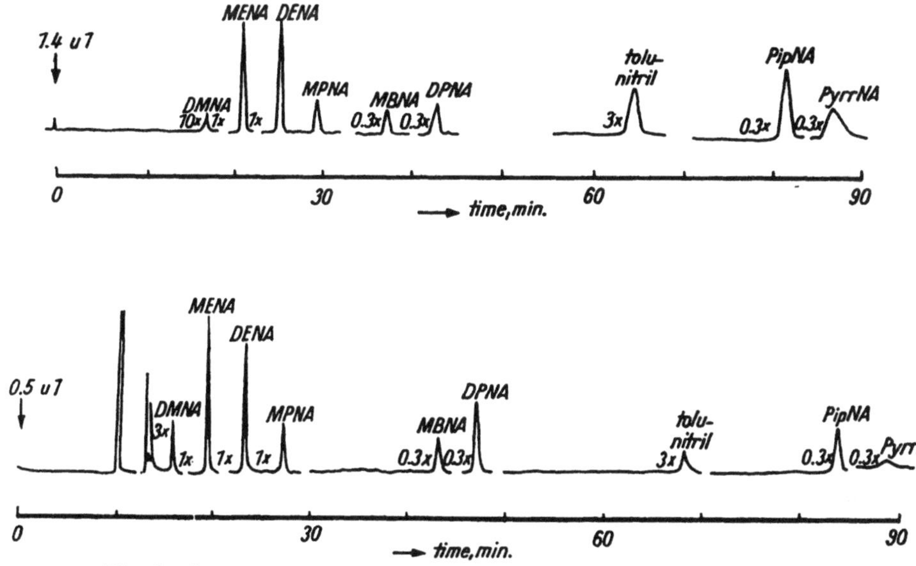

Fig. 6—9
Mass fragmentogram of the molecular ion masses of dimethyl-(DMNA), methylethyl-(MENA), diethyl- (DENA), methylpropyl- (MPNA), methylbutyl- (MBNA) and dipropylnitrosamine (DPNA), nitrosopiperidine (PipNA) and nitrosopyrrolidine (PyrrNA) (24). Gas chromatographic conditions: 150 m × 0.75 mm i.d. stainless steel column, stationary phase Ucon 50 H.B. 280 X; column temperature 110 °C; Varian Mat 731 mass spectrometer, resolution M/ΔM 4000, variable slit separator.

and lower parts respectively, mass fragmentograms are given of the molecular ion masses of the compounds in an extract of cured horse meat, spiked and unspiked with nitrosamines (24).
It is *suggested* to use this method more widely in aroma research; the concentrations of compounds of interest could be determined in a mixture without the need of fully removing others, thus resulting in more reliable results and in simpler sample pretreatment.

6.6. Medical applications

Humans secrete a variety of chemicals in the form of waste products produced by normal body functions. The means of exuding these chemicals may be oral (breath and saliva), skin (sweat and sebum), and in the forms of urine, faeces and flatus. The qualitative and quantitative composition of the mixture of volatiles leaving the body in one way or the other, depends on a person's state of health. It is well known that many diseases cause distinctive, identifiable odours. For example, systemic diseases such as diabetes, Asiatic cholera, typhoid fever, typhus and cholera may impart peculiar odours to the perspiration (25). Organic volatile compounds secreted below their odour threshold values may prove of physiological and diagnostic significance.

The techniques and methods developed in aroma research make it possible to study certain peculiarities of body emissions. An interesting example was reported by LEE et al. (26): three children had a fishy body odour and their urine, too, was smelling fishy. The compounds in the vapour above an urine sample were accumulated by adsorption in a trap containing Chromosorb 105 (see Chapter 2.2.). The adsorbed volatiles were transferred to a large-bore capillary column coated with Carbowax 20 M. A chromatogram of more than 100 peaks was obtained. All compounds were sniffed at the end of the column; one had an intensely fishy odour. GCMS analysis showed this compound to be trimethylamine. Elimination of fish and eggs from the children's diet resulted within a week in the disappearance of the fishy odours.

Several investigators (e.g. 27, 28, 29) have tried to find correlations between the presence of volatile organic metabolites in urines of healthy people and patients, hoping that the results could be used as a diagnostic tool. The most promising and powerful technique at this moment is the combination gas chromatograph-mass spectrometer-computer (GC-MS-Com). Very complex chromatograms are obtained and recognition of anomalies is in most instances possible only by computer analysis. This is particularly true if the disorder causes only small changes in the metabolic profiles or if an extra compound is hidden under a "normal" GC peak and, therefore, apt to be overlooked. JELLUM et al. (30) attempted to produce an automatic procedure for detecting anomalies in complex mixtures; they called their system CASAC (Computer Assisted Search for Anomalous Compounds). In principle, this system works as follows. Spectra from compounds in a normal sample (e.g. urine pooled from 25 healthy persons) and spectra of the compounds in an abnormal sample (e.g. urine of a patient), analyzed under exactly the same conditions, are compared by the computer. Any lacking normal or extra constituent is indicated. Spectra of these compounds can be printed out, and identification is then carried out in the normal way. The CASAC system is not yet working foolproof. Problems derive from the apparent impossibility of exactly reproducing the GC separations. For aroma investigators, further development of the system deserves attention.

Breath samples, too, have been tried as a source in which compounds indicative of a certain disease might be qualitatively and quantitatively analyzed. The concentration of these compounds is generally too low for direct analysis; accumulation by freezing out or adsorption is necessary (see Chapter 2.2.). Knowledge about the composition of breath samples of normal man is necessary before the analysis of respiratory gases of patients can be used as a diagnostic tool. CONKLE et al. (31) tested eight human subjects to establish a "trace compound base line". They found 69 different

compounds originating from human metabolism, from previous exposure (e.g. solvents in the laboratory), and from unknown origin. PAULING et al. (32) trapped human breath volatiles, 10—15 exhalations per sample, transferred them to a GC column by heating the trap, and obtained a chromatogram with 250 peaks.
It is *suggested* to follow closely any further development of computer programmes in medical analysis for comparing the composition of samples. Application of an improved CASAC system will be advantageous e.g. in off-flavour studies.

References

1. MAARSE, H. and M. C. TEN NOEVER DE BRAUW, Catty odour as an air pollutant (in Dutch), Chem. Weekbl. **68** (6), 11—12 (1972)
2. AYLWARD, F., G. COLEMAN and D. R. HAISMAN, Catty odours in food: the reaction between mesityl oxide and sulphur compounds in foodstuffs, Chem. Ind. (London) 1563—1564 (1967)
3. MAARSE, H. and J. SCHAEFER, Quantitative headspace analysis: total and specific group analysis, in "Analysis of foods and beverages. Head space techniques", ed. CHARALAMBOUS, G., Academic Press, New York, San Francisco, London 1978, p. 17—35
4. SCHAEFER, J., J. M. H. BEMELMANS and M. C. TEN NOEVER DE BRAUW, Investigation into components responsible for malodour of pig-feeding sites (in Dutch), Landbouwkundig Tijdschrift **86**, 228—232 (1974)
5. LOGTENBERG, M. TH. and B. STORK, Development of methods for the determination of malodour of exhaust air from swine buildings (in Dutch), C.T.I. report No. 01-4-40130, Central Technical Institute TNO, Apeldoorn (The Netherlands) 1976, pp. 27
6. BEROZA, M., Insect sex attractants, Am. Sci. **59**, 320—325 (1971)
7. JACOBSON, M., M. BEROZA and R. T. YAMAMOTO, Isolation and identification of the sex attractant of the American cockroach, Science **139**, 48—49 (1963)
8. JACOBSON, M. and M. BEROZA, American cockroach sex attractant, Science **147**, 748—749 (1965)
9. PERSOONS, C. J., F. J. RITTER and W. J. LICHTENDONK, Sex pheromones of the American Cockroach, *Periplaneta americana*. Isolation and partial identification of two excitants, Proc. K. Ned. Akad. Wet. Ser. C **77**, 201—204 (1974)
10. PERSOONS, C. J., P. E. J. VERWIEL, F. J. RITTER, E. TALMAN, P. J. F. NOOYEN and W. J. NOOYEN, Sex pheromones of the American Cockroach, *Periplaneta americana*: a tentative structure of periplanone-B, Tetrahedron Lett. 1976, 2055—2058
11. PERSOONS, C. J., Structure elucidation of some insect pheromones: a contribution to the development of selective pest control agents, Thesis, Agricultural University of Wageningen 1977
12. GROB, K., The glass capillary column in gas chromatography. A tool and a technique, Chromatographia **8**, 423—433 (1975)
13. GROENEN, P. J., personal communication
14. GROENEN, P. J. and L. J. VAN GEMERT, Flame photometric detection of volatile sulphur compounds in smoke from various types of cigarettes, J. Chromatogr. **57**, 239—246 (1971)
15. BERTSCH, W., F. SHUNBO, R. C. CHANG and A. ZLATKIS, Preparation of high resolution nickel open tubular columns, Chromatographia **7**, 128—134 (1974)
16. BERTSCH, W., F. HSU and A. ZLATKIS, Heart cutting technique in high resolution gas chromatography applied to sulfur compounds in cigarette smoke, Anal. Chem. **48**, 928—931 (1976)
17. BLOMBERG, L., Gas chromatographic separation of some sulphur compounds on glass capillary columns using flame photometric detection, J. Chromatogr. **125**, 389—397 (1976)
18. HORTON, A. D. and M. R. GUERIN, Quantitative determination of sulfur compounds in the gas phase of cigarette smoke, J. Chromatogr. **90**, 63—70 (1974)

19. GROENEN, P. J. and M. C. TEN NOEVER DE BRAUW, Determination of volatile N-nitrosamines in the vapour phase of the smoke from various tobacco products, Beitr. Tabakforsch. **8**, 113—123 (1975)
20. DOOLEY, C. J., A. E. WASSERMAN and S. OSMAN, A contaminant in N-nitrosodimethylamine: confirmation by high-resolution mass spectrometry, J. Food Sci. **38**, 1096 (1973)
21. GOUGH, T. A. and K. S. WEBB, A method for the detection of traces of nitrosamines using combined gas chromatography and mass spectrometry, J. Chromatogr. **79**, 57—63 (1973)
22. GROENEN, P. J., R. J. G. JONK, C. VAN INGEN and M. C. TEN NOEVER DE BRAUW, Determination of eight volatile N-nitrosamines in thirty cured meat products with capillary gas chromatography-high resolution mass spectrometry: the presence of nitrosodiethylamine and the absence of nitrosopyrrolidine, I.A.R.C. Sci. Publ. **14**, 321—331 (1976)
23. GROENEN, P. J., M. W. DE COCK-BETHBEDER, R. J. G. JONK and C. VAN INGEN, Further studies on the occurrence of volatile N-nitrosamines in meat products by combined gas chromatography and mass spectrometry, Proc. 2nd Int. Symp. Nitrite Meat Prod., Zeist 1976, 227—237
24. NOEVER DE BRAUW, M. C. TEN, C. VAN INGEN and P. J. GROENEN, Identification and quantitative determination of trace amounts of nitrosamines in cigarette smoke and cured meat products, in "Application de la spectrométrie de masse (sm) et de la résonance magnétique nucléaire (rmn) dans les industries alimentaires", XV Symp. Int., Bologna, Italy 1975, 99—111 (1977)
25. DISTELHEIM, I. H. and A. DRAVNIEKS, A method for separating characteristics of odours in detection of disease processes, Int. J. Dermatology **12**, 241—244 (1973)
26. LEE, CH. W. G., J. S. YU, B. B. TURNER and K. E. MURRAY, Trimethylaminuria: Fishy odors in children, New Eng. J. Med. **295**, 937—938 (1976)
27. ROBINSON, A. B., D. PARTRIDGE, M. TURNER, R. TERANISHI and L. PAULING, An apparatus for the quantitative analysis of volatile compounds in urine, J. Chromatogr. **86**, 19—29 (1973)
28. ZLATKIS, A., H. A. LICHTENSTEIN, A. TISHBEE, W. BERTSCH, F. SHUNBO and H. M. LIEBICH, Concentration and analysis of volatile urinary metabolites, J. Chromatogr. Sci. **11**, 299—302 (1973)
29. LIEBICH, H. M., Specific detection of volatile metabolites in urine of normal subjects and patients with Diabetes Mellitus using computerized mass fragmentography, J. Chromatogr. **112**, 551—557 (1975)
30. JELLUM, E., P. HELLAND, L. ELDJARN, U. MARKWARDT and J. MARHÖFER, The development of a computer-assisted search for anomalous compounds (C.A.S.A.C.), J. Chromatogr. **112**, 573—580 (1975)
31. CONCKLE, J. P., B. J. CAMP and B. E. WELCH, Trace composition of human respiratory gas, Arch. Environ. Health **30**, 290—295 (1975)
32. PAULING, L., A. B. ROBINSON, R. TERANISHI and P. GARY, Quantitative analysis of urine vapor and breath by gas liquid partition chromatography, Proc. Nat. Acad. Sci. **68**, 2374—2376 (1971)

Index of authors

Index of subjects

Bei Fragen zur Produktsicherheit wenden Sie sich bitte an:
If you have any questions regarding product safety,
please contact:

Walter de Gruyter GmbH
Genthiner Straße 13
10785 Berlin
productsafety@degruyterbrill.com